Springer Series in Design and Innovation   47

Springer Series in Design and Innovation (SSDI) publishes books on innovation and the latest developments in the fields of Product Design, Interior Design and Communication Design, with particular emphasis on technological and formal innovation, and on the application of digital technologies and new materials. The series explores all aspects of design, e.g. Human-Centered Design/User Experience, Service Design, and Design Thinking, which provide transversal and innovative approaches oriented on the involvement of people throughout the design development process. In addition, it covers emerging areas of research that may represent essential opportunities for economic and social development.

In fields ranging from the humanities to engineering and architecture, design is increasingly being recognized as a key means of bringing ideas to the market by transforming them into user-friendly and appealing products or services. Moreover, it provides a variety of methodologies, tools and techniques that can be used at different stages of the innovation process to enhance the value of new products and services.

The series' scope includes monographs, professional books, advanced textbooks, selected contributions from specialized conferences and workshops, and outstanding Ph.D. theses.

The volumes of the series are single-blind peer-reviewed.

Keywords: Product and System Innovation; Product design; Interior design; Communication Design; Human-Centered Design/User Experience; Service Design; Design Thinking; Digital Innovation; Innovation of Materials.

How to submit proposals

Proposals must include: title, keywords, presentation (max 10,000 characters), table of contents, chapter abstracts, editors'/authors' CV.

In case of proceedings, chairmen/editors are requested to submit the link to conference website (incl. relevant information such as committee members, topics, key dates, keynote speakers, information about the reviewing process, etc.), and approx. number of papers.

Proposals must be sent to: series editor Prof. Francesca Tosi (francesca.tosi@unifi.it) and/or publishing editor Mr. Pierpaolo Riva (pierpaolo.riva@springer.com).

Michela Barosio · Elena Vigliocco ·
Santiago Gomes
Editors

School of Architecture(s) - New Frontiers of Architectural Education

EAAE Annual Conference—Turin 2023

Springer

Editors
Michela Barosio
DAD
Politecnico di Torino
Turin, Italy

Elena Vigliocco
DAD
Politecnico di Torino
Turin, Italy

Santiago Gomes
DAD
Politecnico di Torino
Turin, Italy

ISSN 2661-8184   ISSN 2661-8192 (electronic)
Springer Series in Design and Innovation
ISBN 978-3-031-71958-5   ISBN 978-3-031-71959-2 (eBook)
https://doi.org/10.1007/978-3-031-71959-2

This Springer imprint is published by the registered company Springer Nature Switzerland AG
The registered company address is: Gewerbestrasse 11, 6330 Cham, Switzerland

If disposing of this product, please recycle the paper.

School of Architecture(s)

The Torino EAAE Annual Conference 2023 investigates the plurality of architecture as a discipline and the role of architectural education in training, questioning, and practising this plurality. This plurality is intended in terms of approaches, methods, topics, and values. The conference has been an occasion to think differently, reflecting upon the context of the discipline to understand the knowledge of the future, focusing on the question: what is Architecture in the age often described as post-architecture? This new perspective allows us to call into question some historical grounding principles of architectural education: the schools of architecture as a place where a style, a language is transmitted through the technique of the imitation of the masters, the everlasting character of the architectural artefacts built to last and the role of the architect as individual solely talented interpreter and author of architectural and urban artefacts.

A reflection on the ways of transmitting architectural knowledge, specifically design skills, in the age of post-architecture is needed. Several models of architectural education still coexist in the European context. Some of them still refer to the educational model of the Beaux-Arts. Settled in France at the end of the seventeenth century, this model was the first example of architectural schooling, further developed in the eighteenth century by François Blondel. It is still a pedagogical reference for many architectural schools. The central learning experience was structured around small independent ateliers where students learned directly under a "master", following his direction and imitating his language and practice under a strict hierarchy. Two other activities completed the Beaux-Arts way of teaching: The annual Paris Salon, where the best students' works were selected and displayed to the public, and the Parisian life of cafés, an informal extension of the ateliers, where design tendencies were discussed. Opposite to this model is the Polytechnic approach. Dating back to the Ecole Polytechnique, a military educational institution established in France at the end of the eighteenth century, this way of teaching aims to transmit technical-oriented knowledge, focusing on developing skills and competencies more than styles or tendencies. Contemporary schools of architecture tend to combine these two approaches with different balances. Some schools are still grounded on recognized masters leading the design approach of the school, while others decide to aim for the implementation of strategic topics to be developed through different learning experiences or to focus on specific design methodologies in order to build a school of thought more than a style of the school [1].

Thinking the Acropolis in Athens or the San Vitale di Ravenna in Italy as architectures built to last, together with Moneo we can say that buildings are always alone [2]. The architectures that have come down to us from the past have stood the test of time because societies have absorbed and inhabited them without distorting them. In the contemporary debate, however, we talk about the fragility of architecture and its temporary character. The contemporary question, however, is not the architecture itself but rather the modification of the reasons that determine its production. If Architecture was celebratory in the past and built to restore authority and power, today, Architecture has seen this role reduced. In the past, only those with a role of power asked for Architecture.

On the contrary, today, the demand for Architecture manifests itself through countless possibilities and different objectives. The expansion of users with a demand for Architecture has increased the variety of designed themes. Furthermore, starting from the nineteenth century, the collapse of most travel restrictions has further increased the possibility of contamination, and what once belonged to a specific place is today worldwide spread—just think of the role that International Exhibitions have had in history. We can observe extreme situations in which the Eiffel Tower and the Egyptian Pyramids are rebuilt on a scale in Las Vegas, and in cities worldwide, we can see the same architecture resulting from a globalised culture. If contemporary architectures are often not designed for a specific context, they are more and more designed for a specific lapse, waiving the everlasting ambition of classical architecture.

At the same time, in parallel with the process of globalised homologation that seems to characterise a large part of the material outcomes of architecture in the contemporary condition, the complexification of production processes, the articulation of an ever-increasing number of subjects and demands, and the intensification in the possibilities of exchange, communication, and knowledge are radically transforming the profile of the architect [3]. The mandate that societies assign to architecture is constantly evolving and mutating and, as a consequence, the figure of the architect is also being actualised, leading to the redefinition of the central target of practice in a shift in which the construction and the building, the objects, lose centrality in favour of an ever greater focus on the individual, the community, and the subjects [4].

Furthermore, while it is true that the discipline's interest in community practices, in the participation and inclusion of citizenship in the city's production processes, and the social role of the architect-designer is not new, and that these themes have characterised the debate for a good part of the last century, the scope and the reasons for the rebirth of this interest today have radically changed and transversally reach all professionals, regardless of their civic and political engagement and positioning. It is a transformation of practice that is reflected in the image that architects have of themselves, both inside and outside the discipline, which explains the radical transformation of working methods, the articulated and diversified cultural production of architects' offices and collectives, and the urgent need to rethink and redefine the aims and purposes of the pedagogical proposals offered by schools of architecture, or rather, of architecture(s).

In this context, the conference endeavours to elucidate a contemporary, more expansive, and inclusive definition of architecture by examining six pairs of antinomian concepts. These pairs include architecture as a method and/or as a discipline; architecture of the Masters and/or of the topics; architecture for architects and/or for the community; architecture as avant-garde and/or market-oriented; architecture inside and/or outside the wall; and architecture disciplinary and/or extra-disciplinary.

Michela Barosio
Santiago Gomes
Elena Vigliocco

References

1. Kaps V, Staub, P (eds) (2018) New schools of thought: Augmenting the field of architectural education. Zurich, Triest Verlag Fur Architektur.
2. Moneo, R (1999). La solitudine degli edifici e altri scritti. Torino, Allemandi
3. Gutman, R (1997) Architectural Practice: A Critical View. New York, Princeton Architectural Press.
4. UNESCO-UIA. Charter for Architectural Education (updated July 2023) https://www.uia-architectes.org/wp-content/uploads/2023/08/FINAL_UNESCO-UIA_CHARTER2023.pdf, last accessed 2024/06/04.

Organization

Local Organizing Committee

Barosio, Michela	Politecnico di Torino
Crapolicchio, Martina	Politecnico di Torino
Gomes, Santiago	Politecnico di Torino
Gugliotta, Rossella	Politecnico di Torino
Vigliocco, Elena	Politecnico di Torino

Scientific Committee

Atalay Franck, Oya	Zurich University of Applied Sciences
Barosio, Michela	Politecnico di Torino
Bologna, Alberto	Sapienza Università di Roma
Boutsen, Dag	KU Leuven
Cabrera, Ivan	Universitat Politècnica de València
Cavallo, Roberto	TU Delft
Corradi, Emilia	Politecnico di Milano
Flynn, Patrick	TU Dublin
Roth-Čerina, Mia	University of Zagreb
Santanicchia, Massimo	Iceland University of the Arts
Scala, Paola	Università Federico II di Napoli
Stewart, Sally	Glasgow School of Art
Valente, Ilaria	Politecnico di Milano
Xavier, João Pedro	FAUP Universidade do Porto
Zupančič, Tadeja	University of Ljubljana

Organized by

European Association for Architectural Education
Association Européenne pour l'Enseignement de l'Architecture

Sponsored by

Reviewers

Ali, Ahmed K.	Texas A&M University
Blanco, Manuel	Universidad Politécnica de Madrid
Boano, Camillo	Politecnico di Torino
Bologna, Alberto	Sapienza Università di Roma
Boutsen, Dag	KU Leuven
Cabrera, Ivan	Universitat Politècnica de València
Caliskan, Olgu	Middle East Technical University
Cavallo, Roberto	TU Delft
Corradi, Emilia	Politecnico di Milano
Dessì, Adriano	Università degli Studi di Cagliari
Dhlavacek, Dalibor	Czech Technical University in Prague
Ferrando, Davide Tommaso	Free University of Bolzano
Flynn, Patrick	TU Dublin
Geddes, Ilaria	University of Cyprus
Gullberg, Johanna Sofia	Norwegian University of Science and Technology
Ieva, Matteo	Politecnico di Bari
Lo Turco, Massimiliano	Politecnico di Torino
Macaluso, Luciana	Università degli Studi di Palermo
Maretto, Marco	Università degli studi di Parma
Mariné, Nicolás	Universidad Politécnica de Madrid
Mariolle, Beatrice	ENSAP Lille
Marzot, Nicola	Politecnico di Torino
Massarente, Alessandro	Università di Ferrara
Mattone, Manuela	Politecnico di Torino
Minucciani, Valeria	Politecnico di Torino
Mueller, Felix	FH Münster
Pedersen, Claus	Aarhus School of Architecture
Ricchiardi, Ana	University of Cyprus

Rolando, Diana	Politecnico di Torino
Roth-Čerina, Mia	University of Zagreb
Scala, Paola	Università Federico II di Napoli
Segapeli, Silvana	ENSA Saint-Étienne
Sentieri, Carla	Universitat Politècnica de València
Stewart, Sally	Glasgow School of Art
Trisciuoglio, Marco	Politecnico di Torino
Valente, Ilaria	Politecnico di Milano
Weischer, Martin	FH Münster
Zupančič, Tadeja	University of Ljubljana

Contents

Roots of Architecture: Ways of Research

Branches of Architecture: Ways of Practice

Ways of Architecture(s)

Impact of Architecture(s)

Impact of Architecture(s)

Andrea Čeko[1], Martina Crapolicchio[2], and Rossella Gugliotta[2(✉)]

[1] Department of Architectural Design, University of Zagreb, Zagreb, Croatia
[2] Department of Architecture and Design, Politecnico di Torino, Torino, Italy
rossella.gugliotta@polito.it

The word impact in research highlights the evidence, the results or consequences of a study on some specific contextual factors [1]. There are many other definitions of impact, such as the one used by the University of Scotland to highlight its contribution to the areas of "Innovation, Architecture and Design" [2]. The report presents the impact as benefits to people as architecture and design stakeholders. On the other hand, the AIA (The American Institute of Architects) asserts that research related to the built environment is underfunded, considering its impact on the economy, human condition, and society at large [3].

This contribution analyses different points of view about the impact of architectural studies, frames a specific approach and highlights the importance of open questions to go beyond a strict definition of impact.

Over a decade ago, assessing the impact of research has become an important vehicle for funded studies [4].

Due to the shrinking of public funding for higher education, the allocation of resources needs to be based on solid data; consequently, monitoring and giving a specific measure unit to allocate the budget to specific institutions and schools [5]. As a significant funding institution, the European community, to monitor its assets, decided to define the impact of research. The Commission's proposal for Horizon Europe describes an innovative approach for considering and disseminating impact, "the Key Impact Pathways" [6]. The impact has been evaluated by defining three fields: scientific, societal, technological, and economic [6].

Defining the impact as a factor has, as a consequence, the need to quantify them to allocate funding. The increase in data availability and its quantifiable feature has led to the definition of various metrics and indicators in research that are easy to obtain from numbers (citation counts, h-index, and journal impact factor). However, many metrics must capture the full range of different research activities and are often narrowly focused on what can be measured, forgetting other parts of the research. One among all, citation data provides a limited and incomplete view of research quality, especially in the Art and Architecture discipline [5].

As a starting point for discussion, the EAAE (European Association of Architectural Education) defines the research impact on architecture as a concern of individuals, groups and institutions in its report. It affects all scales in different research areas and timeframes. Its nature and the ways of measuring it depend on the audiences, areas and contexts in which it occurs. With the definition of impact, the EAAE supports and highlights the need to expand the range of impact assessment indicators, including curation, community

© The Author(s) 2025
M. Barosio et al. (Eds.): EAAE AC 2023, SSDI 47, pp. 3–6, 2025.
https://doi.org/10.1007/978-3-031-71959-2_1

engagement, public presentations, practice recognition by peers, awards, funding and publications [7]. This promotes the deficiency of a singular definition of impact in architectural research, opening it up to different research types and encouraging them to exist. From this starting point, several questions arise, leading to the construction of a PhD Workshop in the EAAE 2023 "School of Architecture (s)" framework stimulating a reflection on these multiple perspectives.

The PhD Workshop was dedicated to investigating how and if the impact of research in the architectural field could be defined. With a challenging impact definition of PhD research in architecture in mind and a current debate within the research community on the use of AI, an attempt was made to ask an AI chat (ChatGPT) to "define or describe the impact of a PhD research in architecture studies". The reply given by the ChatGPT listed several possible impacts of a PhD research in architecture studies as rather abstract and generic answers, with impact definition or description still lacking in clarity. While vaguely focusing on the advancement of knowledge, development of new methods and techniques, influence on practice, contribution to public discourse and inspiration for future research—the AI clearly demonstrates a lack of relevant knowledge or perspective when omitting to mention the nature of impact on/in different research audiences, areas, and timeframes—let alone its assessment in the context and field of architecture studies.

The EAAE23 PhD Workshop encouraged a discussion around these subtopics while relying on the knowledge of previous and available EAAE research, especially the experience of the EAAE's Research Academy Workshop that took place in Zagreb in 2019. Its result, framed as a Research Impact Diagram, plainly shows the Workshop's conclusions as visually comprehensible relations between the various aspects of impact definition, nature and evidence. Laying out and opening the questions of potential impact on/in relevant areas defined by scale (local, regional, national and global), or audiences addressed (individuals, groups, institutions), timeframes in mind, as well as the nature of impact relevance (accessibility, engagement and effectiveness) and appropriate evidencing, further resulted in upgrading the EAAE's definition of impact in 2022, especially in terms of impact assessment [7].

In order to offer perspectives and also give potential relevance to the impact in terms of less conventional referential bodies of evidence, some experiments have been conducted in terms of 'design/artistic practice-driven research' to dissect, cut through, and explore the nature of the complex conceptual landscape of PhD by Design (PbD) [8].

Through its experiential learning-through-evaluation model, the recent CA2RE project and conference example focuses on artistic and architectural design-driven doctoral research and its impact. Here and at the same time, developing a collective learning environment through presentations, performances, exhibitions, and critical discussions means (re)building evaluation criteria for the research, as well as building a platform as a 'design/artistic practice-driven research' community [9].

The reflection on impact doesn't aim to archive any new definition of the word. Moreover, it gives another dowel in the discussion.

Frayling's [10] famous tripartite model for practice-related research "into", "for" and "Through" practice classification is upgraded by scholars such as Fraser [11], distinguishing research types by stressing differences in "processes", "outcomes", and "impact". This tripartite classification highlights the presence of the impact definition in the frame of design research [4]. Consequently, other specifications emerged from the EAAE discussions, reframing the evidence of the impact in and on research. However, a big step has already been made in finding different synonyms and interpretations of impact in architectural research; another needs to be taken stressing the importance of opening the research to validate different evidence or body of knowledge (as mentioned in the Research Impact Diagram developed duringEAAE Research Academy in Zagreb RA Workshop 2019) enlarging even more the range of action of architecture. Furthermore, the discussion is enriched by the perspective of doctoral students who raise the issues from below, from a different position than the many starting points.

Acknowledgment. The authors acknowledge the participation of PhD students Diana Salahieh, Šárka Jahodová, Sotiria Inetzi, Carlo Vannini, Erenalp Büyüktopcu, and Hazal Çağlar Tünür. Additionally, the contributions of professors Dag Boutsen, Mia Roth-Čerina, Michela Barosio, Andrés Ros Campos, and Tadeja Zupančič have been pretious in fostering the debate on the impact of architectural research.

References

1. Reed, M.S., et al.: Evaluating impact from research: a methodological framework. Res. Policy **50**(4), 104–147 (2021)
2. Research Impact: I the Year of Innovation Architecture and Design, University of Scotland. https://www.universities-scotland.ac.uk/publications/research-impact-in-the-year-of-innovation-architecture-and-design/. Accessed 17 January 2024.
3. American Institute of Architects (AIA).: https://www.aia.org/pages/5626-architectural-res earch. Accessed 17 January 2024.
4. Aydemir, A.Z., Jacoby, S.: Architectural design research: drivers of practice. Des. J. **25**(4), 657–674 (2022)
5. Gervits, M., Orcutt, R.: Citation analysis and tenure metrics in art, architecture, and design-related disciplines. Art Doc.: J. Art Lib. Soc. N. Am. **35**(2), 218–229 (2016)
6. Horizon Europe programme analysis—European Commission. https://research-and-innova tion.ec.europa.eu/strategy/support-policy-making/shaping-eu-research-and-innovation-pol icy/evaluation-impact-assessment-and-monitoring/horizon-europe-programme-analysis_en. Accessed 17 January 2024.
7. EAAE: (2022) EAAE Charter on Architectural Research. Update approved by the General Assembly 02.09.2022. https://www.eaae.be/about/statutes-and-policypapers/eaae-charter-architectural-research/#:~:text=Architectural%20research%20impacts%20multifarious% 20audiences,of%20impact%2Dmaking%20and%20assessment. Accessed 5 January 2024.
8. Van Reusel, H., Michels, C., Schoonjans, Y.: Dissecting the archipelago: PhD by design concepts in the fields of architecture and urban design. ARENA J. Archit. Res. **6**(1), 4 (2021)
9. CA2RE Zagreb homepage. https://ca2re.eu/events/zagreb-2023/. Accessed 17 January 2024.
10. Frayling, C.: Research in art and design. Royal College Art Res. Papers **1**(1), 1–5 (1993)
11. Fraser, M.: Design Research in Architecture: An Overview. Routledge, London (2013)

Speculating Beyond Academia. A Critical Reflection in the Light of the Experience of the PhD Workshop Held During the EAAE 23 Annual Conference

Sotiria Inetzi[1]([✉]), Carlo Vannini[2], and Diana Salahieh[3]

[1] School of Architecture, National Technical University of Athens, Athens, Greece
inetzi.s@gmail.com
[2] Department of Architecture and Design, Sapienza—Università di Roma, Rome, Italy
[3] Department of Urban Design, Czech Technical University in Prague, Prague, Czech Republic

1 Introduction

Engaging in speculative endeavors beyond the confines of academia entails a deliberate focus on forging novel trajectories within the cognitive and knowledge-production realms of architectural research. It can be understood as an approach where research acts as a continuous path of inspiration and not (only) as a procedural tool for academic advancement. Within architectural doctoral studies, *research by design* and *design driven research* emerge as attitudes towards incorporating the artistic design processes in research, trespassing the scope of architectural research within academic boundaries.

This pursuit of research aimed at enriching applied architecture through the exploration of newly opened cognitive paths calls for recognizing research on architecture as a dynamic sphere. Research and practice enter an interrelation where thinking, practice and social factors go back and forth, contributing essentially to the formation of both research and applied architecture. Similarly, investigating the impact of architectural research necessitates a paradigm shift mirroring the spontaneous and creative nature of design processes and possibilities in the architecture(s) field.

2 The Process: How to Speculate Beyond Academia?

The PhD Workshop commenced with thought-provoking questions which set departure points to ignite the dialectic and reform genealogies, in aim to explore the concept of impact and how it could be interpreted beyond academic boundaries. Amidst delving into these considerations, the dialogue transgressed into several key questions: What impact do PhD students hope to achieve? Can impact be effectively evaluated? How can it be measured? Does it share commonalities with scientific relevance, and what constitutes scientific relevance in the context of architecture?

Doctoral students were key participants in this workshop, however interventions from professors and conference participants enriched the discussion and widened the lens to understanding impact and its implications. Seated in a round-table dynamic facilitated

M. Barosio et al. (Eds.): EAAE AC 2023, SSDI 47, pp. 7–11, 2025.
https://doi.org/10.1007/978-3-031-71959-2_2

the integration of dialectics as a central axis. The discussion reiterated keywords that called for visualization and classification of meanings as a step forward to dissecting how impact unfolds in architectural research. Essentially, the exploration revealed that knowledge in the field of architecture isn't solely contingent on the (potential) impact or impact factor as conventionally defined in scientific fields. The workshop participants recognized the importance of shifting meanings and the dynamic relationships between them (Fig. 1). This realization prompted a reevaluation of the fundamental question: What is ultimately defined as impact in the context of architecture?

In adopting a perspective for a qualitative rather than quantitative framework to describe impact, the workshop participants reached a juncture where the initial questions spurred the generation of new inquiries. Is impact tied to the research's relevance? Would 'influence' better capture the essence of impact? Or is impact more aligned with the practical utility of the research? When does the impact of the research begin?

Furthermore, the discussions questioned how our understanding of architectural research and knowledge affects the forms of impact we, as architects, imagine and seek. These questions gave rise to a consideration of the nature of objective truth in architectural research. Architectural research and knowledge are recognized as dealing and interacting with dynamics, issues or processes that take place outside of a laboratory. It depends on observing and remarking interrelations through architecture, spatial experience and the user. Social and political aspects enveloped in problems of architecture and the city may act as major and contemporaneous contributors to the research path. Therefore, in recognizing the social potential of architectural research, time plays a key determinant in assessing how and when the impact unfolds. It follows then that measuring the impact of architectural research cannot be iterated with quantitative metrics.

Fig. 1. Post-it notes representing the keywords and factors that influence research impact. EAAE 23 PhD Workshop outcomes, September 2023.

Furthermore, when discussing different forms of impact, it was evidently agreed that there are both tangible and intangible dimensions. For example, doctoral research can have a profound methodological impact; not only in producing a novel research methodology but inspiring unconventional perspectives and research tools. Moreover, an equally noteworthy impact, albeit subtle in its definition, is the interpersonal form.

Unanticipated reading encounters have the potential to exert influence over the personal trajectory or scholarly purview of that individual. Architectural research frequently delves into the realms of creative knowledge and the expanses of design possibilities. Consequently, architectural research possesses the capacity to instill inspiration in a manner that is unpredictable and transformative. It has the potential to induce a paradigm shift within the design process, operating within a dynamic, non-linear interactive relationship.

3   Drawing Towards Impact: An Architect's Tool to Speculation

In the process of shuffling, categorizing, and restructuring the post-it notes to visualize the dialectic and the relationship between factors that influence and contribute to research impact, the workshop participants faced a practical challenge. Encapsulating the complex networks into a compact diagram was confronted by the potential limitations of keywords and their associated definitions. As such this necessitated an alternative exploration into the depth of ideas, transcending potential divergences arising from linguistic nuances or contexts, and engaging in different research tools from the architect's toolkit.

This awareness prompted an artistic exercise proposed by Professor Zupančić. The exercise entailed a five-minute sketch to visually represent an individual interpretation of research impact (Fig. 2). The result was a tapestry of symbolic, abstract, and diagrammatic sketches. The diversity of visuals reflected the interconnectedness of ideas and doctoral students' arrays of hopes and forms of impact within their research journeys. Evidently the exercise unveiled the intrinsic language of sketches, the power of drawing as a form of dialogue and the nuances of translation; mirroring the parallels in architectural research that meanders between linguistics and design, artistic expression and language competence.

Fig. 2. Several sketches created in the workshop aimed to encapsulate an understanding of research impact. EAAE 23 PhD Workshop outcomes, September 2023.

4   Reflections: Towards a Nuanced Appreciation of Architectural Research and Its Impact

Within the realm of architecture, diversity assumes a pivotal role in the intricate interplay of imagination, research, and creation. Analogously, the workshop discussions recognized that the impact of architectural research manifests itself at varying stages, scales, and extents throughout the research process. However, the challenge lies in the complexity of articulating the impact of architectural research, greatly influenced by the disparity between how it is envisioned by the researcher and anticipated by the public and academic society, in contrast to the reality of how impact unfolds in diverse and unforeseen ways.

Moreover, the discussions underscored that impact is a qualitatively diverse concept, encompassing both tangible and intangible forms. Undeniably, doctoral research carries transformative potential, influencing the researcher through self-development and self-reflection. Consequently, although the measurement of impact may elude direct quantification, it remains imperative to actively seek and detect impact, recognizing its nuanced and multifaceted nature.

Thus, the conventional definition of impact underwent a shift. In architectural research, scientific relevance is viewed as generating outcomes that contribute to the field's thinking, philosophy, and perception, fostering innovation and advancement. In parallel, a potential qualitative impact may correlate to a societal shift, architecture—either imaginary or implemented—scope to facilitate society and to be inhabited by societal dynamics. Essentially, architectural research catalyzes innovation in both theory and practice, emphasizing the reciprocal relationship. This interplay emerges as the pathway to unveiling complex truths in architecture and architectural processes. This perspective challenges the linear interpretation of impact found in other scientific fields, as impact in architectural research defies straightforward translation into measurable financial results or production processes.

Finally, the Workshop yielded a diverse array of perspectives on "the impact of architectural research," sparking a dialogue that beckons for further exploration. The workshop's overall creative and critical dynamic interactions served as a reflection of the transformative potential inherent in architectural research, extending beyond its immediate academic context to envision a broader societal influence. Recognizing the vigorous interaction with both societal and scientific dimensions in architectural research expands the scope of its impact.

Seeds of Architecture: Ways of Teaching

Seeds of Architecture: Ways of Teaching

Michela Barosio[✉]

DAD Department of Architecture and Design, Politecnico di Torino, Torino, Italy
michela.barosio@polito.it

Within the general frame of the conference enquiring about the plurality of architecture as a discipline, the two sessions dedicated to the ways of teaching specifically addressed the different teaching approaches questioning two of the six dichotomies characterizing the contemporary debate on architectural education. The first tension explored is the supposed opposition between architecture considered as a discipline, with its specific field of knowledge and epistemology, and, on the other hand, architecture is regarded as a method, a mindset, a modus operandi that can be applied to other fields than the built environment. The second dichotomy concerns the origin of architectural education which can be grounded on the study and imitation of the Masters of Architecture, as it was when architecture was taught in the frame of the École des Beaux-arts, or can be grounded in training the student to face specific topics or burning issues of the architectural discipline.

In the sessions, several presentations focused on describing pedagogical experiences tackling contemporary challenges at different levels. They ranged from curriculum innovation and complete educational programs to single teaching activities.

Among those experiences some refer to methodological approaches proposing socially situated practice activities, pleading for interdisciplinary studios to mix cultural backgrounds and working methodologies, or international programs enabling collaboration between different countries and fostering cosmopolitan architecture. Other interventions are focused on urgent topics such as environmental sustainability, fragile territories and rural areas development, ethical approach to environmental and building design as well as the role and impact of artificial intelligence in the architectural pedagogy.

The type of Design teaching experiences presented through the sessions range from design studio teaching practice-based oriented and body-centered learning experiences to theoretical courses, questioning the role of the architect in contemporary society or the approach to the building demolition process. Despite these reflective activities, some interventions complain of a lack of critique, intended in the sense of a gesture that arrests, disorganizes, denaturalizes, and de-hegemonizes [1], in architectural education where new forms of critical theory should be enquired.

The discussion of the presentations seems to conclude that architectural education, both for its topics and its methods, can be considered a general education that can benefit many other fields [2]. This consideration leads to the conclusion that teaching architecture can be part of a non-architecture education, it can be considered as a minor of a different field major, contributing to developing skills such as team building, complexity management, or envisioning capabilities to other professional curricula. Architectural education also develops risk-taking skills, deals with the fragility of the environment

M. Barosio et al. (Eds.): EAAE AC 2023, SSDI 47, pp. 15–16, 2025.
https://doi.org/10.1007/978-3-031-71959-2_3

and the territories, and, through the pedagogy of mistakes, improves students' resiliency. Those specific skills are very valuable in tackling contemporary global challenges and crises that are characterized by a pervaded dimension of uncertainty.

References

1. Foucault, M.: What is critique? In: Lotringer, S., Hochroth, L. (eds.) The Politics of Truth. Semiotext(e), New York (2007)
2. Barosio, M., Boutsen, D., Ceko, A., De Loof, H., De Walsche, J., Gomes, S., Harriss, H., Marcaccio, R., Roth-Cerina, M., Sentieri, C., Sashore, N., Vannucchi, F., Van Reusel, H.: *Architecture's Afterlife. The Multisector Impact of An Architecture Degree.* Routledge, New York (2024)

"Self-efficacy" as a Value in Architectural Pedagogy

Naime Esra Akin[✉]

Aarhus School of Architecture, Exners Plads 7, 8000 Aarhus C, Denmark
nea@aarch.dk

Abstract. The latest research we conducted involved designing a tool to support the pin-up format, which is used to increase the effectiveness of the presenting/commenting activity as the main activity of project-expositions in architectural education. This research provided clues about the roles that students prefer to have as architects. Being a "Master" who has an extraordinary level of proficiency, skills, or performance in the information age can be stressful for students. We questioned the relationship between self-efficacy and the pin-up format, considering the students' experiences with project-expositions. To understand the relationship between students' social intentions, feelings, and the spatial organization during pin-up events, we monitored their emotions and thoughts about presenting their projects through a survey and some random interviews conducted over time. As the stress factor in project-expositions is directly related to the well-being of students, we focused on the emotional aspect of well-being. We were excited to learn that students expect peer-learning from pin-up activities, and that self-efficacy is an outcome of this. The survey was conducted in the Danish context during the Spring semester of 2023. The data collected can be considered informative because the 2nd year students represented a variety of nationalities and genders. This research establishes a strong foundation for asking the question, through the lens of well-being, of how pedagogical approaches in architectural education should be designed to support students in becoming either "Masters" or "self-efficacious architects."

Keywords: Architectural education · well-being · pin-up · self-efficacy · pedagogy · peer-learning

1 Introduction

Master architects as celebrities exist since 1500s [1]. They built up the agenda of architecture through years. Master architects become *hierarchically visible* through *competition*.

Today, visibility is possible for everyone through information technologies. Although there are still celebrities, who is visible on the media having the "wow factor" [2], but today, collaborative initiatives bring diversity in the agenda of architecture instead of one genius move to be followed.

M. Barosio et al. (Eds.): EAAE AC 2023, SSDI 47, pp. 17–28, 2025.
https://doi.org/10.1007/978-3-031-71959-2_4

Well-known collaborative initiatives like Rotor [3]; Rural Urban Framework (HK) [4]; Riwaq (PS) [5] are accessible through internet, so that they can grow, develop and increase the quality and content of their work. This openness also allows the young architects to be a part of an initiative at the early stages of their professional career. They are *equally visible* through *collaboration*. Architectural production engages with the society and architects become more powerful through collaboration.

Attention on the Masters as the figures of power on the big companies might be fostered by the capitalist world. Though, the young generations see the ecological facts and focus on the individual topics, which have urgency in their context. Is the role of architectural education to teach how to follow/become a Master, or foster the students to become self-efficant architects? What is the students' intention about becoming a Master?

This research aims to understand the intentions of the students as the future professionals, if the pin-up events can support the performance of the presenting/commenting activity in a definite direction in architectural education.

2   Architectural Education

Almost all over the world, architectural education is built upon project-based learning. "Learning by doing, was introduced into art and architectural education at the Ecole Nationale et Speciale des Beaux-Arts in Paris in the 1890s" [6]. Senior students were tutoring the junior students. There were "Charrette, French for "cart," refers to the carts in which the finished drawings were placed at the deadline hour for transport to the "Master" for critique [7].

An ongoing tradition of sharing/exposing the project is the pin-up event. Due to the general intentions of being approved, this very important moment of opening a project to discussion is stressful. The moment of encounter with the audience as the Master of their own projects is the pin-up event.

2.1   Pin-Up

The term pin-up may refer to drawings, paintings, and other illustrations as well as photographs. "The term was first attested to in English in 1941 even though the practice is documented at least back to the 1890s. Pin-up images could be cut out of magazines or newspapers, or they could be on a postcard or lithograph. Such pictures often appear on walls, desks, or calendars. Posters of these types of images were mass-produced and became popular starting from the mid-20th century." [8].

Pin-up events are the indispensable component of architectural education as the sharing/exposing moment for the project. A student positioned as the Master, becomes *hierarchically visible*, which triggers the *fear of failure*, causing stress, which affects the performance in a negative direction. (Fig. 1) Is this the only way of exposing the project to an audience as a goal/tasks/challenge?

The topic of pin-up, or crit (criticism in architectural education) has been studied for breaking the conventional power relations in the context of communication by Kathryn Anthony and Thomas Dutton. Crits might become an event of motivation and a tool of exploration through the opinion exchange. Learning about others' perspectives might open another related topic to develop the research about design [9, 10].

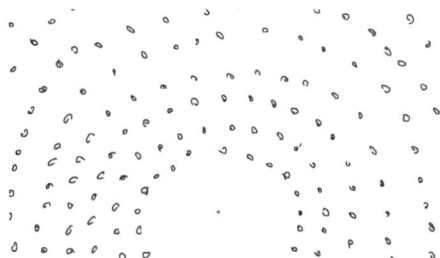

Fig. 1. The red dot is at the center of all attention to a mass of audience, Aarhus School of Architecture, 2023. (Teacher's Training Workshop Project, drawing by Naime Esra Akin)

Lately, Patrick Flynn, Miriam Dunn, Maureen O'Connor, and Mark Price organized an environment to study on changing the dynamic of the crit into a dialogue through experimenting new feedback methods over a full academic year with third year architecture students. Crits named as Round Table Review, Submission: Closed Juries & Open Feedback, Online Learning, and 'Red Dot' Review was made. The main benefits of these different crit events which were held at different phases of the semester was defined in the context of "clarity of feedback, stress reduction and productivity, peer learning, changing the power imbalance" towards reducing the stress of assessments and having a positive impact on design progress. The conclusion is "a reform of the crit can make educators and students engage in an open dialogue, centered on mutually engaged learning and can thereby develop a new pedagogy in architectural education." [11].

These researches and experiments are encouraging to have an experimental approach to pin-up. This paper is about providing data for re-thinking about the impact of spatial organization and format of the pin-up event on the students' self-efficacy in the context of being critical to the image of "master".

2.2 Self-efficacy in Architectural Education

Self-efficacy is an individual's belief in their capacity to act in the ways necessary to reach specific goals. According to Albert Bandura, one's sense of self-efficacy can play a major role in how one approaches goals, tasks, and challenges [12].

Instead of targeting to be perfect as a Master by taking the hierarchically leading position in the pin up event, all the students can be equally positioned to focus on their research and share their approaches in a clear and open way to learn from their peers, who are focusing on the same topic from other perspectives. The pressure of "presenting the genius idea" can be shifted to "sharing the idea", to be developed through an *equally visible* process of discussion (Fig. 2). In other words, instead of positioning as the Master, having self-efficacy as one of the members of the architectural design studio will decrease the stress and increase performance of the student.

Fig. 2. The red dot is one of the other colored dots that sharing the attention by being an element of the non-hierarchical space. Tutor(s) is also involved in the circle, Aarhus School of Architecture, 2023. (Teacher's Training Workshop Project, drawing by Naime Esra Akin)

3   The Research

The main aim of the research is creating a tool which acknowledges the emotional and social effect of pin-ups on students and creates a motivation for the students to consider self-efficacy as a matter of spatial organization of the pin-ups, which may be designed in collaboration with the students.

3.1   The Architectural Design Studio Environment as the Context of the Research

Our pedagogical approach is structured upon developing the research topics and creative skills of the students for empowering them to realise their thoughts through their intentions. The goal is to create a place for bringing out their own capabilities in the world. The students are encouraged to feel free to concentrate on their own knowledge and experiences, to find out their own interest, and be confident with their own insight to search and design. the pin-up events serve as peer-learning workshops with an added value of exchanging the ideas/information/reflection on the common assignment.

Studio culture is considered as an ecology of education where the bottom-up activities are carried out as a local ecosystem. Ultimately, these actions have an impact on the long-term social structure. Students are considered as responsible for their own choices of learning in a framework of sustainable architecture. The tutor is a mentor, guide, and provocateur asking the questions to support each and every student's individual and open-ended process of design. Experiential methods are introduced for triggering the self-awareness, cohesion, respect, inspiration and collaboration. Rights and responsibilities are formed together through weekly architectural design studio meetings to change/add things in the studio process. Starting from the 2023 spring semester, all students take roles in studio organization by the teams of pin-up, exhibition, social meeting, studio care, and studio meeting.

3.2   Well-Being as the Frame of the Research

"Self-efficacy" is a component of "well-being". The World Health Organization describes "well-being" as a situation enabling to function psychologically, physically, emotionally and socially well. According to the Foresight Mental Capital and Wellbeing Project 2008 [13] wellbeing is "enabling people to develop their potential, work productively and creatively, form positive relationships with others and meaningfully contribute to the community". In other words, well-being, is a key for a sustainable society. Individuals with high levels of well-being are more productive at work and are more likely to contribute to their communities.

As the children and young people spend a considerable amount of time at the school during a critical period for the development of their personality and socio-emotional competences, the schools have a key position for developing well-being in the society. "The link between academic and socio-emotional learning has been clearly underlined by empirical evidence, including neuroscientific research, demonstrating that learning is a relational and emotional process." [14] Addressing learners' well-being is therefore the key not only to raising educational outcomes, but also building a society, where the value of well-being is promoted through inclusive, collaborative, creative and self-efficant individuals.

According to European Commission European Education Area, Well-being is about students': Feeling safe, valued and respected; Being actively and meaningfully engaged in academic and social activities; Having positive self-esteem, self-efficacy and a sense of autonomy; Having positive and supportive relationships with teachers and peers; Feeling a sense of belonging to their classroom and school; Feeling happy and satisfied with their lives at school. Specifically, architectural education—like other art and design educations—depends on creativity of the students, adds an extra layer of personal development process although it is higher education. Therefore, well-being needs to have priority in accordance with the performance of the students.

3.3   The Survey as a Tool of the Research

The research depends on a survey documenting the well-being experience of the 2nd year (4th semester) students, in spring semester 2023. Student body consisted of a mixture of worldwide nationalities (Nationality—9/Denmark, 4/Australia, 4 Germany, 2/Belgium, 1/Israel), gender (8/Female, 12/Male), and status (9/Local, 11/Exchange). There are almost an equal number of students from both genders. In other words, the collected data can be considered as a representation of the regular stage of architectural education through the diversity of backgrounds and the level of the studio.

The targeted pin-up events are those that took place in the first half of the semester. Questions of the survey were grouped based on the feelings, actions, self-efficacy, relationship with peers and tutors, and belongingness they felt during the pin-up events they joined. They were inspired by PISA program of OECD [15] (Fig. 3) and revised (with the support of the student counselling psychologist) to challenge the students to think on their own reactions to pin-up events.

The research was introduced openly to the students and requested their support to find out the types of spatial organization through monitoring the semester in the context of the students' well-being through the survey. The results of the survey are achieved through questioning causation, correlation, cross-tabulation of the data.

Design of the survey depends on its graphical effect. (Fig. 4, 5) Color is the connection in-between the abstraction of the choice and the real situation.

. Innovative ideas
. Contextual and global thinking
. High level of motivation for
communication (listening and responding)
. High level of self-efficacy

. Joyful, lively, proud, happy, cherful	. Obliged, scared, miserable, afraid, sad
. Cooperative	. Competative
. Belonging	. Non-belonging
. Self-efficant	. Fear of failure

Fig. 3. OECD's Program for International Student Assessment's (PISA). Aarhus School of Architecture, 2023. (Teacher's Training Workshop Project, Naime Esra Akin)

Indirectness of the numerical expression of grading a specific emotion according to a specific piece of life might be more direct through the psychological effect of the colors. Yellow and orange as warm colours, evoke emotions, such as happiness, energy, optimism or enthusiasm; light yellow and green as slightly cool colours are linked to calmness, sadness and indifference. The colors were used as the indicators of the choices from strongly agree to strongly disagree.

Main topics of the survey were joy, cooperation, competition, belonging, feelings, self-efficacy and feeling of failure which reveal the relevant data with emotional, social and intellectual well-being. The questions were ordered randomly for not to impose a definite intention to answer the questions. Instead, randomness was expected to bring out the real thoughts of the students. It was delivered and received as an e-mail individually (March 20–April 21, 2023) and the names/responses were kept hidden for activating the experiential approach instead of rational choices.

The survey as a tool to think and communicate on self-efficacy is a component of consisting well-being in architectural education. Color-coded data is organized according to the positive/negative statue of the questions and translated into numerical data, as a second step of detailed reading. Some calculations can trigger the curiosity to find more relevant patterns. Survey is designed as a database, which is open to question. You can find out a variety of answers due to the correlations you make. In other words, the results might be read in different ways regarding the questions/correlations made by the reader.

Therefore, through a meta-rationale reading, one can notice relevant factors that others overlooked, asks the key question that no one had thought of, changes the description of the problem so that different solution approaches appear, rethinks the purpose of the work, and combines the multiple contradictory views, not as a synthesis, but as a productive patchwork. In other words, this survey/database is open to be questioned, re-organized and invented for a variety of reading opportunities to inspire the reader through reading. It might be more important to process continuously, instead of taking a decision.

Fig. 4. The delivered un-organized survey that filled out by only one student, Aarhus School of Architecture, 2023. (Teacher's Training Workshop Project, Naime Esra Akin)

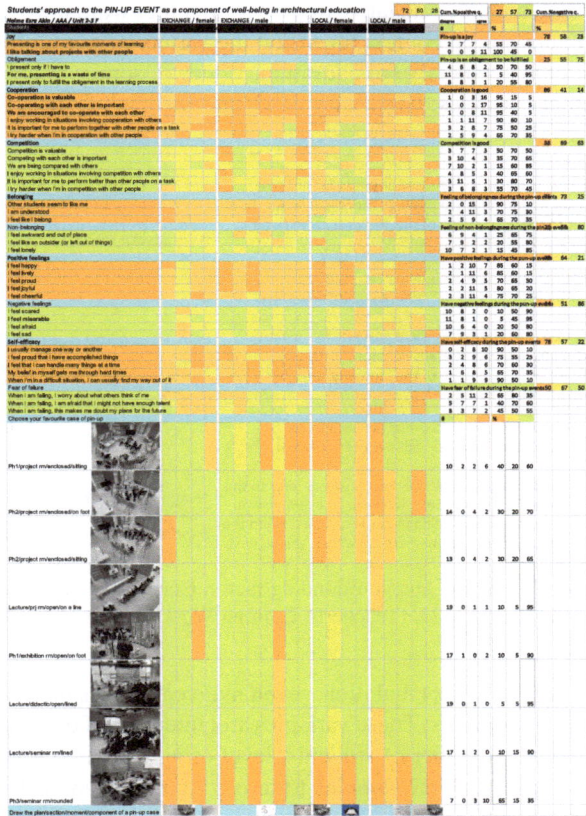

Fig. 5. The organized survey that filled out by all of the students, Aarhus School of Architecture, 2023. (Teacher's Training Workshop Project, Naime Esra Akin)

3.4 Result of the Research

Color coding helps reading the database/survey at one glance. Positively worded questions were mostly agreed with (shown in orange), while negatively worded questions were mostly disagreed with (shown in green).

In the context of the numerical data, the cumulative percentage for negative questions is in correlation with the cumulative percentage for positive questions. %72 of the students feels positive about the positive questions, and %72 of the students feels negative about the negative questions. In other words, %72 of the students feels positive about the pin-up context regarding the variety of pin-up formats.

In the perspective of causation, values of some variables are affected by the change of other variables: (%) (A-agree, B-intermediate, C-disagree) (Fig. 6).

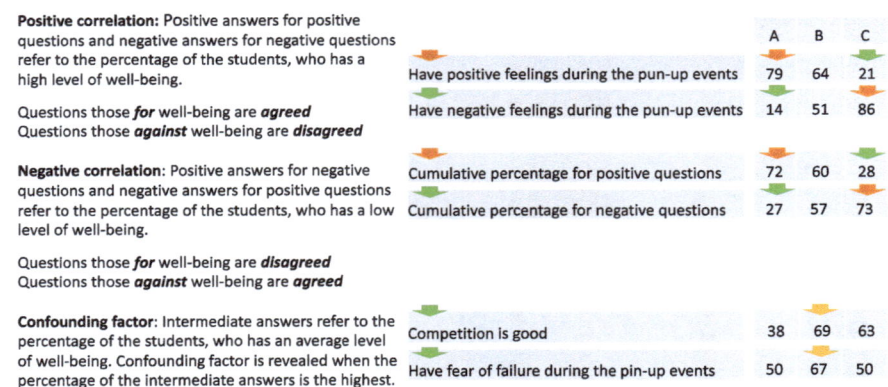

Fig. 6. Correlation and confounding factor. Correlation and confounding factor, Aarhus School of Architecture, 2023. (Teacher's Training Workshop Project, scheme by Naime Esra Akin)

Through cross tabulation, seven main categories involved in the survey opens a variety of perspectives. These variables are; intention of questions, gender of students, status of students, number of students, level of agreement, specific pin-up spaces, suggested pin-up spaces. Some samples for cross tabulation are as follows:

- *What are the topics, which have the maximum number of agreements?*
  Cooperation, positive feelings, self-efficacy, joy, and belongingness.
- *What are the topics, which have the maximum number of disagreements?*
  Competition, negative feelings, fear of failure, non-belonging, obligement.
- *What is the effect of the gender and the status of the students on the topics?*
  More agreement for male students; competition and cooperation.
  More agreement for female students; belongingness and fear of failure.
  More agreement for local students; obligement, competition, non-belonging, fear of failure
  More agreement for exchange students; joy, cooperation, belonging, self-efficacy.

The correlation in-between the feelings and the spaces show that the students agree with having the positive feelings and disagree with having the negative feelings during the pin-up events they have experienced during the first part of the semester. Majority of the students find joy of pin-up in exchanging ideas on projects with peers; strongly agree with cooperation and disagree with competition; agree with the feeling of belongingness; agree with self-efficacy during the pin-up events.

The students' response in overall to the spatial organization of the pin-up event have the characteristics of rounded, small scale, enclosed, and non-hierarchical (Fig. 7). The infographic is the overall visualization of the database without any comments. It has the same color code with the database to establish a connection for further readings through a variety of correlations (Fig. 8).

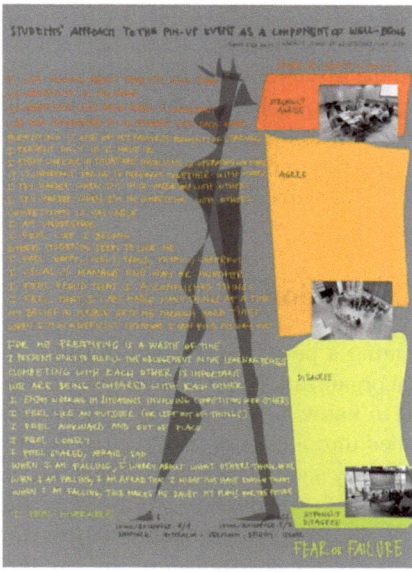

Fig. 7. The pin-up/presentation events experienced by the students, Aarhus School of Architecture, 2023. (Teacher's Training Workshop Project, photos by Naime Esra Akin).

Fig. 8. Infographic explanation of the overall data, Aarhus School of Architecture, 2023. (Teacher's Training Workshop Project, collage by Naime Esra Akin)

Regarding the random discussions and interviews with the students, the pin-up evens are preferred to be closed to big audiences. The students appreciate the value of discussing to learn about each other's skills and share specific issues like drawing techniques, network, experiences with the other students, who are under the same circumstances with themselves, though they don't want to be the focus of the discussion. The discussion is considered useful in case there is an equal and sincere atmosphere. To see each other's faces in a close distance, showing the materials of their projects and being able to take notes/draw sketches on these printed materials make a big sense to them for a fruitful meeting. This kind of a close contact with a small group of students, and a tutor, who chairs the pin-up is the most relaxing and joyful way of developing not only the individual projects but also the social relations at the same time, so that, the communication between the students continues after the pin-up through discussions on individual projects and social matters.

Both survey and the verbal communication indicates that the intention of the students is not positioning themselves as the ***hierarchically visible Master*** in a competitive manner, instead, they prefer to be ***equally visible*** through ***collaboration*** during the pin-up events.

3.5 Side Effects of the Research

Through a close observation during the survey process, it is noticed that the survey itself was considered positive by the students as a tool of communication, because of the direct and open effect of the evaluative discussions on the pin-up formats designed with the pin-up team. Attendance to the pin-up events raised comparing with the Fall semester, due to the discussion on the spatial organization for a better learning environment. The survey was motivating for the pin-up team to think on a variety of spatial organization and reserve time to organize the space. It raised a spatial awareness through experimenting the relationship in-between the user and the space/spatial elements. Students started to observe themselves during not only the pin-up events, but also in the architectural design studio space.

4 Conclusion

Presenting a project is a stressful process in architectural education. The hierarchical spatial organization creates a negative feeling of self-exposure like a Master who is supposed to know all about design. In this research, the moment of exposure/pin-up was unfolded through a frame of well-being for learning more about the students' feelings. Both survey and the verbal communication indicated that the intention of the students is not positioning themselves as the *hierarchically visible Master* in a competitive manner, instead, they prefer to be *equally visible* through *collaboration* during the pin-up events. Specifically, the *peer-learning* and *equality* were pointed as the most important dimension of exchanging the ideas/information/reflection on the common assignment. They preferred to position themselves as a part of a collaborative design process; although all projects were individual, the students felt self-efficant by exposing their projects in the friendly and project-oriented atmosphere of small groups to be developed together. By leaving the judging role, the position of the tutor was opened to discussion as an observer and/or a group member, for enabling the students to create a new understanding of Master in architecture.

Referring to Kate Raworth's idea of sustainable economy, architectural production, education and profession needs to create an ecology enabling others to create value [16]. In other words, cooperating and collaborating is the raising value that is already appreciated by the young generation. Architectural education needs to develop new research methodologies, learning pedagogies for cooperative/collaborative ecologies.

5 Aftermath

5.1 Experimental Pin-Up

After the survey was completed and analyzed, an experimental pin-up was organized with collaboration of the students, which might be called as "science fair". It was designed considering the self-efficacy of the students. The schedule and the spatial organization were designed for a dense, synchronic and repetitive presentation of the project with four instantaneous and quick peer-review of all students in small groups. The aim was shifting the stress of the unique moment of the presentation ritual, towards a normalized communicative activity (Figs. 9 and 10).

The students got it as an opportunity to progress and gain strength in communicating the project and coming over the stress of the moment of encounter at the finals. They also liked the idea of taking detailed feedback from all their peers. As a result of the experiment, all of the students agreed that this was the best pin-up experience they have ever had.

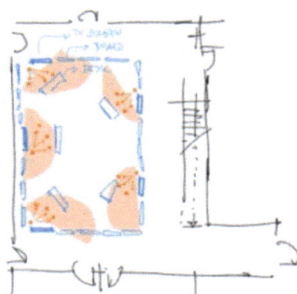

Fig. 9. The "project market" pin-up, Aarhus School of Architecture, 2023. (Teacher's Training Workshop Project, plan drawing by Naime Esra Akin).

Fig. 10. The "project market" pin-up, Aarhus School of Architecture, 2023. (Teacher's Training Workshop Project, photo by Naime Esra Akin)

5.2 Feedback from the Students

Feedback from the students about the pin-up was taken just after the event. The students' response to a few questions referring to the well-being is as follows:

- After a few presentations, presentation process got better.
- The feedback from the peers was satisfying in both quality and quantity. Both the written notes and verbal discussions were detailed enough to develop the design.
- There was the feeling of stress because of the time shortage, but not nervous due to the small groups of audience and friendly dialogue during the presentation.
- Giving feedback was not stressful, instead there was that feeling of giving support through a friendly dialogue.
- There was no feeling of self-exposition, as there were 4 simultaneous presentations sessions.
- The continuous sound in the room was like a background which was not distracting the students.
- There was an energetic feeling during and at the end of the pin-up, comparing with the previous pin-up events.

5.3 Future Projection

Collaborating with the students for designing the pin-up events/spaces may develop the learning environment. A short and clear survey can be used as a tool of communication through open discussion on the process of exposing the projects. "Positioning as a Master" might be used metaphorically or practically for introducing another perspective to ongoing discussion of "the role of architect".

References

1. Gordon, D.J., Orgel, S.: Leonardo's Legend. ELH **49**(2), 220–223 (1982)
2. Peters, T.: The Pursuit of Wow. Vintage, New York (1994)
3. Rotor Homepage. https://rotordb.org/en/about-us
4. Rural Urban Framework Homepage. https://rufwork.hku.hk/
5. Riwaq Homepage. https://www.riwaq.org/support-donate
6. Drawing Matter Homepage. https://drawingmatter.org/the-beaux-arts-tradition/
7. Lackney, J.A.: A History of the studio-based learning model. Educational Design Institute, Mississippi State (1999)
8. Ayto, J.: Movers and Shakers: A Chronology of Words that Shaped Our Age. Oxford University Press, London (2006)
9. Anthony, K.: Design Juries On Trial. Van Nostrand Reinhold, London (1991)
10. Dutton, T.: Voices in Architectural Education: Cultural Politics and Pedagogy. Bergin & Garvey, New York (1991)
11. Flynn, P., Dunn, M., O'Connor, M., Price, M.: Rethinking the Crit: A New Pedagogy in Architectural Education. In: Blythe, R., De Walsche, J. (eds.) 2019 ACSA/EAAE Teachers conference proceedings. Antwerp, Belgium (2019)
12. Luszczynska, A., Schwarzer, R.: Social cognitive theory. In: Conner, M., Norman, P. (eds.) Predicting Health Behaviour (2nd ed. rev. ed.). pp. 127–169. Open University Press, Buckingham (2005)
13. UK Government Homepage. https://www.gov.uk/government/publications/mental-capital-and-wellbeing-making-the-most-of-ourselves-in-the-21st-century
14. EU   Homepage.   https://education.ec.europa.eu/education-levels/school-education/well-being-at-school
15. OECD Homepage. https://www.oecd.org/pisa/pisaproducts/37474503.pdf
16. Raworth, K.: A daughnut for the anthtopocene: Humanitie's compass in the 21st century. Lancet Planet. Health. (2017). https://www.thelancet.com/journals/lanplh/article/PIIS2542-5196(17)30028-1/

Means Oriented or Goals-Oriented Architectural Education

Ahmed K. Ali[✉]

Texas A&M University, College Station, TX 77843, USA
ahali@tamu.edu

Abstract. In a fading material culture, where societies have become far from making and only content with image approximations, architectural education, no exception, is suffering from the absence of visual, haptic, and hands-on knowledge. When pairing the knowledge of materials with its related issues of resource depletion, climate change, and waste, a new seed for teaching architecture is ready to be planted. This paper will focus on a profound distinction between two opposing design methodologies: a conventional method and an unconventional one tested in an architectural design studio environment and can be replicated, reused, and scaled to any architectural education programs worldwide. The aim of this study is to challenge the mainstream model of design in architectural education, provide a resource-based material culture, and highlight the role of design in mitigating climate change. Results shows that the understanding of the properties of materials, evaluating the economic and environmental aspects were critical to achieve the sought learning outcomes.

Keywords: Architectural Education · Material Culture · Haptic Knowledge · Resource reuse · Climate Change

1 Introduction

In today's world, image culture has replaced the material one where everything is becoming digitalized and less experienced. Before the digital revolution, architects, and designers had to physically test their ideas first before implementing them in their work. The current generation of architectural students have become more distant from material culture due to their reliant on digital tools. This separation has created an environment where architects become approximators, not physically analyzing materials, and far less connected with its potentials. This phenomenon seems to exist more in architectural education units that embrace more focus on the digital and less on the analogue. While this study was conducted within an architectural school in the United States, similar observations are shared worldwide. A survey on the status of this phenomena is not part of this study, and therefore left to future investigation.

It might be argued that contemporary digital practices could be influenced by the understanding of materials' properties in a specific way. Since most architects design in a digital environment, designs can sometimes not meet their full potential. When

M. Barosio et al. (Eds.): EAAE AC 2023, SSDI 47, pp. 29–36, 2025.
https://doi.org/10.1007/978-3-031-71959-2_5

understanding a specific material by physically working with it, one can gain a better understanding based on how the material responds to different challenges. For instance, when working with metal, an individual must understand that not all metal has the same characteristics and therefore can't be treated equally. As the case in the automotive industry, sheet metal by-products known as "Offal" when folded, it becomes more difficult to work with due to its stiffening and material memory aspects. Using this material to differentiate between goal-oriented and means-oriented design methods is explained as follows:

Goal Oriented Design. The typical architectural design process within the academic design studio environment has always pushed back materials and methods towards the design development phase. Architecture students are typically guided to start from a given program, site, guidelines, etc. Investigations on materiality, construction, methods, etc. are always left towards the end, and after the design have been finalized. The goal here is to choose the appropriate materials and methods to dress and accomplish the sought design. In an interview, Taeke De Jong, a professor of ecology at the University of Technology in Delft, a leading authority on ecosystems, described two design approaches as "means-oriented" and "goal-oriented." Goal-oriented design is the conventional method in which the goal, or building design, is defined, and every decision is made in fulfillment of that goal. It is not until the design development phase that suitable materials are specified and procured [1].

Means Oriented Design. The means-oriented design methodology, on the other hand, is the opposite process, starting from the means, or materials in our case, available with a less strictly defined end goal. Under this approach, it is necessary to first source and acquires the materials before design starts. Otherwise, uncertainty and potential failure in both sourcing and detailing complicate the process. De Jong stated that most architects are unfamiliar with the means-oriented process and a more structured means-oriented design would be a refreshing change [1]. Similarly, in their book, *Spatial Agency: Other Ways of Doing Architecture*, Awan et al. Made a distinction between the two methodologies and emphasized the role of the architect as "incorporator," the only creative stakeholder in the design and construction process with the potential to transform waste into beauty [2]. Bill Addis, as well, in his book, *Building with Reclaimed Components and Materials*, described the two opposing design methodologies as "normal design" and "design with reclaimed products and materials." He stated that, *"the world of reclamation, reuse and recycling are almost like a parallel universe that is virtually invisible to those familiar only with new construction materials and components"* [3].

The presented study was conducted in a required 15-week graduate level course titled interdisciplinary research-based design studio. The course is intended to small number (10–14) first-year Master of Architecture students who were encouraged to team up and collaborate with a small number (4–6) of students from the College of Engineering registered in an elective sustainable manufacturing course. Both group of students met on a weekly basis along with their two instructors to review the progress of each teamwork. The educational objectives were multi-layered and included but not limited to enhancing interdisciplinary collaboration from design to manufacturing. External experts from the manufacturing industry were involved in midterm and final reviews [4]. The following sections will describe the manufacturing process starting with materials investigation.

The relationship between creative thinking and global problems will be explained in detail using the means-oriented design methodology.

2 Manufacturing Processes

2.1 Materials Investigation

The US manufacturing industry generates approximately 7.6 billion tons of non-hazardous solid waste each year, a large portion of which is either recyclable or reusable [5]. Empirical evidence suggests greater economic, environmental, and societal benefits of reusing industrial waste than recycling it. On average, it costs $30 per ton to recycle industrial waste, $50 to send it to the landfill, and $75 to incinerate it.

The global auto industry generates a steady flow of sheet metal by-products known as Offal [6]. This waste stream produced by its blanking and stamping operations. Offal are consistently sized, corrosion-resistant high-quality irregular shaped sheets of galvanized steel that are produced when windows, doors, and other car body components are stamped out of body panels [7]. Because of their consistent size, shape, and quality, they are valuable for much more than traditional scrap markets. Offal pieces are typically between 0.5 to 3.2 mm thick, have various zinc coatings, and total approximately 1,500 metric tons per year. Promising cost-benefit are expected through the reuse of Offal. One blanking plant in Flint, Michigan generate nearly 40,000 pieces per month in about 11 different shapes and sizes [8].

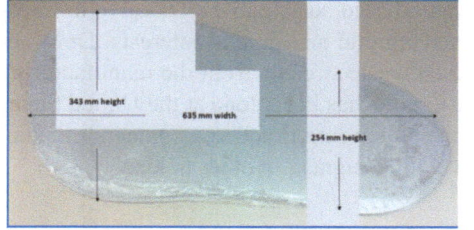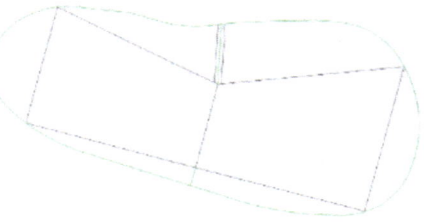

Fig. 1. (model and drawing) 2020 Galvanized sheet metal Offal #8 (left) and the proposed façade panel geometry (right) (Photograph by General Motors and drawing by Jeremy Sims).

Architecture students were asked to study the basic information of Offal and develop a better understanding of its material properties. The irregular shape of the Offal geometry as a by-product of car design parameters, becomes one of the most interesting aspects in this investigation. The transformation of the irregular shapes into a façade-centric paneling system was an educational key moment when a specific application of the building skin became closely tied to the problem of industrial waste. The following case study demonstrates the design process in detail.

In this case study, Offal number eight was utilized to create a faceted paneling system. As seen in Fig. 1, Offal #8 was folded to minimize the overall waste. The design incorporated every square inch of the galvanized sheet metal to maximize the panel size. Using all the surface area of the Offal, helped to hide or secure sharp corners for clean appearance, easier installation, and a safer panel. Figure 2 illustrates the steps associated with the decision made to use more of the material of Offal 8. The A-symmetrical elongated polygon ensured that there was no waste that would have taken place with the more symmetrical shape. The folding diagram to the right shows a simplistic method of folds taken to arrive at the final Offal product.

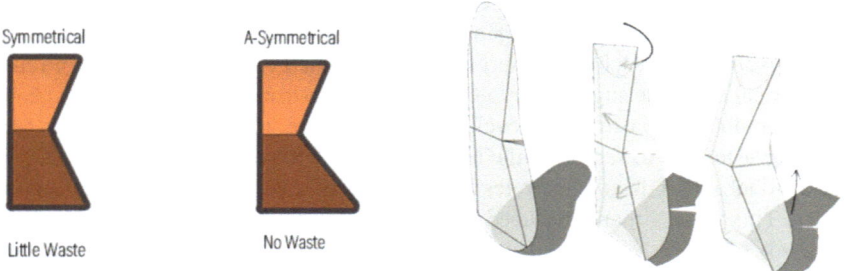

Fig. 2. (drawings) 2020 Maximizing surface area of Offal #8 (left) and folding steps diagram (right) (Drawings by Jeremy Sims)

The next steps in optimizing the Offal and maximizing the surface area to design a façade panel is to test out the design by constructing a full-scale mockup of the panel using the same by-product materials. A seed-planting educational opportunity exists here to introduce, learn, and apply the concepts of Industrial Symbiosis (IS) and circular economy. Circular Economy (CE) is a value-based sustainable alternative to linear economy that connects industries through a symbiotic and mutual interests. Developing building systems and components based on waste streams from the manufacturing industries is a novel approach that very little to none has been done at the United States higher education institutions. Too often in manufacturing practice, engineers may not have the time or opportunity to work closely with designers in other fields.

3   Results

Action-based research, experimental case study, and testing methodology were used during an interdisciplinary research-based design studio setting. The methodology presented here aims to challenge the traditional design process and reverse the materials role in architectural education. The overarching goal through design research is to provide a case study for architects and designers to develop building skin and façade products based on the creative reuse of by-product sheet metal from the auto industry. This approach is sought to help designers evaluate the economic and environmental aspects of their design relative to standardly available market products made from raw materials. While typically the development of a building product is not the responsibility of the architect, alternative materials such as Offal, becomes more convincing when economic savings and positive environmental impacts can be quantified. A holistic life cycle analysis would be necessary to support the evaluation process. Students examined the differences between two design-oriented methods based on the information provided by the industry. The shift in design thinking education occurs when one focuses more on form and less on construction, to deciding on materials and how to turn the waste-flow into real objects. It was critical for the students to first understand the properties of the materials to design according to its limitations. This understanding cultivates a higher sense of responsibility towards resources, the built environment, and the economy of architecture. Additionally, students, with the help of their engineering peers, calculated the total manufacturing process in energy and cost to compare to market products. Fabrication energy used and projected cost included cutting, bending, and punching which are not avoidable in either raw or salvaged materials. Most cost savings reside in the cost of Offal as its value is similar to the value of scrap metal, see Fig. 3. Technical challenges in fabrication led to experiments that provoked new meaning in materials. When challenges arise carefully processed ideas make conversations to solve technical complications. This process can help further understand the challenges within a design. Without challenges there is no cause for creativity if designs are not pushed beyond their bounds.

Total Manufacturing Process

Fabrication Energy

Required Cutting Energy (kJ/m^2)	791.97
Required Bending Energy (kJ/m^2)	17.52
Required Punching Energy (kJ/m^2)	34.21
Total Fabrication Energy (kJ/m^2)	**843.69**

Fabrication Cost

Cutting Cost for labor & machine ($/m^2)	3.18
Bending Cost for labor & machine ($/m^2)	5.17
Punching Cost for labor & machine ($/m^2)	5.84
Total Cost for labor & machine ($/m^2)	**14.19**

Fig. 3. (Tables) 2020 Total Manufacturing process fabrication energy and cost.

4   Creative Thinking Addressing Global Problems

While industry is intimately aware of the disrupting demands imposed by CE, there is a lack of relevant academic initiatives in the US. In 2014, General Motors Company claimed that it generated nearly one billion dollars in annual revenue through reuse and recycling its by-products and avoided releasing over 10 million tons of CO2-equivalent emissions into the atmosphere [9]. This educational-based research intended to accelerate the value that can be added by design to industrial waste-flow and by-products streams. The primary goal of this study was to introduce a new teaching model in architectural education that is based on creative resource reuse of materials. Ultimately, students will be able to apply the acquired knowledge to develop building products, systems, and components with minimal processing of by-products while providing maximum utility, see Fig. 4.

Preliminary data collected from prototypes designed and built by students in the last few years indicates promising energy reduction, reduced heat island effects, water

Expressed Facade Options

Fig. 4. (drawings) 2020 Different façade patterns shown with 35 units (5X7) covering an area of ~ 62 sqf (~5.7 m2). With 1500 Offal generated/month, 43 composition (260 m2) can be produced each month (Drawings by Jeremy Sims).

conservation, and food production. The result of this initiative is expected to positively impact the education, manufacturing, and the building industries through the development of a synergistic closed-loop supply chain of materials through a circular economy approach [10]. The development and testing of this unique educational model for utilizing non-hazardous industrial waste to advance the current knowledge in the fields of green building materials, industrial engineering, and sustainable manufacturing. The prototype approach model develops novel solutions to reuse manufacturing waste by matching its physical and chemical characteristics to the requirements of building elements through student participation. Moving from the individual unit design to a multi-unit façade system is the next step, see Fig. 5. The proposed model can be replicated and applied to other manufacturing industries to open further research possibilities of reusing a wide variety of non-hazardous solid waste.

Fig. 5. (photo) 2022 Fabrication and assembly of multiple metal façade units using real galvanized Sheetmetal by students (Photograph by the Author).

5 Conclusion

Planting a new seed of discovery in architectural education could be achieved through the convergence of material culture and resource awareness. In this new approach, architectural design studios, presents waste-related, theoretical, and real-life challenges to teach creative design thinking to students. While architects typically design and then figure out materials, this approach uses "synergistic means-oriented design" to put the materials first, such as manufacturing waste, and then identify an application to use it. When students are challenged with this type of project-based assignments, they are excited to think about the problem, rather than just the goal of designing a building. When they start with the waste problem, they must learn about things that aren't just architecture. They investigate ecology, manufacturing, steel production, and industrial symbiosis, before they design and start to employ creative design thinking to come up with solutions. The design education allows students to be critical, incredibly creative, and constantly pushing boundaries. As Taeke de Jong iterated *"... You need to do a conversion in your way of thinking to begin to love waste as a material."*

Collaboration between architects, engineers and the manufacturing team can influence the project before the construction process even begins [4]. The initial problems and ideas need to be discussed between every member to make sure that each person from their respected disciplines is on the same page. When the entire team is involved and engaged from the beginning of a project, more likely than not individuals with different learning background will create diverse opinions when it comes to how a project is facilitated. Each profession will bring up different solutions that one discipline might not initially think of until later in the project. Not only does this process save more time but it also increases efficiency. Architects are often viewed as the master of tectonics and add the visual appeal to a project, as where the engineers dive more into the scientific methods of how a project needs to come together. The manufacturing team can implement their knowledge on different desired products and how they will perform. The manufacturing team will also be able to determine whether a product would be a good fit for the assignment. Students in their collaborative experience were able to understand the value of other disciplines to the design and development of products.

Acknowledgment. The author would like to thank General Motors Company for providing the materials used in this study. Thanks also to the Texas A&M University School of Architecture Fabrication Facility at the RELLIS. Special thanks to Jeremy Sims, Dr. Jorge Alvarado, and his Engineering students for their collaboration with architecture students and their contribution to the project presented in this paper.

References

1. Hinte, E., Peeren, C., Jongert, J.: Superuse: Constructing New Architecture by Shortcutting Material Flows. 010 Publishers, Rotterdam (2007)
2. Awan, N., Schneider, T., Till, J.: Spatial Agency: Other Ways of Doing Architecture. Routledge, Abindon, Oxon England, New York (2011)
3. Addis, W.: Building with Reclaimed Components and Materials: A Design Handbook for Reuse and Recycling. Earthscan (2006)

4. Ali, A.K.: A case study in developing an interdisciplinary learning experiment between architecture, building construction, and construction engineering and management education. Eng. Constr. Archit. Manage. **26**(9), 2040–2059 (2019)
5. Ishola, B.: Handling waste in manufacturing: encouraging re-manufacturing, recycling and re-using in United States of America. Procedia Manuf. **39**, 721–726 (2019)
6. Horton, P.M., Allwood, J.M.: Yield improvement opportunities for manufacturing automotive sheet metal components. J. Mater. Process. Technol. **249**, 78–88 (2017)
7. Sheet, S. Car body made from galvanized steel sheet. In: Sheet Metal Forming and Energy Conservation: The 9th Biennial Congress of the International Deep Drawing Research Group, Ann Arbor, Michigan, USA, October 13–14, 1976. American Society for Metals (1976).
8. Ali, A.K., Wang, Y., Alvarado, J.L.: Facilitating industrial symbiosis to achieve circular economy using value-added by design: a case study in transforming the automobile industry sheet metal waste-flow into Voronoi facade systems. J. Clean. Prod. **234**, 1033–1044 (2019)
9. Aguilar Esteva, L.C., et al.: Circular economy framework for automobiles: Closing energy and material loops. J. Ind. Ecol. **25**(4), 877–889 (2021)
10. Shekarian, E.: A review of factors affecting closed-loop supply chain models. J. Clean. Prod. **253**, 119823 (2020)

Architectural Thresholds: Critical Theory as *soglia* in Teaching Architecture and Urban Design

Camillo Boano[✉]

Interuniversity Department of Regional and Urban Studies and Planning, Polytechnic of Turin, DIST, Turin, Italy
camillo.boano@polito.it

Abstract. In Architecture and urban design education, critique it is hard to find anymore: it does seem lost in the call for a renewed disciplinary autonomy, localised territorial interests and the sole artistic and sovereign agency of the maestro: it seems hidden in the sustainability solutionism or reduced to margins, the sole negative with its arrogant tone, stripped from the forms of seminars and discussions and forced to be ancillary to design studios extremisms. The same happen to any form of critical theory, being marxist, relegated to its obsolescence, feminist labelled as activist and anti disciplinar or decolonial, still misjudged as infused in the call of social justice and exoticism. Starting from these assumptions and grounding the reflections in the pedagogical experiences of the author, the paper ask how is then possible to reconfigure critique in the present within a planetary catastrophe in which design is always/already implicated and entangled? What visions of critique are required to intervene into the tangle of ecological, economic, cultural, and sociopolitical conditions of today? What form this could take in architectural and urban design education? Mobilising a Foucault definition of critique as gesture that arrest, disorganise, denaturalise and dehegemonise and expanding it with a partial reading of abolitionist literature.

Keywords: Thresholds · Pedagogy · Destituent · Minor

1 Introduction

This paper was imagined and partially written in the enduring aftermath of Melonism, Orbanism and the transnational rise of authoritarian populism, the emergence of BlackLivesMatter and Fridays for Future, the unrelenting global ecological devastation, the precarities exacerbated by the coronavirus pandemic, the ever present humanitarian crises in war zones like Palestine, Syria, Ukraine and too many more to list here, the visible increase in violence against women, minorities, and in its imperial colonial version the one affecting people on the move across Latin America, Central Africa as well as in the Mediterranean, the intertwined predicaments and pleasures wrought by new technological interfaces, artificial intelligence, and social media, a radical rethinking of the

© The Author(s) 2025
M. Barosio et al. (Eds.): EAAE AC 2023, SSDI 47, pp. 37–46, 2025.
https://doi.org/10.1007/978-3-031-71959-2_6

operations of "care" through trans/queer/crip/feminist lenses, a quite innovative, refreshing and politically engaged Venice Biennale though with a return to the vernacular, the local, the ecological. A landscape that gives architecture, its practice, and its pedagogy (and beyond) a special urgency in reinterrogating positions, conditions, and possibilities of critique and, possibly, a reformulation around its active and situated orientation.

The paper emerges from two interrelated pedagogical experiences. The first one is the "Architecture, Society and Territory B" studio of the Sustainable Architecture MSc Program at the Polytechnic di Torino, where students from all over the world produced architectural research within Anthropocene, beyond its sterile interpretations of 'crisis' focusing on a contested territory and a critical vocabulary fostering the development of a spatial reflection regarding the inhabitation of an uninhabitable world. The second is a Doctoral Course titled "From refusal to abolition. Critical theory and the architecture of livability" developed with Marco Trisciuoglio, a series seminars set out to reflect on the role of architecture and design in the tension between refusal and abolition (of canon, of agency, of author, of site, to the limitless of architecture) engaging directly with an expanded epistemic and geographic with more non-European epistemologies in "the rich fields of global studies…postcolonial, decolonial and settler colonial studies" [1], feminist, political ecologies, more-than-human geographies, new materialism, and black studies.

In the studio we have developed what we called a 'pedagogy of uselessness'[1] an approach that questions how different cultural worlds and realities come into being, how diverse onto-epistemologies encounter one another. The pedagogy of uselessness has two orientations: a) it does not argue for solutions and b) it does not align to the arrogance of utility. Both positions align it to a decolonial practice, and uselessness, approached in the form of a laboratory of imagination, was conceived as political practice that think design through the radical yet taxing power to imagine elsewhere and otherwise, rather than through the evaluative criteria of sustainable development and environmental governance risk prediction calculative figures. Usefulness, was an experiment in critique, was a way to bring critique back in the version suggested by Amy Allen's being something that "refers simultaneously to a tradition, a method, and an aim" [2]. I'm not directly discussing that though.

If it clear that in architecture and urban design education, critique it is hard to find anymore as it does seem lost in the call for a renewed disciplinary autonomy, localised territorial interests and the sole artistic and sovereign agency of the maestro but also hidden in the sustainability solutionism reduced to margins, the sole negative with its arrogant tone, stripped from the forms of seminars and discussions and forced to be ancillary to design studios extremisms. The same happen to any form of critical theory, being Marxist, relegated to its obsolescence, feminist labelled as activist and anti disciplinar or decolonial, still misjudged as infused in the rhetoric of social justice and exoticism. Starting from these assumptions and grounding the reflections in the pedagogical experiences mentioned ask how is possible to reclaim critique in the present within a planetary catastrophe in which design is always/already implicated and entangled? What visions of critique are required?

---

[1] The Pedagogy of Uselessness was a title of a paper developed with Richard Lee Peragine and presented in the Conference Weaving Worlds in TU Delft on the 28th of June 2023.

2   Critique in Crisis

As critique is in crisis in so many disciplines and so many places of public and cultural discourses, more critique is needed everywhere. Banned, rendered superfluous, condemned to irrelevance, to obscurancy and idealism at best. Critique-focused courses have been closed at all latitudes in favour of problem-solving, easy marketable challenges, reading list reduced to the bare twitter-like maximum; intellectual circuits have prefigured even the end of critique beyond its necessity. Critical thinking has limited contrast capacities to the powerful alliance of science, politics, and economy. However, what sort of critical thinking is needed in a time when its very existence seems threatened? When the very essence of liveable futures, existence is at stake? How can we address contemporary issues without repudiating the intellectual legacies of the past and reiterating the very same stultifying language? Very many questions.

Fassin and Harcourt's *A Time for Critique* argue that "the challenges critique faces today call for a reappraisal of its practice, and simultaneously a deepening and a displacement of our own reflection" [3] and appreciate that we are always/already implicated and entangled and how then critique is not an action performed by a neutral subject, philosopher or architect, changing the world, but a more situated position of enquiry and intervene into tangle of ecological, economic, cultural, and sociopolitical conditions of today [4].

The spatiality of globalisation, the continuous redefinition of boundaries between security and insecurity, knowledge and unknowability, certainty, and experimentation, disciplinary and wild, have redrawn what, in science as in politics, has characterised modernity for centuries, making the threshold the most suitable topography to interrogate contemporary spatiality and design. Threshold means many things: the French *seuil* refers to the *solea* of the sandalwood, designating at the same time the movement of passage and grounding; the German schwelle refers to the door lintel and its structural capacity that the verb *schwellen* used by Benjiamin also means 'to swell, swell, rise' and that in the more common English with thresholds also implies the sense of 'to hold back, hesitate, waver' before a territory.

The threshold is not a boundary but an area, an infrastructure, a territory that while contemplating an inside and an outside does not rigidly distinguish them but encompasses them both. With Agamben, it makes them indiscernible. These are not empty, clear, safe spaces but spaces of difference, where we stumble, assemble and clash, are opaque and non-unique. The architectural threshold is therefore a figure to describe and present a reflection on a pedagogy that, borrowing from Deleuze, implies a different way of thinking, as it induces us to identify a potential otherwise production in those spaces that exceed representation, distort cartography while implying a scenography between the sayable and the visible that distorts image and language. An excess that remains unspeakable, suspended almost, interstitial, in-between, with contradictions and aporias. Therefore, the question of thresholds, of the architectural threshold becomes, wherever one poses it, from philosophy to geography, from history to design, a political question and therefore useful for seeking some form of criticism.

Foucault's text *What Is Critique?* do remain to me - and for what I aim to reflect here - central. There, he famously posed the question "how not to be governed like that" [5]. Foucault approaches 'critique' as a technology of the self, a practice involving the subject. In his "very first definition of critique," Foucault characterizes it as "a way of thinking…I would very simply call the art of not being governed or better, the art of not being governed like that and at that cost" [6]. The text then goes on to specify this 'way of thinking' as a 'critical practice' which lies in the 'desubjugation' of the subject itself and is directed against the "movement through which individuals are subjugated in the reality of a social practice through mechanisms of power" [7].

In another text *What Is Enlightenment?* Foucault identified what he called an 'attitude of modernity'. "By 'attitude'" - Foucault explained - "I mean a mode of relating to contemporary reality; a voluntary choice made by certain people; in the end, a way of thinking and feeling; a way, too, of acting and behaving that at one and the same time marks a relation of belonging and presents itself as a task" [8]. According to him, this attitude of modernity would bring together philosophical inquiry and critical thought focused on contemporary historical actuality: "relentless criticism of all existing conditions, relentless in the sense that the criticism is not afraid of its findings and just as little afraid of the conflict with the powers that be" [9]. The idea was that critique must serve as the way to awaken a new sense of human dignity and bring about social change. Critical theory and critical praxis are not activism, and neither are contemplation. They are not abstraction, and certainly are not simple action [10]. Does seem that for my argument, take a definition or an attribute of critique as 'curios activity', as a practice that arrest, disorganize, denaturalize, dehegemonize, both propositions and institutions.

2.1 Critique in the Anthropocene

In the Anthropocene any critical though cannot insist exclusively on solution-based and technical approaches toward solving climate disruption. What happens when things are beyond repair? We can only stay with the brokenness of things. As Thieme puts it, "staying with the trouble is a disposition that gives permission for things not to work (necessarily) and not to be fixed (right away), at least not in the way that adheres to familiar and mainstream metrics of expertise" [11]. Thinking on the planetary technologies of domination, colonial neocolonial, securitarian, extractive, is key to transcend the crisis as in imposition of exception, allow to think and practice a different gaze, perspective and optic from the ordinary people (as the southern city do not correspond to the modernist biased form) that live in such conditions where makeshift forms of life and creative infrastructures have to be coupled with the cumulative dynamic of exclusion, expulsions and deep inequality. How is then possible to reconfigure critique for the present within a planetary (non-innocent) condition where we know that we are always/already implicated and entangled; What visions of critique are required to intervene into the tangle of ecological, economic, cultural, and sociopolitical conditions of today? [12].

3 Sketching an Architectural *soglia*: Minor, Adjacency and Destituent

How can the monumental task of abolishing the present conditions be accomplished without creating a newly terrifying monster in its turn? A minor design. An inversion. It is not another planet, another future, distant or not, but an act of an inverse nature: a reconfiguration of the conditions of possibility. It is an effort of unmaking, of redefinition to resignify territories; ultimately, to undo or deactivate an established territorial order of modernity, security, and escape. A possible avenue, imperfect, and provisional is an appreciation of an architectural thresholds a *soglia* between spaces, figures, and territories, around three gestures of such critique: the minor, the adjacency and the destituent. I'll address them very quickly and in a sketched manner.

3.1 A Pedagogy of Minor Practices, Spaces, and Discourses

A critical theory as *soglia* in teaching architecture should be minor and should be focused on minor practices, spaces, and discourses. Minor is an adjective that qualifies an action. It is a tonality. A differential tone, a reduction. It is a difference of status, of recognition, of measure, of position. Deleuze and Guttuari writing in *Kafka, Toward a Minor Literature* state that 'minority' becomes the category to which one turns in order to "subtract politics from all contradictions and regressions" and that, while remaining in language, it takes on the most disruptive meaning precisely because "major and minor, rather than opposing peculiarities or qualitative indices, are intensive processes that interact reciprocally, modifying the relations of force from time to time prevailing in a given language. The minor language does not tend to replace a major one, nor does it claim its own status, but it acts within it" [13].

Deleuze and Guattari argue that a minor literature means writing in a major language in ways that subvert it from within. Major and minor are not different languages so much as different 'uses' of the same language, not a translation but more of a subversion, escape, or transformation. Not a limited or imperceptible position, nor a language spoken by a minority, rather a metamorphic becoming, a vibrant wave around a transgressive movement. Deleuze and Guattari call this deterritorialization. A term with which they designate a movement of leaving the habitual in a continuous flight from itself, in the oscillation between possibility and impossibility which is never completely resolved.

Erin Manning's *The Minor Gesture* marks the idea of minority as a "gestural force that opens experiences to its potential variations" [14]. The tense relation of the minor to the major is not only because the major is dominant, but because its rhythms are not controlled by a pre-existing structure. The minor is thus resisting to be set aside, neglected, or forgotten in the interaction of the major arrangements. The minor invents new forms of existence and with them, in them, we come to be [15]. These temporary forms of life travel through the everyday, creating untimely structures, activating new modes of perception, inventing languages that speak, as Cindi Katz suggests in *Towards Minor Theory* [16] to reconfiguring the production of knowledge in geography and reposition it as 'situated' and 'interstitial'. Katz explicitly critiques the dominant position suggesting a minor approach that is framed outside the exact dichotomies "of the local versus the global, structure versus agency, class versus gender, culture versus economy" [17]. With Katz minor is not a theory of the margins, but a different way of working with the material. It is about making, finding, elaborating, inhabiting ruptures: a tension out of which something else could happen.

A minor theory can erode the major one with a series of different positions, remaining with its assertions, interstitial "[…] a minor theory is not about mastery but […] its intent is to mark and produce alternative subjectivities, spatialities and temporalities"

[17]. What Katz denounces, is the way theory is made. There is an arrogance to the way major theory is established. "Minor theory" is not a distinct body of theory, proper to specific authors and specific disciplines, so much as a gesture, "a way of doing theory [...] backwards, made up of fugitive moves and interstitial emergent practices [...] Minor theory challenges major theory and celebrates its distinctions, but its movements alter the constitution of major theory as well" [18]. Precisely because "the intention is to recognize and release a multitude of 'other histories vibrating within' the affirmations and arguments of the major theory" [19].

3.2   A Pedagogy of Adjacency and Commitment to Difference

A critical theory as *soglia* in teaching architecture should develop a sort of adjacency due to its power of commitment to alterity.

Tina Campt in her *A Black Gaze* introduce a powerful concept of adjacency. She does so, discussing black artistic works and their visual language. Specifically looking at the 'autoportrait' of Diamond Reynolds, she said "it demands we partake of the labour of adjacency, which requires us to listen attentively to her quietly enthralling image and feel accountable to it. Reynolds's act of defiance was her refusal to remain silent. Yet autoportrait renders her neither speechless nor silent" [20]. Adjacency is "the reparative work of transforming proximity into accountability; the labour of positioning oneself in relation to another in ways that revalue and redress complex histories of dispossession" [21]. And she continues "it is not a gaze restricted to or defined by race or phenotype. It is a viewing practice and a structure of witnessing that reckons with the precarious state of Black life in the twenty-first century. A Black gaze transforms this precarity into creative forms of affirmation. It repurposes vulnerability and makes it (re)generative". Ultimately, the quiet, still-moving-image of Diamond Reynolds's refusal to embrace silence is, to me, a powerful means of reckoning with the impact of the ongoing war currently being waged against Black bodies. It is a still-moving-image of refusal—a quiet refusal to explain, a refusal to capitulate, a refusal to be anything else than who we are, even at the cost of death" [22]. The engagement with art as form or refusal.

We can paraphrase Campt question what constitutes an architecture of refusal? A design of refusal. An attempt to reclaim architectural practice from the capitalist ideology of production. It proposes, instead, that the possibilities of architecture are not determined by the performance-and-deadline-driven excesses of post-industrial society. An architecture of refusal is not a refusal of architecture, but rather a refusal to see building as the only valid architectural response to the question of space, place, and occasion; a refusal to understand architecture as always 'problem-solving'; a refusal to view money as the bottom line. It is an opportunity to slow down and re-evaluate the terms of the discipline, consider the possibilities of an expanded architectural practice, and to participate in the creation of an architectural commons.

An architecture of refusal is close to the 'I prefer not to' voiced by Bartleby as but framed as a mode of engagement that creates the possibility for what Camps describe as a process of 'reassemblage in dispossession': everyday micro-shifts in the social order of racialization that temporarily reconfigure the status of the dispossessed. A fugitivity is not an act of flight or escape or a strategy of resistance.

Adjacency allow us to think agency, and relations as redemption and control, as we are never fully in control of our own processes. Encounters, episodes, inclinations, patterns, become "another way of talking about mediation" as Laurent Berlant [23] poignantly writes. Adjacency allows for different rationalities, moving from linearity, determinism, and functionalism to something more ambivalent, embracing the situation in which we live and from which we must imagine a future of coexistence in and with "the inconvenience of other people" [23]. For Berlant, inconvenience is the affective sense of the familiar friction of being in relation and continuously adapting to these relations. The central element is to acknowledge one's implication in the pressures of coexistence. This condition suggests the importance of "the evidence that no one has ever been sovereign, only that most operate according to an imaginable, often distorted image of their power over things, actions, people and causality" [24]. The inconvenience of other people becomes a pragmatic political topic for any disaster reflection: With whom can you imagine sharing the world's sidewalk? What do you do with the figures of threat and dread that your own mind carries around? Berlant's book is a reflection on 'over-closeness' in the world and how we live with it. Inconvenience is a key concept of this book: the affective sense of the familiar friction of being in relation with no reduction, sustainability, recovery possible. At a minimum, inconvenience is the force that makes one shift a little while processing the world. The important thing is that we are inescapably in relation with other beings and the world and are continuously being adjacent to them.

A possible ambivalence in the pedagogy of architecture maybe forces research and knowledge production to lose its innocence and change its rationalities moving from linearity, determinism, functionalism to something less hard more ambivalent embracing the situation in which we live and from which we must imagine a future of coexistence in an adjacency with 'the inconvenience of other people'.

3.3 A Destituent Pedagogy One that Radically Breaks with the Modern Logic of Sovereignty and Realisation

A critical theory as *soglia* in teaching architecture radically break with the modern logic of sovereignty, realization and try to repair souls in the same act of repairing the world. Destituent was a term/concept used for the first time by the Colectivo Situactiones in Buenos Aires to describe the original features of the Argentine *piqueteros* movement of 2001 which was capable of bringing about real change in Argentina by delegitimizing the existing political forces [25]. More recently, the concept is found in Agamben, who expresses the full force of its political meaning. In his last instalment of the *Homo Sacer* project, *The Use of Bodies*, Agamben [18] suggests that a destituent power is one that "deactivates something and renders it inoperative – a power … without simply destroying it but by liberating the potentials that have remained inactive in it in order to allow a different use of them" and that "while remaining heterogeneous to the system, had the capacity to render decisions destitute and suspend them" [18].

Destituent power is configured as a way of practising and thinking about politics that radically breaks with the modern logic of sovereignty. Consequently, it can be seen as a radical alternative to constituent power in a time that is not about control and sovereignty, but of immanent permanence, albeit in a potential form, of multiple and plural instances of liberation that do not find a solution in institutions (state and non-state), and somehow remain in *écarts* with respect to dominant forms of societal control.

Destituent subjectivities are linked to the idea of politics without foundation (a politics without *arché*). A destituent politics has a limited but precise task: to create the conditions, that is, the vacuum, so that another politics, the one that today seems impossible, can happen. Destituent is a politics not founded by power. Destituent is a "power that deposes power without setting a new one in its place" [26]. It indicates a movement to be made: to unleash a politics of the event. Recently, Agamben suggested that, only if "it is subjected to a decisive critique and if we free ourselves from a concept that has dominated and continues surreptitiously to dominate Western thought and politics: the concept of realisation" [27]. Agamben asks "is it possible to free from its central tenets of realisation?" where realisation is intended as "the idea that political action consists in realising, in facts or deeds, a doctrine, a philosophy, an ideal, a plan, or whatever else one wants to call this sort of obscure presupposition to every political praxis?" [28]. Destituent thinking is something removed from the model of realisation. Realisation is not related to construction, to the building process, or to its materiality, but rather a "matter of rendering it inexecutable" [29]. Desituent here is understood more as a withdrawal, a more radical and certainly more visible response in the present global condition of dispossession. What Stefano Harney and Fred Moten call abolition: "not so much the abolition of prisons but the abolition of a society that could have prisons, that could have slavery, that could have the wage, and therefore not abolition as the elimination of anything but abolition as the founding of a new society" [30]. As The Invisible Committee observes, "what is at issue is nothing less than the repairing of our souls through the very act of repairing our world" [31]. The nature of such a revolution "is no longer merely political or cosmopolitan but anthropological" [32]. For Aarons (and others) the logic is not abstract, rather is the actual intimate relationship between place-making and the politics of destituent power through "affirming another idea of living, which presupposes—literally coincides with—the affirmation of a fragmentary experience of collective dignity" [33]. Weather in the form of desertion, withdrawal, new form of alliances, new form of narrative, support of struggles, the logic of destituent potential is essentially inconsistent with plausible and well-established principles of the grammar of political change. It does not fit with the modern political and juridical canon that sees any new accomplishment in the broader sphere of human social life as the realisation and constitution of orders.

4   Towards Abolition: Hope for Critical Theory

As Tony Fry and Mladina Tlostanova have repeatedly argued, "we must acknowledge the damage done by human designs and how our current design infrastructures keep designing even though we have little knowledge of the ongoing agencies of our designs and the values and knowledges and how they keep designing after we have designed and made them" [34]. To be sure, refusal should not be mistaken as simply passive withdrawal or retreat; rather, they are the active forms of a radically different mode of being and doing. There is something prophetic about abolition; some element of the elsewhere that marks its practice, and its discourse. In the work of undoing, there is a crack. In the refusal, a moment of imagination. When conceived as abolitionist, critique becomes different, not the road to enlightenment, not a route to a new politics but the end as such. A much-needed direction in architecture that consists in the construction of a space, a *soglia*, of "struggle against the arbitrary diktats of autocracy" [35] against absolute truth.

References

1. Challand, B., Bottici, C.: Toward an interstitial global critical theory. Globalizations 1–27 (2022).https://doi.org/10.1080/14747731.2021.1989140
2. Allen, A.: The End of Progress. Decolonizing the Normative Foundations of Critical Theory. Columbia University Press, New York (2017)
3. Fassin, D., Harcourt, B.E.: A Time for Critique, p. 2. Columbia University Press, New York (2019)
4. Thiele, K., Kaiser, B.M., O'Leary, T.: The Ends of Critique. Methods, Institutions, Politics. Rowman & Littlefield Publisher, Lanham (2022)
5. Foucault, M.: What is Critique? In: Lotringer, S., Hochroth, L. (eds.) The Politics of Truth, p. 44. Semiotext(e), New York (1997)
6. Foucault, M.: What is Critique?, p. 45 (1997)
7. Foucault, M.: What is Critique?, p. 47 (1997)
8. Foucault, M.: What is Englightment? In: Rabinow, P. (eds.) The Foucault Reader. An Introduction to Foucault's Thoughts, pp. 34–50, p. 39. Penguin, London (1984)
9. Foucault, M.: What is Englightment?, p. 41 (1984)
10. Harcourt, B.E.: Critic and Praxis. Columbia University Press, New York (2021)
11. Thieme, T.: Beyond repair: staying with breakdown at the interstices. Environ. Plan. D Soc. Space **39**(6), 1095 (2021)
12. Thiele, K., Kaiser, B.M., O'Leary, T.: The Ends of Critique (2022)
13. Deleuze, G., Guattari, F.: Kafka. Pour une Litérature Mineure. Les Édition de Minuit, Pairs, p. 4 (1975)
14. Manning, E.: The Minor Gesture, p. 7. Duke University Press, Durham (2016)
15. Manning, E.: The Minor Gesture, p. 10 (2016)
16. Katz, C.: Towards minor theory. Environ. Plan. D Soc. Space **14**(2), 495–510 (2016)
17. Katz, C.: Towards minor theory, p. 498 (2016)
18. Agamben, G.: The Use of Bodies, p. 274. Stanford University Press, Stanford (2016)
19. Katz, C.: Revisiting minor theory, p. 598 (2017)
20. Campt, T.: The Black Gaze. Duke University Press, Durham (2022)
21. Campt, T.: The Black Gaze, p. 113 (2022)
22. Campt, T.: The Black Gaze, p. 104 (2022)
23. Berlant, L.: On the Inconvenience of Other People, p. 22. Duke University Press, Durham (2022)
24. Berlant, L.: On the Inconvenience of Other People, p. 3 (2022)
25. Laudani, R.: Il Movimento della Politica: Teorie Critiche e Potere Destituente. Il Mulino, Bologna (2016)
26. Carvalho, L.: A violence other than violence. South Atl. Q. **122**(1), 73–86 (2023)
27. Agamben, G.: Destituent potentiality and the critique of realization. South Atl. Q. **122**(1), 9–17 (2023)
28. Agamben, G.: Destituent Potentiality and the Critique of Realization, p. 10 (2023)
29. Agamben, G.: Destituent Potentiality and the Critique of Realization, p. 15 (2023)

30. Harney, S., Moten, F.: Undercommons. Fugitive Planning and Black Study. Minor Composition, Wivenhoe (2013)
31. Comitato Invisibile: L'Insurrezione che Viene – Ai nostri Amici, translated by Marcello Tarì. Nero, Roma (2019)
32. Aaron, K.: Exile and fragmentation: the new politics of place. Philos. Today **67**(2), 395–404 (2023)
33. Aaron, K.: Exile and Fragmentation, p. 401 (2023)
34. Fry, T., Tlostanova, M.: A New Political Imagination Making the Case. Routledge, London (2021)
35. Eagleton, T.: The Functions of Criticism. Verso, London (2005)

How to Tackle Crisis in Architectural Education? Truth or Dare

Hazal Çağlar Tünür$^{(\boxtimes)}$ and Göksenin İnalhan

Istanbul Technical University, Istanbul, Turkey
caglarhazal@gmail.com

Abstract. Architectural practice is influenced by all the dynamics of daily life, such as climate change, pandemics, political changes, economic issues, globalization, and social inequalities, so the roles and responsibilities of the architect have also changed. As a result, the practice and architecture discipline have been compressed into a narrow field, and it can be called a crisis. Some educators act more rationally and dynamically, taking structural and spatial initiatives to work out these crises on the spot by overcoming economic, political, or regulatory conjunctures like design-build studios. Nonetheless, in traditional studio practice, professionals produce intellectual content and projects based on architectural knowledge by addressing these issues in a more philosophical or political on paper. By comparing these two design studio modalities through literature and case studies, this paper will explore how an inclusive, socially engaged, anti-crisis design practice in architectural education can be addressed in the curriculum, how it can find an answer in design pedagogy, and how we can make it sustainable in the future of the discipline. This research will take these two avenues of approaching these crises and compile their potential contributions toward developing a responsive, resilient, and inclusive habitus-of-learning approach for architectural education.

Keywords: Architectural education · design studio · crisis · design-built studio

1 Unveiling Crisis and Architectural Education

Architectural practice is influenced by various dynamics of daily life, including climatic emergencies, pandemics, political transformations, economic issues, social inequalities, globalization, and its challenges. However, in the Anthropocene era, where human-made elements dominate, there is a contradiction in the exclusion of society and other biotas from constructing the urban environment. Architects play a crucial role in re-establishing the relationship between the city and society. The social aspect of architecture has been emerging since the early 1900s, with different causes and solutions. In this Anthropocene age, the roles and responsibilities of architects have significantly transformed. Instead of solely adhering to traditional design practices, architects now find themselves at the intersection of complexity.

© The Author(s) 2025
M. Barosio et al. (Eds.): EAAE AC 2023, SSDI 47, pp. 47–55, 2025.
https://doi.org/10.1007/978-3-031-71959-2_7

They must have proficiency in understanding intricate systems and collaborating with professionals from diverse disciplines. The profession requires continuous self-updating as architects respond to ever-shifting needs, conditions, and the broader socio—economic dynamics shaping our societies. Architects have become proactive learners, constantly seeking ways to enhance their practices in light of the evolving environment. To thrive in this era of rapid change, architects have embraced the philosophy of perpetual learning, recognizing that becoming digitally literate is not just a choice but a necessity They must adapt to emerging technologies, incorporate sustainable practices, and address dynamic user needs [1, 2].

The contemporary architect is a multifaceted professional with creativity, adaptability, and a profound sense of responsibility. This changing role of architects must be addressed from the very beginning, starting with architectural education. The global issues mentioned earlier have compressed architectural practices into a limited field, giving rise to a "crisis" A crisis, in its etymological definition, is a moment when inevitable change prompts a reevaluation of the current situation. It is crucial to address this emphasis on change from the outset, starting with architectural education. The design studio presents an opportunity to incorporate influences from everyday life into the curriculum while adhering to specific criteria.

Moreover, it can deploy practical tools and chart a meaningful course for society, contributing to local spatial production processes. Therefore, it is imperative to consider the transformative potential of education on society as an opportunity and ask the question: How should architectural pedagogy be adapted to restructure the responsibilities of architects in a world undergoing crisis?

In the aftermath of the pandemic, there is a paradigm shift in all realms of education. This study focuses on the role of the design studio within this rapidly changing crisis environment and reexamines the roles that architectural education assumes. As the circumstances of our era demand a reconsideration of design education and pedagogies, this paper explores two distinct approaches to addressing these crises and consolidates their potential contributions towards fostering a responsive, resilient, and inclusive learning environment for architectural education.

2 Educator Approaches: A Spectrum of Responses

In the realm of education, where responses to the ever-evolving challenges of climate change, pandemics, political shifts, globalization, and social inequalities are imperative, this chapter navigates through the diverse landscape of educator approaches, focusing on two distinct paradigms within architecture education: the pragmatic, hands-on ethos of design-built studios and the academically engaged conceptual exploration within conventional studios. Design-built studios are characterized by spatial initiatives rooted in overcoming real-world constraints. In this experience, rational and dynamic ways of building a bridge are chosen by bringing theory and practice closer together. On the other hand, conventional design studios draw upon a diverse spectrum of academic and philosophical insights, guiding spatialization processes. Within this framework, design problems are systematically addressed, undergoing structured resolution as a result of the conceptual discourse.

While acknowledging the nuanced nature of design studios, it is evident that instances arise where these models converge. The study was initiated to select studios from two different models concerned with one of the problems called crisis, to compare and compile their solutions, and to draw certain conclusions from this. The objective is to illuminate their potential in addressing crises and to contemplate the possibilities of a novel approach. This examination aims to shed light on moments of alignment between these distinct paradigms and explore prospective avenues of operation.

In this chapter, it is embarked on a comparative analysis of two distinct design studio paradigms. Through an examination of both literature and practical case studies, we delve into the nuanced differences between design-build studios and traditional studio practices. Our investigation encompasses a detailed exploration of their respective effectiveness, particularly within the context of the current crisis in architectural education.

2.1 Embracing Real-World Challenges: Design-Build Studios

Design-built studios emerge as more than a curriculum—they embody a philosophy. This immersive learning-by-doing approach transcends the traditional boundaries of the studio with a dynamic process. This approach carries an occasionally unpredictable nature due to the multitude of variables inherent within its contextual framework, extending well beyond the confines of the studio. Students not only gain technical knowledge on a tangible and haptic level—encompassing aspects such as site, structure, materials, and building details—but also undergo a genuine building experience involving skills such as communication and negotiation. These processes go beyond the student-teacher binary and include encounters with people outside the school such as user groups, NGOs, other professionals, and contractors. With technical tours and site visits, the construction process goes through many stages such as material selection, fundraising, costing, negotiations with users, marketing, and collaborating when necessary [3].

The foundations of the first 1:1 scale design studio practice in an institutionalized architecture school were laid in the Bauhaus. Considering the Bauhaus' educational mediums and strategies, it can also be claimed that it was the first design-and-build studio [4]. While it began to spread with the Yale Building Project in the late 1960s as a reaction to the aesthetically-oriented, elitist, two-dimensional approaches of the Ecole de Beaux-art, it can be said that many schools today are striving to create new pedagogies and programs in this regard [5]. Based on this view, one of the most iconic of these is the Rural Studio in Alabama which prioritized addressing community issues through environmentally and socially sustainable solutions. Mockbee takes action not only out of a conscientious reflex but also with the idea that the profession should challenge the status quo to make responsible environmental and social changes. He proposes expanding the architectural education curriculum from paper architecture to creating real buildings and instilling a sense of community service [6]. Similarly, Bennett and Reynolds [7] highlighted this studio's possibility of threatening the status quo via social and cultural intervention.

The concept of design-build studios and the pursuit of disrupting the norms can be juxtaposed with Ivan Illich's revolutionary notion of "unschooling" Just as Mockbee assists hands-on architectural experiences, Illich advocates for a paradigm shift in education, encouraging us to question conventional learning methods and embrace experiential, self-directed pathways to knowledge. In this context, design-build studios can be considered as a practice of unschooling, inspiring new and radical ways to tackle the crisis. Illich presents the idea of liberalizing the process of teaching and learning, advocating for a broader spectrum of individuals to take on the role of educators, and advocating for an increase in the availability of educational opportunities [8]. Furthermore, Illich's perspective shares certain resemblances with design-build studios. These studios encompass avenues such as learning through civic engagement, and learning from each other/peer learning, which are notably prominent in design-built studios [9]. These characteristics serve to distribute the power and authority of the instructors to base, thereby reflecting the principles advocated by Illich. The concept of ambiguity, which defines design-built projects as both a weakness and strength, contrasts with the more structured nature of conventional studios. Illich's criticism revolves around the idea that educational systems often prioritize skill acquisition and development through repetitive processes. However, this adherence to predictability becomes problematic when conditions shift, rendering the acquired skills less effective.

Unlike the controlled environment of a studio, where outcomes can be managed, the design-built project atmosphere is marked by a multitude of inputs that introduce unpredictability. It offers a platform for students to navigate and embrace uncertainty, thereby cultivating skills beyond the boundaries of expectables. While traditional institutional education may be criticized for its potential limitations in preparing students to face the ever-evolving challenges of the professional world, design-built studios stand as a testament to the effectiveness of learning through ambiguity. On the other hand, the open-ended nature of these projects can make it difficult to define clear goals and criteria for success. This ambiguity might pose challenges when assessing student performance and evaluating project effectiveness. Moreover, the iterative design and construction process inherent in such projects might extend project timelines, potentially clashing with academic schedules and resource constraints. Risk management also becomes critical, as ambiguity can lead to unexpected design changes, unforeseen challenges, and construction delays. When all of these challenges are addressed, the will to make this system maintain for many years may break down [10]. According to Illich, the school combines the student's growing up with a sense of weakness stemming from ignorance with the obligation to make a humiliating commitment to the teacher. This can actually be linked to Chris Argyris[11]' mystery—mastery term, which has been much debated in architectural education. On the other hand, creating a tangible product, seeing it being used, or solving a problem, can distract students from this feeling. In design-built studios a top-down hierarchical scheme of education is replaced by the side-by-side that collective productions require. Therefore, the unquestionable positionality of the educator disappears.

Power issues in education can be reviewed not only in terms of educator roles but also in terms of cultural trends, iconic figures, and even East-West examples and approaches. Golzari et al. [12] says they have developed various teaching tactics to raise awareness of this issue. Instead of the usual master plans, there should be encouragement to get closer to the local, 1:1 scale, contextual research instead of form, and study of the neighborhood are some of them. They also advocate low-cost, low-technology environmental and economic design. In this way, they have taken a step to relieve their concerns about climate change. In essence, the benefit of focusing on the environment and ecology is the research of the local conditions of the design site and the development of a design that respects them. They add that while rejecting global icons through collaborative work on site, they emphasize the cultural realities and socio-economic needs of silent or invisible urban communities.

One of the most important achievements of such projects is that the academy creates points of opening up to the community, where not only the student but also the community meets and reproduces knowledge. Working with a diverse group can be seen as breaking of environmental uniformity for the student's world, enriching perspective and natural development of critical thinking. As cited by Guaita and others [13], the process of 3 Swiss architecture schools invited to work on a building project in Open City El Portico in 2014 is described as follows: drawing and construction are connected through the construction process. They are creating tacit and embodied knowledge at this moment, obliging students to critically engage with all the realities of the building (material, human, and temporal), creating a circular process that also connects the community and the environment. Another essential achievement emphasized in their article could be the reinterpretation of techne and poesis through the design-built project. Interpreted as a fusion of techne (technical skill) and poiesis (creative expression), students' time and effort in construction underlines the synthesis of practical and creative elements in the learning process. The ongoing construction process serves as a platform for continuous learning and knowledge production. This implies that the act of building itself contributes to the design process and provides opportunities to refine and redefine the project's potential. Hence, there is an understanding of the continuous learning inherent in construction processes and the integration of technical skills with creative expression for a holistic educational experience [13].

2.2 Tradition Evolved: Inquiring the Traditional Studios

In education, each discipline possesses its unique characteristics, requisites, and principles. Architectural education, for instance, places a strong emphasis on observation, tactile experience, and physical perception. A cornerstone of this type of education involves learning by doing within a studio environment, where practical application is the key. Historically, architectural education evolved from the master-apprentice relationship into more formalized structures. The Ecole de Beaux Art is an early and influential example, infusing architectural education with a corporate identity and laying the groundwork for the studio culture. This model and its associated culture have been disseminated globally, occasionally transforming. Many subjects specific to the design studio, such as the unquestionable authority established by the critics, long working hours, and juries, are Ecole De Beaux-Art's legacy [14]. In the 1920s, critiquing the Ecole de Beaux Art for its detachment from the human dimension, excessive emphasis on aesthetics, and pedagogy focused on competition victories rather than learning prompted a new European approach to education, culminating in establishing the Bauhaus. The Bauhaus curriculum prioritized collaborative work, innovation over imitation, and the practical design and creation of tangible structures within workshops and laboratories. The school's holistic design approach offered students an academic atmosphere and an immersive living environment, blurring the lines between studio and life [15]. These two foundational institutions have evolved yet retain relevance in various aspects of architectural education.

To grasp the essence of the design studio, one must go beyond surface-level definitions and narrow interpretations. Two pivotal institutions from history, the École des Beaux-Arts, and the Bauhaus, form the foundation of the modern design studio, each with its unique approach. These institutions continue to exert significant influence, shaping aspects like curriculum organization and preserving traditions. Institutions like the École des Beaux-Arts and the Bauhaus have a lasting impact on the pedagogical approaches of the modern design studio. The Bauhaus, for instance, encouraged hands-on problem solving and practical experimentation, as reflected in Donald Schön's depiction.

Similarly, the concept of the hidden curriculum, as highlighted by Dutton [16], is seen in the perpetuation of traditions and rituals within the design studio, maintaining connections to broader cultural, social, and power dynamics. It also touches upon the impact of the hidden curriculum on relationships within the studio and its hindrance to learning, a viewpoint shared by Jeremy Till [17] and Garry Stevens [18], who criticize the hierarchical and elitist norms in architectural education.. In the evolving landscape of architectural education, there's a growing recognition of the need to address pressing social and humanitarian challenges. The following examples show pioneering design studios committed to making a meaningful societal impact through innovative and socially conscious architectural practices.

It could be argued that the pedagogical approach of the educator and many decisions regarding the project are, in a sense, interconnected by an invisible thread. Elitist and masculine attitudes in education can play a decisive role in every aspect of the studio, from the choice of the project topic to the selection of the project area and the target audience. For this reason, the infiltration of radical pedagogies into the design studio, which has become highly visible, especially in the previous few years, has had a very positive impact. Within these concerns, the global housing project at Tu Delft, as reported by Mota and Gameren [19], is a multicultural project that covers specific challenges addressing the housing crisis and socio-spatial inequalities. The teaching method employed in the Global Housing studio aims to address the widespread uniformity in thinking by fostering the growth of critical awareness concerning evolving social dynamics, challenges, and experiences in an ever-changing world. One of the approaches that both reinforces inclusivity and proposes an alternative by subverting the ongoing power relations within the studio is the autobiographical spatial narratives that Aykaç [20] uses as a radical pedagogy in the design studio. Studio work of this nature, conducted through participatory processes that prioritize inclusivity over dominance, holds the potential to challenge prevailing power dynamics. Simultaneously, it addresses pressing social issues by deeply comprehending and assimilating the nuances of the local context.

In today's Anthropocene age, a project that can develop a solution via social/humanitarian problem at an urban scale and within a specific time limit should be recognized as the success and sensitivity of the design studio. Conventional design studio provides a controlled and focused environment where students can immerse themselves fully in the design process. This focused setting allows for deeper exploration and experimentation, fostering a more comprehensive understanding of design principles and strategies. Particularly those grounded in ethnographic narratives have the potential to spark self-reflection and stimulate fresh inquiries into the students' core identity. In doing so, they can lay the foundation for a new portrayal, dismantling the conventional archetype of the architect—a figure often burdened with lofty titles like star or leader because this traditional identity is increasingly under scrutiny, particularly within the context of radical pedagogies. The studio format encourages iterative design processes, where students can continually refine their ideas and solutions. This iterative approach enables students to explore various design options, leading to more robust and refined outcomes. Studios often integrate various skills, from concept development and visualization to presentation techniques. This holistic approach equips students with a well-rounded skill set that applies to various aspects of architectural practice. And maybe the most potent part as the conventional studio encourages the exploration of conceptual ideas and theories, allowing students to delve into design philosophies and ideologies that may not be as readily addressed in live projects with practical constraints. Students in a conventional studio have greater freedom to experiment and take creative risks without the immediate real-world consequences that live projects may entail.

3 Speculating for Building an Inclusive, Socially Engaged, Anti-crisis Design Practice

The traditional studio and the design and build studio, often perceived as divergent educational models, possess a nuanced and interrelated nature that challenges categorical distinctions. They share common ground in experiential education, enabling both to engage with real-world problems. While design-build studios operate amidst uncertainty and actual variables, traditional studios function within more controlled environments and established frameworks. Both models impart valuable skills, yet a distinction arises regarding knowledge transfer: traditional studios emphasize a defined, direct, and hierarchical approach, while design-build studios foster multidirectional and unpredictable information flow. Active participation remains integral to both models. Upon comparing these two educational approaches, it becomes evident that design-build studios play a vital role in addressing crises. Therefore, recognizing the distinctions between vocational schools and academies, it is essential to incorporate design-build studios into the curriculum. However, implementing such projects in a curriculum necessitates thorough administrative preparation. Communication, technical details, site arrangements, and resource allocation are among the initial prerequisites. Consequently, educators must undergo training and establish networks before students engage in these endeavors. Furthermore, defining assessment criteria poses a challenge in this context. Methods such as observation, self-assessment, and peer assessment gain prominence. However, the inherent uncertainty of the process complicates the objectivity and fairness of assessments. Effective architectural practice necessitates strong communication skills. However, within the educational setting, communication typically occurs primarily between students, instructors, or peers. Live projects introduce an element of unpredictability, fostering unexpected encounters and enhancing understanding of diversity. This sense of community engagement can contribute to addressing specific issues related to crises. In contemporary society, globalization's rising tide of individualism has led to passive citizenship. Collaborative endeavors that bring individuals together to create not only benefit students but also educate communities about social inequalities. It is essential to leverage local knowledge without romanticizing or fetishizing it.

Today, with the advent of artificial intelligence, access to academic knowledge has become more attainable, thanks to the ability to access information in a synthesized and categorized manner, provided the suitable script is applied. Therefore, incorporating personal narratives and indigenous wisdom into the studio environment extends beyond technical and theoretical realms, enriching students'learning experiences with diverse perspectives. The role of academia extends beyond the education of its students. Indeed, one of the most crucial dimensions of live projects, particularly within underdeveloped societies, is the transmission of architectural culture to the broader community. These projects facilitate the public's comprehension of the distinction between exemplary and substandard architecture, promote mutual learning, and enable society to engage with this culture, even from its periphery. Consequently, live projects hold significance in fostering a society that actively seeks superior, inclusive, humanitarian designs globally.

In establishing an educational practice resilient against crises, one of the most crucial means of fostering awareness of these issues involves the early internalization and unwavering ommitment to professional ethics within the educational domain. Ethical considerations must be addressed unequivocally and comprehensively. An all-encompassing approach to ethics is imperative. For instance, it is considered unacceptable for a building to prioritize clean energy while neglecting accessibility for individuals with disabilities. Similarly, a project cannot claim humanitarian values while exploiting the labor processes involving stakeholders and architects. The acceptance of a merely superficial commitment to ethics should be discouraged.

Contemporary discussions underscore the prominence of radical pedagogies within the educational landscape. Integrating these approaches into the studio allows for the introduction of stimulating topics and mini-projects. It is imperative to acknowledge that the era of a monotonous architectural culture has passed. Instead of adhering to a single narrative, canon, or myth, embracing diversity by respecting individual uniqueness, worlds, lives, and stories becomes essential. From this perspective, valuable lessons can be derived. Allocating space within the studio for students' narratives and fostering a pluralistic attitude may yield more attuned and socially aware graduates.

References

1. Nicol, D., Pilling, S.: Changing Architectural Education: Towards a New Professionalism. Taylor & Francis, London and New York (2005)
2. Meyer, M., Norman, D.: Changing design education for the 21st century. J. Des. Econ. Innov. **6**(1), 13–49 (2020)
3. Morrow, R.: Live Project Love: Building a Framework for Live Projects, Architecture Live Projects: Pedagogy into Practice. Routledge, London and New York (2014)
4. Lonnman, B.: Constructing design in the studio: projects that include making. In: 98th ACSA Annual Meeting Proceedings, pp. 67–76. New Orleans (2010)
5. Canizaro, V.B.: Design-build in architectural education: motivations, practices, challenges, successes and failures. ArchNet-IJAR Int. J. Archit. Res. **6**(3), 20 (2012)
6. Oppenheimer, D.A.: Rural studio: Samuel Mockbee and an architecture of decency, pp. 34–38. Rural Studio Samuel Mockbee and an Architecture of Decency (2014)

7. Bennett, B., Reynolds, R.: A Pedagogical Gap Architecture Live Projects: Pedagogy into Practice. Routledge, London and New York (2014)
8. Illich, I.: Deschooling Society. Harper & Row, New York (1971)
9. Harriss, H., Widder, L.: Architecture Live Projects: Pedagogy into Practice. Routledge, Oxford (2014)
10. Harriss, H.: Architecture Live Projects: Acquiring and Applying Missing Practice Ready Skills. Royal College of Art, London (2015)
11. Argyris, C.: Some limitations of the case method: experiences in a management development program. Acad. Manag. Rev.Manag. Rev. **5**(2), 291 (1980)
12. Golzari, N., Fraser, M., Sharif, Y.: Learning by praxis: rethinking architectural pedagogy through hybrid cross-cultural design research. In: The Routledge Companion to Architectural Pedagogies of the Global South, pp. 343–355. Routledge, Oxford and New York (2022)
13. Guaita, P., Baur, R., Jolly, D., Jolly, V.: "El Pórtico de los Huéspedes": exploring other ways of building at the Open City in Valparaiso, Chile. In: The Routledge Companion to Architectural Pedagogies of the Global South, pp. 394–406. Routledge, Oxford and New York (2022)
14. Fisher, T.: In the Scheme of Things: Alternative Thinking on the Practice of Architecture. Minnesota Press, Minneapolis (2000)
15. Anthony, K.: Architecture school: three centuries of educating architects in North America. MIT Press, Cambridge (2012)
16. Dutton, T.: Design and studio pedagogy. J. Archit. Educ. **41**(1), 16–25 (1987)
17. Till, J.: Architecture Depends. MIT Press, Cambridge (2009)
18. Stevens, G.: The Favored Circle: The Social Foundations of Architectural Distinction. MIT Press, Cambridge (2002)
19. Mota, N., Van Gameren, D.: Dwelling beyond cultural differences: architectural education for peripheral urbanization in Bangladesh, Ethiopia and India. In: The Routledge Companion to Architectural Pedagogies of the Global South, pp. 419–432. Routledge, Oxford and New York (2022)
20. Aykaç, G.: The possibility of radical resources and participation in architectural education: autobiographical spatial narratives. Archnet-IJAR: Int. J. Archit. Res. (2023). https://doi.org/10.1108/ARCH-05-2023-0130
21. Leach, N.: Architecture or Revolution. Routledge, London (1999)
22. Online Etymology Dictionary Homepage. https://www.etymonline.com/search?q=crisis
23. Tschumi, B.: Architecture and Disjunction. MIT Press, Cambridge (1996)

Demolition(s) in Question: A Pedagogical Approach Case Study: Toulouse Le Mirail

Isabel Concheiro Guisan[✉]

Haute école d'ingénierie et d'architecture de Fribourg, HEIA-FR Joint Master of Architecture Fribourg – JMA-FR, Fribourg, Switzerland
isabel.concheiroguisan@hefr.ch

Abstract. The transformation of non-heritage protected mid-twentieth century European architecture, especially public housing, is a key contemporary issue that raises several architectural, environmental and social questions. While several examples show the potential for transformation, demolition is unfortunately still considered commonplace, leading not only to the loss of architectural heritage and affordable housing, but also to the destruction of communities built over decades. In this context, architects have an important role to play in proposing alternatives. Within the framework of architectural schools, and in addition to design-based strategies, it seems necessary to confront students as well with a critical analysis of complex demolition processes in theoretical courses. The course "Transformation Strategies" at the HEIA-FR proposes collective research aimed at questioning demolition processes, and at developing a sensitive approach and methodology to reveal architectural, environmental and social values as a basis for the transformation of mid-twentieth century architecture. The first case study aims to question the ongoing demolition process in the Toulouse Le Mirail district, designed by Candilis, Josić and Woods in the 1960s.

Keywords: Demolition · Transformation · Heritage · Value · Public Housing

1 Demolition(s)

"[The book's] concern was with why society does not value its good buildings, to try and think out why good buildings, even when they—like new housing—are something society needs, are subject to senseless destruction." Peter Smithson [1]. The destruction of significant buildings without heritage protection is not a new phenomenon in the history of architecture, ranging from the demolition of public buildings such as John Soane's Bank of England in the 1920s, Victor Baltard's Halles in Paris in the 1970s, to the ongoing demolition of mid-twentieth century architecture, especially public housing, such as Allison and Peter Smithson's Robin Hood Gardens or the *ville nouvelle* de Toulouse Le Mirail by Candilis, Josić and Woods, to name just a few.

Beyond the specific reasons behind the destruction of a given building, we can argue that demolition often implies a lack of temporal distance to recognize the architectural and cultural value of buildings [2], coupled with an apparent *need* to replace somehow

M. Barosio et al. (Eds.): EAAE AC 2023, SSDI 47, pp. 56–63, 2025.
https://doi.org/10.1007/978-3-031-71959-2_8

obsolete buildings with *up-to-date* architecture that responds to contemporary functional, aesthetic, or financial needs [3].

In the case of post-war public housing, we could add a strong stigmatization of certain neighborhoods related to systemic social problems, together with a lack of maintenance and the weakening of public investment in favor of the private real estate market since the 1980s. This combination of factors leads to the devaluation of a heritage generally subject to strong real estate pressures, and fuels political discourses that justify demolition as a "solution" for urban renewal [4].

Although many post-war housing projects in Europe embodied innovative and experimental approaches, and several examples demonstrate their potential for transformation through different strategies of energy renovation, typological reinterpretation, enhancement of original qualities or residents' participation—such as Tour Bois le Prêtre (Paris, 2011), Kleiburg (Amsterdam, 2016) or Le Lignon (Genève, 2022)—, demolition is unfortunately still considered commonplace. This results not only in the loss of architectural heritage, much-needed affordable housing, and a major source of embodied carbon, but, most importantly, in the destruction of communities built over decades.

In this context, architects have an important role to play in countering the lack of architectural and social values that generally underpin demolition policies, particularly in relation to public housing. Far from a nostalgic or purely preservationist approach, or a lack of recognition of failures to improve, the interest in this question—both "aesthetic and ethic" [5]—, lies in recognizing the architectural, social, and environmental value of what already exists, and the potential of architecture to be transformed.

In addition to design-based strategies, architects can also contribute to challenging demolition processes and building new narratives through research, publications, photographic essays, films [1, 4, 6, 7], or other forms of collective action, bringing the "not to demolish strategy" [8] to the forefront of the architectural and political debate.

2 Toulouse Le Mirail and France's Urban Renewal Policy

"Le Mirail was an incredible architectural project, studied by universities all over the world. To demolish it would be catastrophic. We can restore, reimagine. It's done elsewhere, but why not here? " Cathérine Beauville [9].

In 1961, the project by Candilis, Josić and Woods won the international competition to build Le Mirail, a new district of Toulouse for over 100,000 inhabitants. Three neighborhoods of the original project were built—Le Mirail with the university, Reynerie and Bellefontaine.

The project was defined by the relationship between three different housing typologies—the *villas patio*, the *petits collectifs* and the *tripodes*—and by the articulation of various types of public spaces—the *dalle,* a continuous pedestrian space connecting the three neighborhoods, the *coursives,* open-air elevated corridors connecting different buildings, and a system of natural public spaces between the buildings [10].

Although since the 1980s, the increase in social problems and the lack of maintenance have led to a strong stigmatization [11], Le Mirail bears witness to the modernist idea of "housing for all" [12] and of "a planning culture with a social idea as its generative core" [13]. Far from being obsolete, the original qualitative value of the project, both at the architectural scale (housing typologies) and at the landscape scale (common green spaces), is more than relevant today [14]. Both the quality of housing and the positive environmental impact of natural spaces in the city are at the core of today's so-called "sustainable neighborhoods". (see Fig. 1).

Fig. 1. The Messager and Poulenc buildings and the natural public space threatened with demolition, Toulouse Le Mirail, 2023 (Photograph by Adam W. Pugliese, 2023).

Since 2003, the district has been part of France's National Urban Renewal Program (PNRU) developed by the National Agency for Urban Renewal (ANRU), which aims to transform 600 so-called "fragile neighborhoods" through the rehabilitation, demolition, and construction of new housing, by both public and private actors. By 2030, this program will have financed the demolition of 270,000 social housing units nationwide [15].

In the case of Le Mirail, the Reynerie and Bellefontaine neighborhoods have been facing the demolition of a significant part of their housing and public spaces for almost twenty years and are currently facing the planned demolition of 1,421 dwellings, along with 780 trees and 3.7 hectares of green space [16] (see Fig. 2). Le Mirail university, designed by Candilis, Josić and Woods in 1964, was demolished and replaced by a new building. Although this demolition affects the unprotected heritage of Candilis, Josić and Woods, it was done independently of the ANRU's urban renovation policy and is therefore not the subject of this paper.

While these so-called "urban renewal" processes could be an opportunity to improve the existing neighborhoods based on their architectural and social qualities, prioritizing the preservation of existing buildings and communities, questioning certain aspects of the original projects and transforming them in collaboration with the inhabitants, they are rather an opportunity to implement a top-down demolition policy with little regard for the loss of heritage, environmental impact, and social consequences.

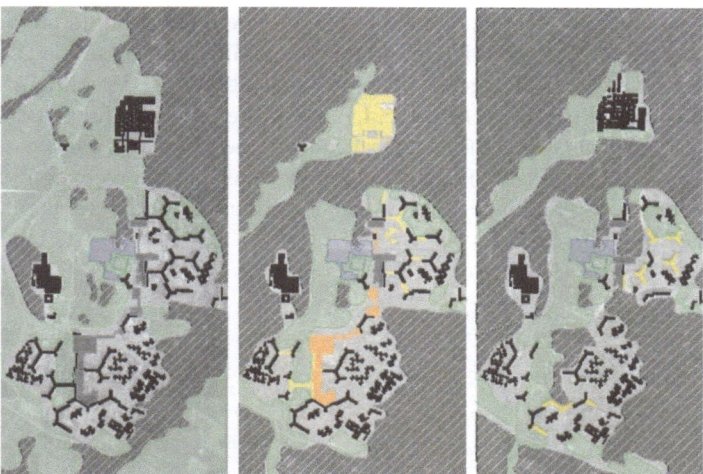

Fig. 2. Natali Céspedes, Estelle Delavy, Gabriel Dos Santos, Vincent Dumont, Alvina Ferrera, Samara Zuber, Toulouse le Mirail built project, demolitions since 2000s and planned demolitions, 2023 (drawings by students © 2023 HEIA-FR/JMA-FR, Tracés).

In addition to Le Mirail, several social housing projects in France of recognized architectural quality but not listed, are now threatened with demolition, such as La Butte-Rouge garden city in Chatenay-Malabry by Bassompierre, de Rutté, Sirvin and Arfvidson (1931), the Époisses in Beçanson by Maurice Novarina (1967), or La Maladrerie in Aubervilliers by Réné Gailloustet (1975), to name just a few, illustrating the main impact of this policy on the destruction of the social housing heritage of the twentieth century.

In this context, several local collectives and associations of residents, architects and social actors, among others, have come together to prevent further demolitions, creating a national collective called "Stop aux demolitons ANRU" and launching a petition calling for a "moratorium on demolitions", opening a necessary debate on the urgent need to move beyond *tabula rasa* approaches.

3 Pedagogical Approach

The course "Transformation Strategies" at the HEIA-FR aims to address this issue in architectural education by proposing a collective research, exploration and analysis to challenge the demolition and identify the potential of transformation of mid-twentieth century architecture, particularly social housing, by recognizing its architectural, environmental and social existing values.

Through documentary and field research, the students' work, developed since the spring semester 2023, aims, as a first case study, to make a sensitive inventory of threatened buildings and public spaces in the Reynerie and Bellefontaine districts of Toulouse Le Mirail.

This pedagogical approach is based on the following six principles that aim, on the one hand, to broaden the analytical scope and tools for envisioning transformation projects based on the highlighted qualities, and, on the other hand, to provide the students with a methodology aimed at questioning demolition processes and contributing to counterbalancing biased images that are often used to justify demolition.

Revealing the cultural and historical value of the project. Through documentary sources, the students analyzed the historical context, the competition process, the architectural and urban principles, the construction systems and the reception of the work at the time, leading to highlight the architectural principles of the project that are still present today. (see Fig. 3).

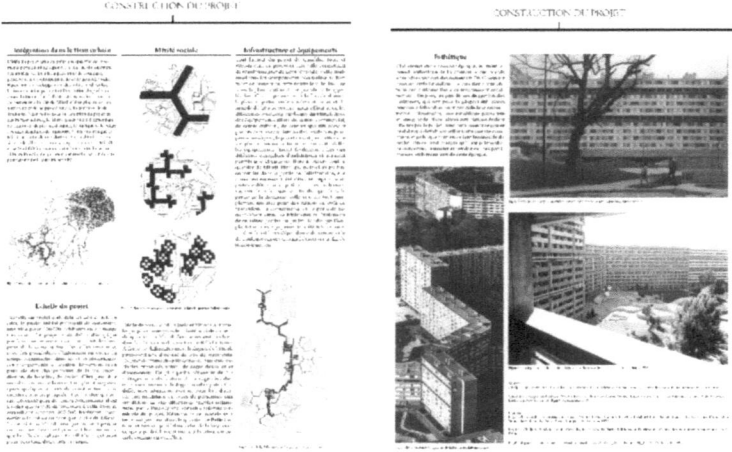

Fig. 3. Fabian Billon, Coline Bonnafous, Marie Donzé, Issa Kashivagui, Francis Labhard, Maxence Launay, Rémi Mauduit and Stéphane Vallon, Analysis of the cultural and historical value of Toulouse Le Mirail, 2023 (work by students © 2023 HEIA-FR/JMA-FR).

Revealing the landscape value of the neighborhood. Through the graphic analysis of the evolution of the project, the students analyzed the landscape principles of the original project in relation to the natural conditions of the site, and their potential in the transformation of the neighborhood. (see Fig. 4).

Revealing the architectural and typological value of the dwellings. Through the redesign of the buildings, the students identified the different housing typologies, understanding their spatial and structural qualities and principles that allow for potential transformations.

Revealing the *lived experience* [16] value of the inhabitants. Through visits, photographies and video interviews with residents, the students highlighted on the one hand, the perception of the quality of life in the apartments and the neighborhood, and on the other hand, the impact of twenty years of demolition, disinvestment and degradation, and the form of institutional social violence induced by this situation. (see Fig. 5).

Revealing the environmental value of structures and natural areas. Through LCA (Life Cycle Analysis), the students will be able to quantify the embodied grey

Fig. 4. Marine Hayoz and Killian Piguet, Analysis of the existing landscape and landscape concept of Toulouse le Mirail, 2023 (work by students © 2023 HEIA-FR/JMA-FR).

Fig. 5. Pierre Crevoisier, Alexandre Olszak and Corentin Weber, Revealing the *lived experience* value of Toulouse le Mirail during a video workshop with Maxime Faure and Adam. W. Pugliese, 2023 (work by students © 2023 HEIA-FR/JMA-FR).

energy of the building structures to be demolished and the environmental impact of demolition policies.

Uncovering the key stakes in the demolition process. Through the analysis of the public inquiries and institutional communications, the students begin to analyze the political, economic and legal issues and the role of different actors behind the demolition process.

4 Conclusion

Questioning demolition and redefining the notion of value should become central themes in architectural education and in the work of architects in the years to come.

In the absence of heritage protection, the architecture of the second half of the twentieth century is potentially under threat, especially that which has been stigmatised and considered worthless. In this context, some of the outcomes of this pedagogical experience will be the following:

To develop a critical approach to complex and inextricable demolition processes as future professionals, by being confronted with the contradictions and the consequences of these processes through an in-situ and in-depth approach.

To reverse biased representations by raising awareness of the intelligence of the original projects, developing a sensitive and holistic view to highlight the existing values on which to build on contemporary transformation projects.

To build alternative narratives and publicly challenge demolition-oriented decision-making processes by placing architecture and social issues at the center of the public debate on urban regeneration processes.

To contribute to the development of an expanded notion of heritage, that embraces the architectural, environmental and social values of both extraordinary and ordinary architecture.

Historically, the European city was built by reinterpreting the existing architecture in the light of the issues of a given era. In the face of climate change, it is essential to build the city on its architectural and social heritage, revealing its qualities and imagining new projects, discourses and levers for action, to transform cities in a truly sustainable way.

References

1. Smithson, A., Smithson, P.: The Euston Arch and the Growth of the London, Midland & Scottish Railway. Thames and Hudson, London (1968)
2. Pevsner, N.: Foreword. In: Smithson, A., Smithson, P. (eds.) The Euston Arch and the Growth of the London, Midland & Scottish Railway. Thames and Hudson, London (1968)
3. INA. https://www.ina.fr/ina-eclaire-actu/video/caf93022331/baltard-j-achete. Accessed 28 Oct 2023.
4. Thoburn, N.: Brutalism as Found Housing, Form and Crisis at Robin Hood Gardens. Goldsmiths Press, London (2022)
5. Smithson, A., Smithson, P.: The new brutalism. Archit. Des. **1957**, 113 (1957)
6. Faure, M., Pugliese, A.: Les insulaires. France (film) (2021).
7. Miah, K.: Lived Brutalism: Portraits at Robin Hood Gardens. London (photographs) (2016)
8. Arc en rêve. www.arcenreve.eu/rencontre/ne-pas-detruire-est-une-strategie. Accessed 28 Oct 2023.
9. actuToulouse. https://actu.fr/occitanie/toulouse_31555/au-mirail-des-habitants-de-l-immeuble-le-plus-vetuste-de-toulouse-on-fait-tout-pour-nous-faire-partir_47892373.html Accessed 28 Oct 2023.
10. Candilis, G., Josic, A., Woods, S.: Toulouse-Le Mirait Geburt Einer Neuen Stadt La Naissance d'une ville Nouvelle. Birth of a New Town, Karl Krämer, Stuttgart (1975)
11. Fernández Salgado, C.: La construction de la stigmatisation: effets de la postmodernité sur l'image des derniers grands ensembles de logements sociaux au travers de l'étude de cas de Toulouse Le Mirail. Doctoral Thesis ETSAM, UPM, Madrid (2020).

12. Candilis, G.: Bâtir la vie Un architecte témoin de son temps. Stock, Paris (1977)
13. Asabashvili, L.: No title. In: Bodrožić, N., Šimpraga, S. (eds.) Motel Trogir: It Is Not Future that Always Comes After. Slobodne veze/Loose Associations, Zagreb, Onomatopee, Eindhoven (2016)
14. Girometti, L., Leclercq, F.: Rapport de la mission sur la qualité du logement. In: Référentiel du logement de qualité, pp. 41–55. Ministère du logement, Paris (2021)
15. ANRU. https://www.anru.fr/le-programme-national-de-renovation-urbaine-pnru. Accessed 28 October 2023.
16. Collectif d'architectes en défense du patrimojne de l'équipe Candilis au Mirail. Le Mirail, un patrimoine à réhanilier au sens propre et figuré. In: Obsolescence et programmes en danger. Journée d'études de Docomomo France, Paris (2023).

Studio Life: Mechanisms of Competition and Collaboration in Architectural Labour Processes

Camilo Vladimir de Lima Amaral(✉)

Politecnico di Torino, 10125 Turin, Italy
camilo.delima@polito.it

Abstract. The aim of the current study is to explore creative labour production and reproduction processes in architectural design studios in Brazil, UK, Belgium and Italy. In contrast to free autonomy narratives, participant observation has evidenced three conflicting mechanisms in creative labour processes, namely: subjectification, distinction and hierarchical expropriation. This issue is pivotal in a transition period-of-time when architecture confronts the myths of geniuses and focuses on knowledge exchange paradigms. Architecture was not herein approached as substance nor as form (an immutable essence and ideal), rather, it was explored as processes engaged with discipline and dialectics. Discipline was investigated based on the Grounded Theory used to code conflicts and recurrences by focusing on how it reinforces subjectivities and practices. In addition, Action Research was used to explore architecture's social dialectics by focusing on collaborative methodologies and on how architecture (re)produces ways of seeing by revealing (visualizing hidden properties), imagining (conceiving future scenarios) and refunding (articulating virtual seeds for shared social realities). Results have indicated proposals to result from collaborative work, subjectivities to be enclosed in hegemonic narratives, fantasies to hide the actual collective process of production, and allegedly individual creations to be forms of fetish. This finding suggests a paradigm transition still in course, with overlapping conflicts between invention and labour, competition and collaboration, distinction and collective dialogue, as well as seductive narratives and negotiated practices.

Keywords: Creative Labour · Design Process · Architectural Studios · Discipline · Collaboration

1 Approaching Creative Labour Production in Design Studios

The current study explores creative labour production and reproduction in architectural design studios, by analysing experiences lived in Brazil, UK, Belgium and Italy. This investigation was inspired by Bruno and Woolgar's book "Laboratory Life: The Construction of Scientific Facts" [1], who approached scientists' practices from the perspective of an anthropologist who arrived in a remote tribe and tried to understand how social representations and relations interacted within the production of facts. In our case, it was done to develop a similar approaching distance to help better understanding theoretical device types adopted by people to reflect on their own practices.

M. Barosio et al. (Eds.): EAAE AC 2023, SSDI 47, pp. 64–76, 2025.
https://doi.org/10.1007/978-3-031-71959-2_9

This engaging-distancing movement enabled focusing participatory investigation on ideas supporting different actors involved in creative labour production. The investigated scenario has evidenced a structured process of both disciplining subjectivities and dialectical exchanges (among subjects and context). On the one hand, disciplinary aspects set different subjects into different roles and actions, by both creating distinction among them and promoting creative surplus value-extraction processes. On the other hand, dialectic social conditions and collaborative processes were traced in revealing and imagining processes associated with social reality refunding.

This finding has evidenced an ongoing paradigm shift, according to which, subjectivities are still enclosed by hegemonic narratives, individual creation fantasies and power relations. In contrast, results have shown that creation is a collaborative process with scarce narratives available, since architects tend not to see their work as labour. This issue is pivotal in a transition period-of-time when architecture confronts the myths of geniuses and focuses on paradigms of knowledge exchange and inclusion. Furthermore, this is an overall spread narrative system that encompasses different human activity fields. Thus, a new approach to it can help better understanding other phenomena taking place in the neoliberal context. For instance, we tend to say that Thomas Jefferson "invented" the lightbulb and to hide the fact that he had a laboratory structured as assembly line, where hundreds of scientists worked in different parts of the whole assemblage of that invention [2]. Similarly, the narrative goes that Steve Jobs "invented" Apple and the Iphone, although economist Mariana Mazzucato [3] has provided evidence that most Iphone-related inventions were produced by state funded research and appropriated by individual companies, latter on. Not surprisingly, the most significant shift in Frank Lloyd Wright's style took place with the Broadacre City project. Funded by his wife's fortune, the project took place at Taliesin Fellowship, which was a school that gathered hundreds of architects to exchanging knowledge in a messianic community in the middle of the desert [4].

Therefore, the current study focused on investigating architectural production as process, in contrast to narratives about free individuals acting with free autonomy to develop new architectonic things. Gilbert Simondon [5] calls this process of creating things 'ontogenesis' and he explains it based on the example of a brick 'taking form'. According to him, the 'clay'—as matter—is not just passive raw material,it has multiple possible transformations, as well as aptitudes and tendencies. However, clay is already a processed material whose production is based on selected grains added with the right amount of moisture. Moreover, it was collected in, and transported from, a known specific spot. Its identification and properties' description were only possible based on a long knowledge-acquisition process. The 'mould', in its turn, is not just an abstract shape previously defined through intellectual processes. It plays a procedural role in limiting clay transformation. It performs an active action, more precisely, a reactive action presenting equal and opposite force to the one exercised by the clay in it.

The brick individuation process comprises—already so far—a dynamic interaction system, according to which, potentialities and forces interact with each other to enable a final stability state. However, one should add to this scenario the actual work of artisans, who separate, discharge and press the clay, while using complex and subtle artifices to open and close the mould, and to provide it with perfect geometrical limits. Nevertheless, Simondon continues, the individuation of physical and technical objects happens (usually) only once, whereas living individuations take place in 'metastability', i.e., in the continuous process of becoming.

According to Simondon [6], seeing reality as becoming transforms the finite being of 'substantialism' into a being that is instead 'limited'. Being, as such, can be understood in more dynamic terms, as an undefined being bearing potential energy bigger than its factual actualisation. Thus, the idea of a 'limited being' acknowledges how being can relate to outside matter and how it can incorporate, reorder and transform itself in interaction with external elements. According to the aforementioned author, this is the only way transformation (creation and invention) can be understood in a consistent manner. Finite and eternal beings would not be able to change because what is said to be 'fundamental' is assumed as 'being as such' (the pure ultimate and supreme undifferentiated reality, which is, therefore, inaccessible and immutable). However, Simondon [7] advocates that the limited aspect of a given being is not fixed. This limit is a structuring process, the process of structuring a relational space between the inside and the outside.

Thus, looking back to the current object of study, it is possible to see that architecture is not a substance. It is not an immutable essence, something based on some specific features regardless of its actual production process. Moreover, architecture is not a set of formal properties abstractly defined based on a set of ideal features. Rather, the current study explored how the discipline of architecture limits its being and how the dynamic process of dialectical transformations works.

Discipline was herein approached as a social practice capable of framing ways of behaving, seeing and acting in the world. Therefore, this term refers to no distinctive boundaries, yet the phenomenon encompassed by it is fragmented, porous and interpenetrates different social life domains. The herein adopted approach was mainly inspired in Bourdieu's [8] sociology of the rules of art and expanded Foucalut's [9] argument towards an open disciplining process that do not completely lock subjects in an 'iron cage' [10], but it rather limits practices in a dynamic circumstance of exchanges.

Based on these terms, architecture should be seen through its social dialectics and disciplined production practices, and as a social product that, in its turn, disciplines habits. Thus, the current research started with fieldwork, which was carried out as a starting point to trace the logic-epistemological elements set in motion in concrete social practices and to investigate how they reproduce and reinforce social relationships. Paradoxically, based on these terms, what is socially understood as 'architecture' matters exactly because it guides concrete practices and is what we focus on to deconstruct.

Therefore, the next sections will briefly explore a way of opening architectural studios' "black boxes" to analyse the disciplinary aspects manifested in them, the dialectical dynamics likely to be explored and, finally, to explore how these aspects may suggest an ongoing paradigm shift. Discipline will be herein explored based on the way distinction established in the classroom gives voice to some individuals and to others gives a duty to obey, as well as on how it produces specific subjectivities and results in a hidden creative surplus-value estrangement process. Dialectics will be explored through action research based on flipped classroom, collaboration techniques and a concept of micro-utopias. It will be done to explore the way it might create different ways of seeing things by revealing hidden properties of both society and space, and by imagining future possibilities based on how it can articulate new shared social realities. Finally, the current study aims to explore how we live within conflicting paradigms and between concepts such as invention and labour, competition and collaboration, distinction and collective dialogue, as well as seductive narratives and negotiated practices, thus, pointing towards the need of finding new collaborative approaches to creative processes in the architecture field.

2   Discipline: Subjectification, Distinction and Hierarchical Expropriation Framing Subjectivities

Field investigation comprised a series of "constant comparative analysis", as proposed in the Grounded Theory approach [11, 12]. Overall, it is a qualitative sociological method aimed at developing theories based on rigorous observations. In order to do so, four one-week live-project workshops (two in London, one in Belgium and one in Italy) and two four-month studios in Brazil were approached through participant observation of 100% of their activities, which were recorded in field journals. In addition, interview forms were applied to participants. All collected data were coded to enable retrieving key aspects and repetitions, secondly, they were thematized in concepts presented and discussed in conferences, before being finally reviewed in the present report.

Furthermore, Tedlock [13] set the grounds for an 'ethno-sociology' capable of developing a self-reflexive ethnography (Tedlock, 1991, pp. 78–80). It means that 'participant observation' became a more personal 'observation of participation' by keeping the crucial dilemma between "participation" (which entails emotional involvement) and "observation" (which requires detachment). Thus, its findings result from a unique and specific dialogue that dilutes the mediations shared by those "who observe" and those "who are observed". According to Tedlock, this knowledge 'belongs neither to the realm of objectivity nor to that of subjectivity, but rather to "human intersubjectivity"'. To preserve participants' privacy, we will only mention that the field investigation involved studio experiences from 2014 to 2021.

2.1   Opening Architectural Studios' 'Black Boxes'

The overall aim lied on understanding how architecture is seen as 'things' rather than as processes, thus turning 'things into persons, and persons into things' [14], during this journey. The argument approaches architectural discipline as a non-trivial abstract machinery capable of framing subjectivities [15]. Therefore, this discipline cannot be seen as a closed system no one can escape from. Thus, we traced architecture's non—trivial machinations and limitations rather than seeing this discipline as an 'apparatus'. Subsequent According to Von Foerster [16], a trivial machine is featured by a one-to—one relationship between input and output; this invariability is precisely what defines the machine. However, the non-trivial machine presents varying input-output association, wherein the input, once processed, leads to changes in machine structure, and wheresoever the output also creates a new context that further changes the machines' internal structure. Thus, the non-trivial machine operates with conflicting internal structures.

Therefore, the concept of non-trivial machines can explain how a given social space operates in a complex field of historical and conflicting relations.

Based on that account, it is possible to explore a renewed approach to Marx's 'Fragments on Machines' [17]. According to Marx, machines are objectified knowledge, and the increasingly complexity of machines creates frames that capture collective labour and fragments it into individual efforts. In Marx's condition, he was talking about actual machines capturing collective productive forces. However, authors such as Virno [18] and Lazzarato [19] have updated this idea for contemporary abstract machines capable of capturing intelectual labour, such as mass culture, advertisement, ideologies and institutions. The same process of objectifying abstract scientific knowledge into fixed capital takes place in these abstract machines by articulating inter-subjective dead (past) labour and live labour, in such a way that dead labour both controls and exploits live labour.

To approach architecture as a such complex abstract machinery, the current study focused on investigating how the architectural discipline creates 'black boxes' and hides operations behind the process to produce architectural things and truth, as well as methods implied in this production process. Three black boxes-opening procedures played a relevant role in this process. On the one hand, Bruno Latour [20] has investigated how whenever a given scientist uses an apparatus to observe certain phenomena, what he/she sees on the other side of the apparatus' black box is framed by past theories and hypothesis that account for the production of that black box in first place. According to the aforementioned author, phenomena seen by this scientist only existed through the mediation of this machine, whereas the machine only existed through the past labour reified on it (theories inscribed in this material basis). On the other hand, based on Vilem Flusser's [21] argument about photography, the camera is the one performing the operation of turning reality into codified signals of visual communication, whereas the photographer is manoeuvred by the few potentialities inscribed in the apparatus. Therefore, the photographer actually looks inside, rather than outside, the apparatus; thus, he/she 'reveals' rather than creates things. However, according to Cabral and Baltazar [22], it is not a matter of destroying the 'magic' of the black box or of making its devices predicable and dull, but rather of opening its internal mechanisms to enable potential interactivity processes.

This is the only way we can approach the architectural discipline system in a way— other than being the product of a sort of 'Big Brother conspiracy'—to develop an approach to this metastable field where conflicts and tensions can dialectically emerge. Thus, architecture is not just the reflection of a pre-established status quo system. Although it operates in reproducing specific 'traces' and in conserving specific social structures, there is also room for conflict and transformation in it.

2.2 Distinctions: Who Has a Voice in the Classroom?

The first aspect noticed in architectural productions developed in these educational experiences lies on the distinction among several subjects involved in them. Distinction plays a key role in placing agents in different social positions, so both teachers and students' voices are structured in a hierarchical manner among themselves.

According to Bourdieu [23], pressures mediating the production of different works are both internal and external to the associated field. Moreover, they relate to the symbolic capital associated with participants' previous creative experience. In addition, the routine imposed by institutions on agents establishes a symbolic order, a valuation circle, a publicity level and a definition of legitimacy.

Distinction is the first step in the subjectification process, since it establishes the one(s) accounting for guiding common knowledge production. This process is not straightforwardly objective; but it results from social struggles for dominant positions. However, the distinct centralities of the classroom are those entitled to create habitus, to foment desires, as well as to establish references and hierarchies to properly mobilise individuals' libido and acceptance of order. Precisely because other subjects crave for the same distinction in the field, agents in distinct positions take leading roles and define what is good or bad, in a continuous circular reinforcement.

2.3 Subjectification: The Production of Architectural Subjects

A classical essay written by Judith Butler [24] aimed at denaturalizing the understanding of our own body and sought to understand gender as performance, or as socially constructed phenomenology. Helene Shugart [25] also explored how femininity in women and masculinity in men can be understood as social performance by de-fetishising their mechanisms, based on parody studies. The aforementioned author sees the possibility of denaturalizing the nature of gender and, consequently, of reconstituting desire, by doing so. Arguably, we could also denaturalize architecture and the creative labour involved in its production to set the ground for seeing architecture as it performs.

According to Bronwyn Davies [26], who analysed Butler's relevance for education, subjectification is the process that simultaneously establishes mastery and submission. To master a given topic, students submit themselves to the perceived order of distinctions. It is herein possible tracing a link to Althusser's [27] concept of 'interpellation' (an identity mirror). Interpellations create the subject because they create an image of the world, and place the individual in a relationship with this world, and in a position based on which he/she can act in this world. Interpellations functions through ordinary objects address a subject setting ideological expectations and interaction rules, incorporating ideological discourses (such as the placement of chairs in a studio establishes institutional roles). Thus, these ordinary objects become embodied ideologies shaping internal subjectivities. Therefore, the material condition of a given production mode has a dialectical relationship with subjects' cognitive awareness of their place in this condition. Accordingly, discipline and distinctions produce subjectivity by framing the way architects see the world, their position in it, as well as how they understand the way they can act in the world.

Following Jason Read, the idea of the 'production of subjectivity' implies a double meaning: it is both something 'productive' and something 'produced'. Subjects emerge always in the context of a collective of subjects, i.e. there is no subject detached in an abstract empty space. This means that subjectivity is formed by elements a priori (language, culture, structure, social expectations, among others) 'externalised in machines and internalised in concepts, habits, and ways of thinking' [28]. The dialectic operation of these abstract machinery of subjectivities creates a problem that is specifically political, because they act across the whole society, seizing for individuals what was once formed by collective efforts.

Jodi Dean [29] advocates that the representation of the individual is, itself, a fantasy that places what one can only do among 'others' into an imaginary ego; consequently, the concepts of genius result from a point of view that imagines an abstract individual detached from any context. Dean mentions that Freud's theory produces this type of 'enclosure of the individual'. It is worth emphasizing two additional points in Jodi Dean's account of the enclosure of the subject. She inverts Althusser's famous formula by stating that it is not the case that ideology interpellates individuals as subjects, rather, capitalism interpellates subjects as individuals.

Nonetheless, although the traditional architecture discipline encloses subjects in individual creativity and invention narratives, the observation of participation suggests that the most creative moments in architecture emerge from long processes of exchange, discussion and maturing of alternatives. Thus, collective subjectivity is gradually produced, and it allows eventual individual creation to emerge from this transversal experience. Furthermore, according to Simondon, subjects can change, therefore, they are in a metastable condition that includes non-actualised potentialities. These are non-actualised potentialities that exist in all subjects, and are formed of internal tensions and multiplicities, besides being in constant exchange with their surroundings. It is through these tensions that the architectural discipline operates its abstract machinery, enabling the development of both a given architectonic object and the production of architectonic subjects, themselves.

2.4   Creative Surplus-Value Estrangement

Case studies helped investigating how architecture is not just a practice, but also a subjectification process based on narratives capable of reproducing estrangement before requiring submission and providing mastery. Thus, these narratives stand on a previous acceptance of instrumentalisation.

Notably, instrumentalisation resulted in a continuous authorship appropriating process that can be called 'surplus of creation' estrangement. The dominant group not only proposed ideas to be followed by students, but also appropriated students' work. In a given situation, a door was to be made by subtracting modules from the external wall of the pavilion. The tutor continuously shouted that she had already solved the problem and that students' only task was to 'draw it'.

But a series of issues the tutor was unaware of emerged as students engaged in the design process, namely: symmetry of his sketches was unfeasible; position of the door was impossible because it would make the cantilever bigger; among others. However, the tutor would return to the table many times, commenting with irony as if the students were incapable of drawing his solution; however, there was no solution, yet. Students were aware they were being observed and maintained a subservient attitude agreeing that 'the problem was solved'. After the tutor turned away, they worked on a viable solution, which was only attainable by changing the shape of, and rotating, the entire structure by a few degrees. However, once the solution was found, the tutor joyfully asserted that they had finally understood it.

In another case, the tutor repeatedly changed the ground floor plan designed by students, although his changes made no practical difference. Better solutions proposed by students were re-sketched until the tutor felt to own the solutions. Again, in a paradigmatic case, a design was structured with larger modules in the base and with smaller modules at the top. The tutor took the model and inverted it, asking a fellow teacher, 'Should we do it like this? It will create an optical illusion that the structure is bigger'. Although the intended optical illusion would technically be delivered by the original proposal, students (tired of continuous conflicts) accepted the tutor's suggestion as the 'stroke of a genius'. Their initial design was magically reworked by the genius (a change without changes), and the whole thing now belonged to his act of 'creation'.

Thus, the working subjects became instruments through a double step: first, they became estranged from their own background to fit the distinction structure in the classroom; later on, they were estranged from the product of their own work. This process not just fulfilled the desires of the group in power, but also genuinely created new solutions, which were later assumed to result from their obedience. Thus, the disciplinary black box of architecture not only produces distinctions, but the way subjects see the world, their place in it, and how they can act in it. Those are the cogs and gears of this abstract machinery.

3 Dialectics: The Production and Reproduction of Ways of Seeing

Several authors (including David Harvey and Boaventura Souza Santos [30–32]) have argued that we are experiencing a great time in history, when great changes in social structure, culture, technology and production, as well as the threat of an eminent environmental collapse, force us to imagine new social alternatives. As we started opening the black boxes of architecture, we could recognize how the architectural discipline reproduces social relations, both internally and externally (framing the possible life of ordinary people). Nonetheless, besides acknowledging that design and architecture have strongly contributed to the way people see the world, their place in it and how they can act in the world, if we open that black box to scrutiny, it can also become a way of visualizing alternatives, of exploring different places potentialities and of envisioning new futures.

It means that architecture not only reproduces ways of seeing things, but it also has the potential to explore new ones by revealing hidden properties of social reality through new imaginative tools of representation and critical narration; by imagining other possible future scenarios based on using its design-thinking methodologies; and, finally, by refunding social realities in itself based on articulating virtual seeds and on building shared knowledge. Thus, we explored these issues in action research experiments (these experiments were analyzed in other perspectives in previous studies, see [33]). Just as a chemist would mix different compounds in a lab, we mixed ideas in the studio practice, such as hacking objectified social relationships and developing micro-utopias (understood as the virtual power of concrete potentialities of inexistent worlds).

In order to do so, we adopted 'flipped classroom' methodologies, according to which hierarchy and knowledge production in the classroom was inverted, so the studio became a place of shared knowledge production. In addition, collaboration techniques, such as workshops, kanban and assemblies, were explored to emphasize the collective construction of the debates. Finally, the studios aimed at developing 'live projects' to overcome the traditional 'simulation studio'—where inputs are imagined and assumed—by focusing proposals on the strategies to put them into practice.

In a workshop held in London, UK, a given area was facing a Regeneration Action Plan that proposed the privatization of a square and the demolition of social housing to open room for new luxurious flats. Because the consultation process was a simulacrum, the local community and one association reacted to it by developing an autonomous process. The University engaged in the process by developing an event with students, when a spatial device was built in the square and the community engaged in the resistance to the initial proposal. Students from the local school helped build the design pieces and participated in structure assembly testing. Leaflets were distributed and the local cafeteria provided food for the event. Proposals by locals were collected and kids were encouraged to contribute to the project with their drawings. Moreover, an internet platform was launched to collect stories describing the importance of those spaces to the community and, consequently, the need of protecting them—in compliance with the British law.

A series of interviews regarding alternative practices also helped to see those paradoxes. According to Sarah Wigglesworth [34], participation is a fundamental element in her architectural design process. In fact, she spends most of the time on any presentation she gives for a new project talking about participation. Nevertheless, participation appears to be a 'consulting' process, wherein users are invited to provide input and steer decisions. The architect works as guide in this process. However, the resulting product is a conventional design, an object that hides the means of its own production in its cleanness. By contrast, participation in the architectural studio 'MUF' becomes part of the elements being 'represented' in architecture and participation traces become a set of elements in the final products' aesthetics. Moreover, according to the multi—disciplinary collective Assemble, participation not only enters the process as input, but it also becomes an output. According to an interview [35], the aesthetics of products keeps a certain openness, as if users could see the way the object was produced and think 'I could do that myself'. Thus, the product becomes an empowering object, rather than a form to mask processes with appearances. Furthermore, based on the Assemble's Granby Four Streets project, the design process is actually the means of bringing a community together to actively produce their own spatial conditions.

One of the workshops held in Belgium serves as counterexample of how this line is blurred. The workshop aimed to identify local desires and elements that could be realized, but students struggled to determine the limits of their agency and were continually concerned with meeting their teachers' expectations. This experience brought further questions to light. It was paradoxically unresolved to what extent the resulting objects were more than sculptures representing specific community interests and to what extent they were merely educational activities. Additionally, because the objects symbolized community participation, they were no less constrained by the prevailing ideology than the average 'good citizen' This raised concerns about whether architecture was effecting change or merely reinforcing the status quo, ultimately questioning whether it had become an empty community game.

Departing from these experiences, several studios in Brazil worked with the idea of micro-utopias by taking into consideration immanent potentialities and by exploring their development with the aim to break up established ways of seeing social structures.

Nonetheless, besides the utopian intent, it was quite common for students to rather develop dystopias, a fact that can be read as sign of our times: it is easier imagining society's collapse than any alternatives to current scenarios.

Accordingly, our social-environmental issue is also an aesthetical issue: how can we collectively see the solution for unsustainable development? Cities' aesthetics goes far beyond simple 'beautification', since it establishes a way of perceiving and establishing relevant issues for a given community. Thus, urban space creates an "aesthetic field", i.e., a setting that expands, or limits, the vision shared by a group of citizens.

A studio in Goiânia City (Brazil) worked in partnership with the local environmental agency to investigate and propose solutions to help recovering its main river. As the analysis went by, the collective conclusion was that it was not possible to develop one single project to solve issues faced by the river, since they involved all tributaries of the aforementioned river and the whole city's relationship with them. Therefore, the proposal lay on developing a series of actions to gain public attention to these issues; thus, architecture was no longer a means of designing things, but of making visible the hidden city-local nature relationship. In addition, a collaborative web platform and a NGO were launched as a means to permanently address and share proposals and visions to build new paths.

As a counterexample, the workshop 'Into the Green' held in London also implemented an intervention in a public space during an event that mobilised the local community. As a consequence, an office was opened at the university to develop interventions in that space, so that the ephemeral community activities held in it could become permanent. However, the office was dismantled due to the university's bureaucracy, the community became disengaged, and it put an end to the process.

Finally, a form was applied to participants to help better understand students' experience. The 'labour condition' issue was assessed as the best dimension of their experience, with 80% considering it positive. This topic assessment involved the role played by participants in the production and sharing of ideas, the openness of debates, and the democratic decision-making in comparison to previous academic experiences. Surprisingly, students did not consider this experience significantly different from other past practices. This finding suggests that the actual work experience is always collective in practice. Although the narrative in the field is that the production is done by the main architect in the office, in practice the involvement of trainees in the design process resembles the collaborative methodologies explored in the workshops. This might be the case of any office, where the collective work is anonymous, and the main architect is the reference. Again, Assemble functions on the opposite pole, where the work is done in a collective process, according to which, all activities—from design process to everyday reproduction—are shared.

Form results for the 'delivered products' dimension were mostly paradoxical given the Live Projects' intentions. Based on students' perception, products had weak positive impact on the place (only 19% above neutrality), the connection between creativity and construction was limited (19.8% above neutrality) and the product was quite similar to that of a conventional architectural practice (3.4% above neutrality). However, the analysed interventions were considered a 'valuable project' (31% above neutrality).

This condition points out that the architectural studio has a long tradition and that even counter-practices remain entangled in this long tradition. Thus, it is not the case of advocating for a fully free practice, but, as Hanna Arendt [36] suggested in a different context, we should focus on the continuous fight for freedom, on the continuous deconstruction and critical awareness of limits imposed by the discipline, and on potentialities enabled by a dialectical attitude towards design and social structuring.

4   Design is Always Collaborative: A Dangerous Idea in the Form of Conclusion

The Competition [37] is a documentary movie focused on telling the story of several star architects competing for a project in Luxembourg. Among several similar scenes, one shows the moment when a team of architects explains their proposal to the main architects, and one of them takes the model, turns it upside down and asks whether that was not the best solution. Thus, several hours of hard work by the team were "magically" estranged from the collective effort to become the genius stroke of one single man.

The main argument that can be built in the current research is that architectural production is a long process, just as the previously used brick example, since it involves traditional knowledge and several hours of creative and design-thinking experimentations. Nonetheless, both narratives and field structure set our perception to see individuals as creative drivers. Thus, what is essentially collaborative is captured in an abstract machinery that is reproduced by the discipline and it ends up being perceived as a single-person invention.

Although this machinery operates in quite abstract and subtle terms, the aim of the current study was to investigate how distinction establishes a hierarchical structure in creation processes, how subjectification establishes a thought pattern by placing different subjects in different positions where they can act in and by limiting how they think and act in the world. Finally, this machinery produces the estrangement of collective products, in such a way that subjects involved do not see the product of their work as their own.

This is only possible because field narratives depict architects as individuals in competition, thus enclosing their subjectivities. That hides the actual collective production of architecture. If hegemonic narratives create this process of estrangement, we need new counter-narratives to oppose them. If fantasies of individual creation seed competition between architects, we need a new paradigm for envisioning the collaborative dimension at play. If we should care more about architects being explored in unpaid extra-hours, we need new perspectives to envision power relations inside the discipline and a new consciousness of architects as labourers. Although the collective production is captured by this abstract machinery, contradictory forces are at place. In this sense, there is space for a dialectical paradigm to be built to rethink our practice as fundamentally collaborative.

In this dialectical practice, architecture plays the social role of revealing, reimagining and refunding social structuration. Through representation, diagrams and spatial analysis, architects can make aspects of reality that were not seen by common people before visible. By using design-thinking methodologies, architecture can radically reimagine the foundations of social space. And by setting these ideas into practice, we can collaborate to create more sustainable social structures.

Ultimately, this contradiction between disciplining forces and dialectical processes represents a transition time of conflicting paradigms. Between invention and labour, architects do not see themselves as workers; between competition and collaboration, architects see themselves as individuals in fierce fight; between distinction and collective dialogue, internal hierarchies give voice to few and estrange others; between seductive narratives of great masters and hard-time assemblies of negotiated practices, architects still tend to use the few symbolic power they still have.

References

1. Latour, B., Woolgar, S.: Laboratory Life. Princeton University Press, New Jersey (1986)
2. Mazzucato, M.: The Entrepreneurial State. Demos, London (2011)
3. Simondon, G.: L'individuation à la lumière des notions de forme et d'information, pp. 39–58. Millon, Grenoble (2013)
4. Friedland, R., Zellman, H.: The Fellowship: The Untold Story of Frank Lloyd Wright and the Taliesin Fellowship. Harper Collins, New York (2007)
5. Simondon, G.: L'individuation à la lumière des notions de forme et d'information, pp. 39–58. Millon, Grenoble (2013)
6. Simondon, G.: L'individuation à la lumière des notions de forme et d'information, pp. 39–58. Millon, Grenoble (2013)
7. Simondon, G.: L'individuation à la lumière des notions de forme et d'information, pp. 39–58. Millon, Grenoble (2013)
8. Bourdieu, P.: The Rules of Art. Polity Press, Cambridge (1996)
9. Foucault, M.: Discipline and Punish. Vintage Books, New York (1979)
10. Jameson, F.: The Political Unconscious, pp. 75–78. Routledge, Oxon (2002)
11. Guest, G., MacQueen, K., Namey, E.: Applied Thematic Analysis. SAGE, Thousand Oaks (2011)
12. And Glaser, B., Strauss, A.: The Discovery of Grounded Theory. Aldine, Chicago (1967)
13. Tedlock, B.: From participant observation to the observation of participation. J. Anthropol. Res. **47**(1), 69–94 (1991)
14. Marx, K.: Capital, vol. 1, p. 209. Penguin, London (1990)
15. Lazzarato, M.: Signs and Machines. Los Angeles, Semiotext (e) (2014)
16. Von Foerster, H.: Perception of the future and the future of perception. Instr. Sci. **1**(1), 31–43 (1972)
17. Marx, K. Fragments on machines. In Marx, K. (1973) Grundrisse. Penguin Books and New Left Review, London (no date [1857])
18. Virno, P.: General intellect. In: Zanini, A., Fadini, U. (eds.) Lessico Postfordista. Feltrinelli, Milan (2001)
19. Lazzarato, M.: Signs and Machines. Los Angeles, Semiotext(e) (2014)
20. Latour, B., Woolgar, S.: Laboratory Life. Princeton University Press, New Jersey (1986)
21. Flusser, V.: Filosofia da Caixa Petra, p. 43. Hucitec, São Paulo (1985)
22. Cabral Filho, J., Baltazar, A.: Magic beyond ignorance virtualizing the black box. In: Festival de Arte Digital Essays, pp. 12–23. Instituto Cidades Criativas, Rio de Janeiro (2010)
23. Bourdieu, P.: Distinction. Harvard University Press, Cambridge (1996)
24. Butler, J.: Performative acts and gender constitution. Theatr. J. **40**(4), 519–531 (1988)
25. Shugart, H.: Parody as subversive performance: denaturalizing gender and reconstituting desire in Ellen. Text Perform. Q. **21**(2), 94–113 (2001)
26. Davies, B.: Subjectification: the relevance of Butler's analysis for education. Br. J. Sociol. Educ. **27**(4), 425–438 (2006)

27. Althusser, L.: Lenin and Philosophy and Other Essays. Monthly Review Press, New York (1971)
28. Read, J.: Production of subjectivity from trans individuality to the commons new formations. A J. Cult. Theory/Polit. **70**, 113–131 (2010)
29. Dean, J.: Enclosing the Subject. Political Theory **44**(3), 363–393 (2014)
30. Harvey, D., Whachsmuth, D.: What is to be done? In: Brenner, N., Marcuse, P., Mayer, M. (eds.) Cities for People, Not for Profit, pp. 264–274. Taylor and Francis, Hoboken (2011)
31. Santos, B.: Renovar a teoria crítica e reinventar a emancipação social. Boitempo, SP (2007)
32. D Lima C Amaral.: The Reproduction of Architecture. PhD Thesis, University of East London (2017)
33. De Lima, A.C.: Production of project: a subversive guide to the subject of innovation. Ardeth **5**, 56–77 (2019)
34. Wigglesworth, S.: Interviewed by the author, 11th December, personal archive (2015)
35. Jones, L.: Interviewed by the author, 4th November, personal archive (2015)
36. Arendt, H.: On Revolution. Penguin Books, New York (2006)
37. The Competition.: Directed by Angel Borrego Cubero [film] (2013)

Socially Situated Pedagogies as a Strategy to Innovate Architectural Curricula: The Case Study of SArPe and Its Design Studio Experimentation at the University of Pavia

Ioanni Delsante[1]([✉]), Tabassum Ahmed[2], Maddalena Giovanna Anita Duse[1], and Linda Migliavacca[1]

[1] University of Pavia, 27100 Pavia, Italy
delsante@unipv.it
[2] University of Huddersfield, Huddersfield HD1 3DH, UK

Abstract. The authors discuss a pedagogical experiment that arises out of an ongoing Erasmus+ project titled "Socially Situated Architectural Pedagogies" or SArPe, which involves the Universities of Pavia, Istanbul, TU Delft and Malaga. SArPe project situates itself in between three areas of inquiry: critical/radical pedagogies [1–4]; situated knowledge [5, 6] and commons-oriented knowledge and pedagogy [7–9].

On that basis, the paper aims to critically analyze architectural pedagogy through authors' positioning in respect to the wider debate and a case study-based approach. Namely, a second-year architectural studio held at the University of Pavia in a.y. 2022–2023. Here the studio is reimagined as a site for common-ing of knowledge through collaborations with non-academic actors; challenging hierarchical position between learners and educators; and experiment practice of dialogue.

By doing that, it seeks to broaden reflections on how architectural studio can reconnect to the outside world and, more particularly, how learners and educators (and their mutual positioning) engage with non-institutional stakeholders.

As such a transformative-relational pedagogy is experimented, which activates the traditional studio towards a socially situated pedagogical practice that promotes self-organization, encourages active participation and destabilases hierarchies.

Keywords: Critical/radical pedagogies · situated knowledge · commons-oriented knowledge

1 Introduction

The authors discuss a pedagogical case study originating from an ongoing Erasmus+ KA2 project known as "Socially Situated Architectural Pedagogies" (i.e., SArPe). This project involves a consortium of the Universities of Pavia, Istanbul, TU Delft, and Malaga. SArPe's primary objective is to enhance the responsiveness of the higher education sector to societal challenges through inclusive and participatory activities. It is built

© The Author(s) 2025
M. Barosio et al. (Eds.): EAAE AC 2023, SSDI 47, pp. 77–86, 2025.
https://doi.org/10.1007/978-3-031-71959-2_10

upon an integrated approach to teaching, learning, and training activities, combined with deep engagement, outreach, and dissemination targeting both educators and learners.

Within the broader academic context, the project addresses three distinct target groups (see Fig. 1) and, in particular, educators and learners (within and beyond academia), but also active groups including non-institutional organizations and grass-roots groups. A part of the methodology of the project involves interaction with multiple active groups, allowing the consortium to listen to multiple voices, including those of local civil society organizations and communities.

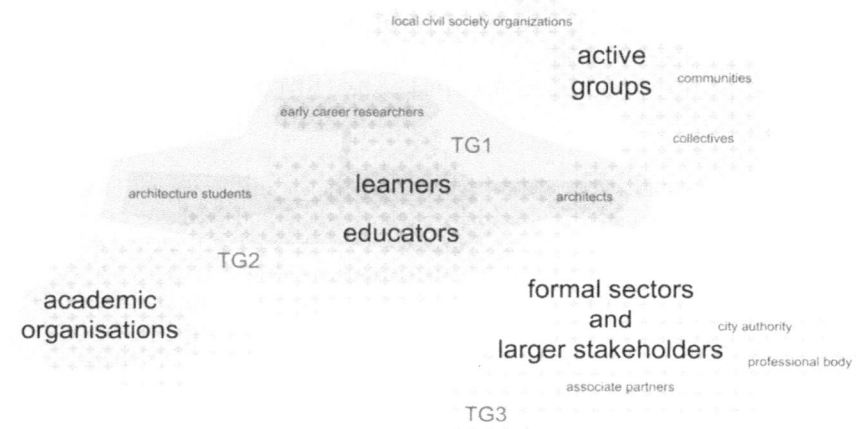

Fig. 1. Visualization of target groups involved in SArPe, Pavia 2022 (Image by Aslihan Senel, produced for 2022 KA2 Erasmus+ Coordination Partnership proposal named SArPe).

2   Setting the Scene

There is a growing debate surrounding current architectural pedagogy [3, 8], a sense of dissatisfaction with studio teaching [10], and criticism of architectural professional training that emphasizes star-architecture [11]. Many of these teaching methodologies follow a hierarchical approach in which the teacher possesses absolute power and knowl-edge, and students depend on them for all facts, ideas, and perspectives. Such one-sided traditional pedagogical models, referred to as the 'banking model' by Freire (1968), portray students as empty vessels who receive information without engaging their crit-ical faculties, creativity, subjectivities, or capacity for questioning. Particularly in the context of the interconnected crises of the 21st century, there is a need to reconfigure the teaching and learning process to make informed and responsive choices in architectural pedagogical practice.

In this context, SArPe joins the ongoing debate and advocates for a socially situ-ated pedagogy in architecture, where knowledge is collaboratively produced through dialogues between learners and educators, as well as between universities and non-academic institutions. Critical pedagogies have long called for critical positioning that recognizes the political nature of teaching and learning and argue for social transformation [3]. The prism of critical pedagogy illuminates the hidden subtleties of social, cultural, political, and even economic conditions within the complexities of teaching and learning [12]. Various formats have been conceived and advocated for, such as the transformative model [10], the dialogical model [13], tutorless studio teaching [14], and even the commoning of architectural pedagogy [8]. More relevant to the scope of this paper, transformative pedagogy challenges the democratic configuration of the studio and seeks to redistribute power [10]. The dialogical problem-posing model, seen as a key tool to challenge and transform the power imbalances and relations between educators and learners [13], allows for the sharing of equal roles, acknowledges the learners' perspective of the world, and nurtures a relationship between the educator and learner in which they learn from each other's perspectives. Opening the studio as a site of co-production, exchange, and dialogue generates new forms of knowledge and dialogue that are crucial for transforming educational practice and for education to be considered a valid political tool.

Specifically related to this paper, the authors have drawn inspiration from the dialogical and transformative approaches of critical pedagogy to approach the module differently and introduce the use of tools and methods that can support the aforementioned ethos and learning approach.

3   Methodology of the Paper

This paper builds on activities performed during the SArPe Project between October 2022 and July 2023. It includes the initial phases of setup and discussion, the initial literature review primarily conducted between October 2022 and January 2023, the preparation phase of a teaching module held in February 2023, and its unfolding in the following months.

Hence, this paper includes an initial section that both identifies key themes and unfolds concepts and ideas that are developed in the following sections (e.g., tools and methods).

It is also case study-based, represented by the module named "Architettura e Composizione Architettonica 1" at the University of Pavia (2022-2023). The module is structured on the basis of a general methodology (research-led teaching), which includes interviews, fieldwork, design activities, as well as a variety of other tools and methods.

These activities are reflected upon through the authors' critical reflections, formed as a team of researchers and participants (authors' positioning in respect to the object of study). Feedback from learners has been collected so far through means such as self-evaluation forms, informal interviews, and other opportunities for engagement that have arisen after the end of the module.

4 Case Study

The module "Architettura e Composizione Architettonica 1" is a 2nd-year module within the Course of Building Engineering and Architecture at the University of Pavia (an integrated BSc + MSc 5-year course). It consists of 9 CFU (both front lectures and studio activities) and 8 h per week over a period of 3 months (12 weeks), with the teaching hours divided between two staff members. Additionally, 4 h of tutoring are scheduled every week (tutor-led). Ten tutors (3rd to 5th-year students) were selected, with 4 of them continuing from previous years and 6 newly appointed. In the 2022/2023 academic year, the module had 42 students, including 5 Erasmus students. Therefore, the current staff-to-student ratio is 1 to 42, while the tutor-to-student ratio ranges from 1 to 8–16, depending on the session and the pool of tutors in attendance. The syllabus was slightly revised at the beginning of the academic year, but it has been regularly amended since the academic year 2018/19.

From the literature review, the possibility of innovating the studio environment has also emerged. This innovation has been conceptualized spatially and socially as non-neutral spaces and democratic environments [10] that generate new knowledge through dialogue and serve as a site for interacting with the social, political, economic, and cultural dynamics of the city [3]. Large tables and centrally organized spaces for sharing, feedback, and peer-to-peer reviews encourage shared learning and critical thinking. Additionally, the ongoing interaction of the studio with stakeholders and the community, both within and outside the studio, situates the design process within the context, needs, and local circumstances [5, 8]. Evaluation has been extended to include peer and self-evaluations to enhance self-awareness and establish a power balance between learners and educators [10, 15]. More broadly, the shift from the traditional studio model towards a commons-oriented studio [16], guided by shared principles and guidelines established as "manifestos" between educators and learners at Compo1, transforms the studio process into a collaborative and non-hierarchical format.

4.1 Semi Structured Interviews with Learners

One of the initial steps taken before drafting the course program is the analysis of the results of the mandatory university questionnaire regarding the modules taught in the previous semester and years. However, the limitation of these results is the lack of specificity in relation to the pedagogical aspect of the module.

The requested feedback primarily concerns the themes of the learning process related to urban commons and communities. As a result, a series of general questions were formulated, followed by more specific questions based on the year of enrollment in the course.

The first choice was to conduct semi-structured interviews [17] (rather than to distribute questionnaires) in order to ask questions in an "open" manner and leaving some room for conversation aimed at visualizing a broader perspective. These were conducted with eight learners, two for each academic cohort, from 2019 to 2023. Engagement was very strong, and all learners were open to sharing their ideas regarding community engagement and the importance of the design process.

"By carrying out work directly on-site and involving the citizens, I might be able to find a more suitable location or a more useful solution for the residents." (A 2019 course attendee).

"Talking to external people is very challenging; it already raises the bar. That's why it is difficult, but also very stimulating." (A 2020 course attendee).

Some of the key findings include: the need for a stronger focus on practical project work, associated with increased stakeholder involvement; positive feedback on the community engagement workshops associated with the project and a push to cultivate a relationship with the city (e.g., Pavia).

4.2   Approach and Methodology of the Module

Approaching the new term, the teaching staff and tutors agreed on certain educational objectives. These objectives included bringing in innovative tools and methods to inform the module's pedagogical approach, creating opportunities to engage with local communities and groups, especially non-institutional ones, and innovating the studio environment and its organization. At the same time, they aimed to build on the positives developed during previous academic years, specifically from 2018/19.

Engaging with the SArPe research project (and international team) provides an opportunity to shape the module as a research-led teaching activity. Consequently, a general methodology (defining what, how, and when of tasks/activities) is designed by the module leader and discussed with the team of educators.

This methodology unfolds through four main areas: secondary and primary data collection, setting a vision and research-by-design, which are not meant to be separate but intertwined. Moreover, critical positioning is integrated at various points during the term. Theory and architectural precedents are considered as a background and run parallel to studio activity.

As mentioned, the module builds on activities that have already been tested in previous years (e.g., psychogeographical dérive), but it aims to examine them in relation to each other and also in relation to the general pedagogical aim, which aligns with the syllabus (Fig. 2).

The general pedagogical approach to the module has been explained to the students on the first day of studio. It has also been reiterated and expanded on during the term. In particular, when some activities were unfolded these were also explained in relation to the general plan.

4.3   Towards a Pedagogical Manifesto

At the beginning of the module, a "Manifesto" is drafted with the purpose of critically positioning in respect to the existing literature and the local context/content of the module. It establishes the key principles, values, and ideas that will be used throughout the term. It is conceived as a dynamic and open document, initially drafted by a small group of educators, but inherently flexible, modifiable, and expandable by the learners themselves.

The learners are introduced to the Manifesto (Fig. 3) during the initial lessons through a presentation outlining the educational objectives of the module, related to the tools and

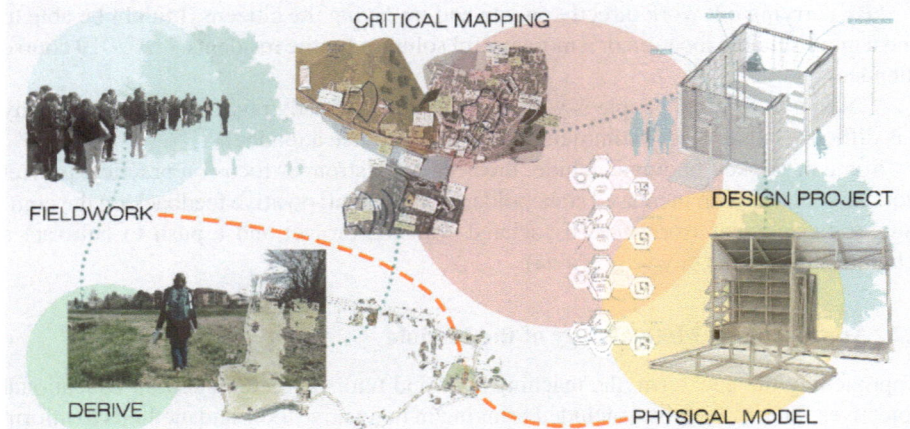

Fig. 2. The collage, featuring a variety of sample images created by learners, effectively illustrates the step-by-step development process of the project. It encompasses the initial fieldwork, the evolution of critical mapping, and the site-specific design process, serving as a representation of the outcomes achieved within the design studio, Pavia 2024 (Image by Linda Migliavacca, produced for 2022 KA2 Erasmus+ Coordination Partnership proposal named SArPe).

methods that the authors intend to adopt, with reference to the relevant literature review. This emphasizes the aspiration to create a horizontal and non-hierarchical environment within the studio, involving the adoption of the terminology of "educators" and "learners," creating uniformity in the "teacher-student" and "student-teacher" relationships. The teaching staff takes on the role of conveying critical knowledge [13]: recognizing the learning dimension of everyone involved in the module does not mean roles and responsibilities are denied. The statement of wanting to use a peer-to-peer dialogue [18] presupposes the idea that interactions between educators and learners, as well as between learners themselves, hold the same value; learners become critical investigators in dialogue with their teachers. The Manifesto also contemplates the "architect's role" and the goal of developing projects embedded in social contexts within communities. The central idea is not the final project but the process through which participants can develop a critical position [19].

Drafting the manifesto is an exercise that enhances awareness of the educational process for the educators. It allows them to develop an understanding of the topics to be addressed during the module and the tools to be used.

4.4 Tools and Methods

Among the methodologies proposed for the general organization of the module, tools derived from the literature review are being applied. They are conceived as a set of exercises to subject learners to, with the aim of creating a learning process.

The authors begin by questioning the course's structure at both the hierarchical level, dismantling traditional models. They are also modifying the study environment, changing the affordances [20, 21] of the classical studio, aiming for a structure that

Fig. 3. Towards a Manifesto, shared with the learners at the beginning of the module, Pavia 2023 (Image by Linda Migliavacca, produced for 2022 KA2 Erasmus+ Coordination Partnership proposal named SArPe).

fosters more open work and discussion. The case study knowledge process is designed as a collection of situated knowledge for the learners. The exercises proposed to achieve this goal include interviews with previously mapped stakeholders, fieldwork, guided tours of the study neighborhood, and a psychogeographic dérive [22] within the same area. Another educational objective is to facilitate dialogue among learners and between learners and educators. The selected tools include role play, allowing for continuous work on the initial analysis of the case study. Other peer-to-peer comparison exercises are used with the concept of reflection-in-action [23]. Finally, efforts are made to reduce aspects of critical design within the learning process and fully engage the learners. Therefore, self-evaluation [15] modules are provided during the final exams.

The application of various tools within the course allows for an understanding of the positive aspects, difficulties, and outcomes. The most innovative approach in this pedagogical process is the option to combine various tools.

5   Conclusive Remarks

The objective of the paper is to convey the process of shaping a module brief with innovative pedagogical intentions while also sharing the authors' critical analysis of what is being done.

Starting from a literature review, the authors were able to ask themselves the 'right questions' so as to try to answer them through such a pedagogical journey. This journey leads them to formulate a process that involves the use of specific tools. These resources should then guide educators and learners through a more effective and engaging learning process.

The most innovative outcome of semi-structured interviews is the importance of listening to the learners, as the feedback received directly from them pertains to important pedagogical themes. In fact, the analysis of the interviews provides valuable insights into the learners' perspectives on pedagogy. The starting point for planning the methodology underpinning the new academic year's brief is a summary of such feedback. Efforts are made to design a schedule that emphasizes learners' interactions with the communities they will collaborate with. This involved introducing innovative tools to initiate the learning process and, most importantly, organizing various events beyond the university setting to engage non-academic partners in the entire project. The authors also attempt to suggest exercises aimed at fostering a deeper connection between the university space and the city, and encouraging learners to become more involved in the environment in which they operate.

The methodological pathway, structured as a research project, enables the creation of a sequence of tasks while remaining open to amendments and integrations, e.g., based on direct feedback from learners. The authors think this has been clearly communicated and well received by them.

Throughout the module, certain activities were implemented more successfully than others, prompting consideration of alternatives for the upcoming pedagogical process. The manifesto makes educators more aware of the pedagogical matters in relation to specific cohorts of learners and identifies priorities among the set of tools to use.

In the specific case, the learners have perceived a change in the educational model; however, a fuller adoption of the principles outlined in the manifesto, diverging from the conventional pedagogical model already in use across the faculty, has not been viable. This could be attributed to various factors among which: current institutional frameworks and their limitations; cultural barriers (of various nature); and also, the reluctance by learners to implement them. The latter might stem from the lack of a collaborative process in designing the Manifesto (co-design), which might have left learners disconnected from the logic behind the course ethos. However, it should also be considered that this module represents (to learners) such a radical innovation, in relation to a system based on conventional approaches in studio teaching such as 'master' led tutorials and top-down project crits.

Regarding the use of different tools and methods, the authors actively aim to promote learners' critical thinking. The main experimental aspect includes the use and combination of various tools, and it is quite evident that the combination of these tools, along with a departure from conventional classroom teaching (e.g., critical discussions), is significant and worked well towards the final outcomes.

In conclusion, the methodology applied for the module has proven effective. A positive judgment has emerged from various perspectives. Objectively, it can be stated that approximately 75% of the participants were able to pass the course in the first three examination dates. Among these learners, over 50% passed the exam with distinction, (grade band A, or $\geq 27/30$). Furthermore, the quality of the outcomes was considered very high both by educators and stakeholders, who had the opportunity to observe the results during the exam sessions and via a public exhibition organized in October. From the subjective perspective of the authors, the pedagogical process developed and then implemented was carried out with great positivity. The module, in general, is positively received by the learners. The critical thinking skills of the learners were cultivated and excellently expressed in their self-assessment forms (filled on the day of the exam, before educators' marks are shared with them) because they are highly beneficial in raising awareness among the learners about their learning and design process. Most of the time, their personal reflection aligns with their outcomes.

However, as the authors delve deeper into the analysis of results, it is important to recognize that the shortcomings and challenges that lie ahead will also come from learner feedback. These reflections will be fundamental in refining the authors' pedagogical approach and enhancing the overall module effectiveness. For now, the reflections are still provisional, and more time is needed to process and evaluate the results. Therefore, the definition of final ideas, methodologies, and tools for structuring the learning process aims to be further validated.

By constantly improving the module brief and the module's pedagogical approach, one further step (as planned by SArPe) will be to amend next year's module syllabus. This will have, potentially, an even stronger impact on students' learning objectives and skills. This is precisely why the process described in the article is part of a three-year long process within the SArPe project, which sees the collaboration of various international partners from academia as well as from the not-for-profit sector.

"Socially Situated Architectural Pedagogies" or SArPe: it is a project that has been funded with the support from the European Commission. This publication reflects the views only of the author, and the Commission cannot be held responsible for any use which may be made of the information contained therein.

References

1. Colomina, B., Galán, I.G., Kotsioris, E., Meister, A.: Radical Pedagogies. MIT Press (2022)
2. Crysler, C.G.: Critical pedagogy and architectural education. J. Archit. Educ. **48**(4), 208–217 (1995). https://doi.org/10.1080/10464883.1995.10734644
3. Dutton, T.A., Mann, L.H. In: Dutton, T.A., Mann, L.H (eds.) Reconstructing Architecture: Critical Discourses and Social Practices (N—New ed.). University of Minnesota Press (1996). https://doi.org/10.5749/j.ctttmg1
4. Hooks, B.: Teaching to Transgress. Education as the Practice of Freedom. Routledge, London (1994)
5. Haraway, D.: Situated knowledges: the science question in feminism and the privilege of partial perspective. Fem. Stud. **14**(3), 575–599 (1988). https://doi.org/10.2307/3178066
6. Rendell, J. Sites, Situations, and other kinds of situatedness. Log (2020) pp. 27–38
7. Bourassa, G.N.: Towards an elaboration of the pedagogical common. In: Means, A., Ford, D.R., Slater, G. (eds.) Educational Commons in Theory and Practice, pp. 75–93. Palgrave Macmillan, New York (2017)
8. Deamer, P., Deeg, L., Metz, T., Tursky, R.: Design pedagogy: The new architectural studio and its consequences. Architecture_MPS 18(1) (2020). https://doi.org/10.14324/111.444.amps. 2020v18i1.002
9. Korsgaard, M.T.: Education and the concept of commons. A pedagogical reinterpretation. Educ. Philos. Theory **51**(4), 445–455 (2019).

10. Dutton, T.A.: Design and studio pedagogy. J. Archit. Educ. **41**(1), 16–25 (1987).
11. Till, J.: Architecture Depends (Illustrated Edition). MIT Press (2013)
12. Wink, J.: Critical Pedagogy: Notes from the Real World (1996). https://www.academia.edu/69930342/Critical_Pedagogy_Notes_from_the_Real_World
13. Freire, P.: Pedagogy of the Oppressed. Seabury Press, New York (1968)
14. Hill, G.A.: The 'Tutorless' design studio: a radical experiment in blended learning. J. Prob. Learn. High. Educ. 5(1) (2017). https://doi.org/10.5278/ojs.jpblhe.v0i0.1550
15. Sara, R.: Live project good practice: A guide for the implementation of live projects. CEBE Briefing Guide No. 8 (2006).
16. Charlot, J., Ertas, H., Pak, B. The Architecture Design Studio as a Commons. City as a Commons, Pavia (2019). https://kuleuven.limo.libis.be/discovery/search?query=any,contains,LIRIAS3397074&tab=LIRIAS&search_scope=lirias_profile&vid=32KUL_KUL:Lirias&offset=0
17. Groat, L.N., Ahrentzen, S.: Reconceptualizing architectural education for a more diverse future: perceptions and visions of architectural students. J. Archit. Educ. **49**(3), 166–183 (1996). https://doi.org/10.1080/10464883.1996.10734679
18. Emam, M., Taha, D., ElSayad, Z.: Collaborative pedagogy in architectural design studio: a case study in applying collaborative design. Alex. Eng. J. **58**(1), 163–170 (2019). https://doi.org/10.1016/j.aej.2018.03.005
19. Sara, R., Jones, M.: The university as agent of change in the city: Co-creation of Live Community Architecture. Int. J. Archit. Res.: ArchNet-IJAR, **12**(1), 326–337 (2018)
20. Gibson, J.J.: The Ecological Approach to Visual Perception. Psychology Press (1979)
21. Greeno, J.G.: Gibson's affordances. Psychol. Rev. **101**(2), 336–342 (1994)
22. Debord G.: Théorie de la dérive, in Les lèvres nues, 9, Décembre (1956).
23. Schön, D.A.: The architectural studio as an exemplar of education for reflection-in-action. J. Archit. Educ. **38**(1), 2–9 (1984). https://doi.org/10.1080/10464883.1984.10758345

Architectural Pedagogy in the Age of AI: The Transformation of a Domain

Mustapha El Moussaoui$^{(\boxtimes)}$ and Kris Krois

Free University of Bolzano, Bolzano, Italy
mustapha.elmoussaoui@unibz.it

Abstract. The introduction of computer-aided design (CAD) programs marked a significant shift in architectural practice, with professionals transitioning from hand drawing to digital tools. This evolution sparked debates on the implications of technology for the quality and authenticity of design. Today, artificial intelligence (AI) presents a similar challenge, raising questions about its advantages and disadvantages in architectural education. This paper aims to investigate the impact of AI on architectural pedagogy, exploring both its potential benefits and the concerns it raises. The research will propose strategies for integrating AI tools into architectural curricula, emphasizing their role as an aid to human creativity and problem-solving rather than a replacement. The study will argue that adapting to AI technologies is crucial for preparing students for the future of architectural practice and ensuring they are equipped to utilize the full potential of these tools while maintaining ethical and responsible design approaches, to use the tools by being pro-active rather than passive.

Keywords: Architectural Pedagogy · Artificial Intelligence · Architectural Practice · Eco-social · Technologies in Design

1 Introduction

The architectural landscape has long been a dynamic field, with its pedagogical methods and design philosophies evolving alongside technological advancements, however sometime late [1]. From the earliest usage of drafting boards and protractors to the more recent reliance on computer-aided design (CAD) tools, each technological shift has carried far-reaching implications for both practicing architects and their educators. Today, we stand on the cusp of another monumental change—the integration of artificial intelligence (AI) into the architectural realm. This integration is not without its controversies, and it urges a renewed investigation into its potential impact, particularly on architectural education.

The introduction of AI into various industries has been both celebrated and critiqued. In medicine, AI offers advanced diagnostic capabilities; in the automotive industry, we see the advent of self-driving cars. In architecture, AI presents avenues for computational design, resource optimization, and predictive modeling, among other uses. However, like any disruptive technology, AI also raises ethical and philosophical questions. Concerns

M. Barosio et al. (Eds.): EAAE AC 2023, SSDI 47, pp. 87–93, 2025.
https://doi.org/10.1007/978-3-031-71959-2_11

range from job displacement due to automation to the elusive, sometimes unsettling, question of what constitutes creativity in the era of machine learning [2]. These issues become even more complex when translated into the educational ecosystem.

In this study, we explore the evolution of architectural pedagogy in response to technological integration. We place emphasis on assessing the extent to which AI has influenced the architectural domain, both in terms of its advancements and the challenges. Given the capabilities of these emerging digital tools, it is imperative to understand their transformative potential within the architectural landscape. As the profession grapples with the integration of AI there arises a pressing need to address the concerns and implications it presents for the future of architectural practice.

Finally, we recommend incorporating both social science and computer science courses to ensure architects remain actively engaged in the architectural process. This approach will equip architects with the skills to address challenges that AI, in its current state, cannot fully grasp. By utilizing the capabilities of AI, architects can take a proactive role rather than merely adapting to technological advancements.

2 Architecture and Technology

2.1 History

Throughout architectural history, there has been a noticeable hesitancy in integrating technological advancements [3, 4]. This inclination towards traditional methods persisted, even as the Gothic era showcased remarkable construction feats [5]. Despite the profound societal changes of the Industrial Revolution, some architects often utilized modern materials to mirror antiquated styles [6]. Twentieth-century luminaries, including Le Corbusier, highlighted the disparities between architectural conservatism and progress in other industries. The unique challenges of architecture, emphasizing individuality in design, often seemed in conflict with technological progress [7]. Yet, with the advent of the digital age in the 1990s, visionary architects began harnessing digital tools, recognizing their potential to redefine traditional architectural paradigms [8, 9].

Moreover, the introduction of Computer-Aided Design (CAD) into the architectural realm was met with a mix of anticipation and skepticism. As noted by [10], many architects expressed concerns, fearing that CAD might obliterate the human touch in design. The hand-drawn sketches, which were seen as the soulful expressions of an architect's vision, were thought to be in jeopardy. The initial CAD systems, with their rigid lines, seemed to lack the fluidity and of manual drafts.

This sentiment was echoed when Building Information Modeling emerged. While BIM promised an integrated approach, streamlining design, construction, and management processes, it was seen by some as an overly mechanized system, potentially undermining the artistry of architecture [11].

2.2 Architecture and AI

The architectural domain, a melding of human creativity and practicality, is currently at the cusp of a profound transformation propelled by technological advancements. Highlighting this shift is an insightful experiment led by DamiLee [12].

This study pitted three human architects against AI in a design competition, with both entities striving to conceptualize a 28 m² house on a challenging sloped landscape. As designs emerged, showcasing varied interpretations members of the Archibeans Discord community were entrusted to cast their judgments, unaware of each design's origin. To many's astonishment, not only did AI designs garner a majority of the votes, but they also demonstrated an ability to produce these designs at a pace humans couldn't rival. Moreover, AI's triumphs extended to rendering and textual description segments of the competition, further amplifying its role in the new architectural age.

To understand better the magnitude of change that is taking place in the domain, we will examine into the processes and stages of architectural design and execution and how AI's influence is already apparent and at which stage (Table 1).

Table 1. Artificial Intelligence in Architectural Practice

Architectural Process Stage	Technological Tools
Schematic Design and Design Development	Prototyping and visualization tools, including Midjourney, DALL-E, and Enscape. AI-driven platforms such as Autodesk Forma and UrbanFootprint for spatial analysis and site optimization [13–17]
Generative Design Solutions	Software like Autodesk's Dreamcatcher, and Rhino's Grasshopper automate design exploration [18, 19]
Construction Documentation	Kreo, Swapp, TestFit and Revit make hold the promise of seamless documentation [20, 21]
Bidding and negotiations	AI-driven estimators like Destini Estimator help with the financial aspects of projects [22]
Permitting and Compliance	AI-empowered tools like UpCodes AI offer proactive compliance checks [23]
Construction Administration	Digital platforms, such as Procore and PlanGrid, bring forth enhanced monitoring capabilities [24, 25]
Robotics and Construction	Robotic entities like SAM and Hadrian X introduce precision-oriented construction methodologies [26, 27]
Inspection, Closeouts, and Post-occupancy Evaluations	AI tools, including SiteAware and Canvas, transform building evaluations [28, 29]

Given the profound impact of these technologies, future architects should be trained not merely to use these tools but to pioneer their evolution, ensuring that technological advancements serve societal needs optimally. The above-mentioned tools, are very limited and a little part of what is really in the market. A revised curriculum should balance technological proficiency, societal implications, and the enduring essence of human-centric design rather than ocular-centric [30].

3 Talks, Surveys, and Suggestions

To understand the impact and willingness of users, academics, and students to adapt to changes, we conducted a survey, receiving feedback from 144 participants, which spanned students, educators, and practicing architects. The goal was to gauge attitudes towards the role of AI in architectural education and to identify perceived challenges and opportunities. 68% of these respondents either strongly agreed or agreed that integrating AI is crucial for the future of architectural education. 17% held reservations against this view. However, concerns emerged too; 71% feared the loss of traditional skills, 32% raised ethical concerns, 62% worried about a lack of human touch in designs. Furthermore, the willingness to adapt to a new curriculum incorporating AI was evident, with 59% being willing.

Despite the support for AI's integration and its benefits, discussions with architects, and academics all converge on a singular thought: AI can automate and expedite many architectural processes, but it remains limited in understanding how impactful it will be on the industry and the domain itself. What is clear is that the first to adapt to technologies are developers rather than academic institutions, and that's to the apparent efficiency, and economic benefits. Afterwards the pedagogical system starts its engines to follow up with the market [31].

4 Architecture Pedagogy

The architectural education landscape has been historically rooted in traditional pedagogy. Central to this approach is the "Architectural Design Studio." In these courses, students have been guided by the design professors, immersing themselves in design projects deeply influenced by architectural history, styles, and movements [32]. However, as we transition into a digital age, the call for integrating AI-assisted concept generation into these studios becomes increasingly evident [33]. Such integration provides architects with the tools to craft initial designs that are both innovative and efficient. For example, by training architects how to present their designs by using AI tools, that will cut the time spent on visuals, and make them focus on the essence of those spaces.

Alongside design, the study of architectural history and theory has played a pivotal role in shaping architects. This exploration, which looks into the cultural and societal contexts of architectural evolutions, equips students with a deep understanding of design implications through time [34]. But with AI becoming an omnipresent force, it's vital to expand the horizons of these courses. While AI may not directly interpret history or theory, its influence necessitates a broader inclusion of the social sciences, in the sense that, future architects will have much free time to draft and visualize their designs, hence, these repetitive tasks done by AI-tools, architects should be equipped more with humanities courses to be capable of touching bases with what is important for human beings, what is essential, and existential. With that in mind, courses of politics, sociology, theory, philosophy, should all be included in the curriculum. For example, a course focusing on "Politics" can dissect the symbiotic relationship between architectural trends and political narratives. Moreover, "Sociology" would provide a lens into societal structures and dynamics, ensuring that designs are deeply rooted in community contexts. While

"Philosophy" propels architects to introspect on the existential aspects of their creations, melding existentialism, aesthetics, and ethics. Additionally, a course on "Micro/Macro Economics", would help students understand the real world implication of design, and how economy has a direct implication on how and why buildings are built. In essence, the proposed courses are aimed at molding architects who are not only technologically adept but also deeply insightful about the eco-social fabric of our societies.

Field experiences, like site visits and internships, have traditionally acted as the bridge between theory and real-world application. These hands-on exposures have been further enriched by technological infusions. In the current moment, we have many technological tools, that would aid construction workers and contractors in understanding the intentions of the engineers/architects/designers – like VR, AR, post-occupancy evaluation tools, BIM coordination, etc. – hence, it would be beneficial if architects already have hands on on these tools, and understanding how they can bridge their creative concepts and ideas into reality in the easiest way possible.

Additionally, courses such as Computer-Aided Design (CAD) have ushered in a paradigm shift in design methodologies [34]. But with the advent of AI, now we see tools that can aid in construction documents, and preliminary design [35]. Furthermore, as visualization processes evolve, traditional representation courses must adopt AI-driven tools, by equipping students with the correct methods of prompt writing, and editing furthermore the designs to fit. These advanced tools, when paired with intricate prompts, can revolutionize the way architects envision and communicate designs.

However, with the digital realm expanding its footprint in architecture, the introduction of courses like "Programming for Architects" becomes imperative. Such courses ensure architects are not just passive users but understand the underlying intricacies of the tools they employ, enabling them to understand the mechanics behinds such tools, and teaching them how to create their own tool one day.

5   Conclusion

This study explored the evolving landscape of architectural practice in the age of artificial intelligence. Through a review of literature, surveys, talks and interviews, it highlighted both the potential and the pitfalls of integrating AI into architectural pedagogy.

Our suggested courses serves as a general format of how AI can be integrated into architectural education in a manner that is both technologically progressive and ethically sound. We assert that future architects should not be limited to traditional construction and design roles, as architecture as we know it, seems to be changing forever. Therefore, architectural education should strive to produce versatile individuals, capable of applying their unique skill sets across various domains—from entrepreneurship to social reform. The courses aim to prepare future architects to be proactive users with new technologies, being pioneers rather than passive users, and late to understand the impact of such technologies on the domain. While if future architects took the leading role, they will be able to shift the market's direction towards their visions, by taking into consideration the historical, social, economic, philosophical, and political, ahead of starting the design, rather than jumping on on the train of construction and market's needs so late, that they will themselves only be tools in the hand of developers/stakeholders.

As AI continues to evolve and permeate various aspects of human life, and architectural process, architectural education cannot afford to remain static. Our study provides an exploration of the challenges and opportunities that lie ahead. Educational institutions must adapt to this changing landscape to equip the next generation of architects with the skills, ethical awareness, and versatility they need to navigate an increasingly complex world.

It is clear that the integration of AI into architectural education is not a question of mere technological adoption, but a complex interplay of pedagogy, ethics, and accessibility. The need for a strategic, thoughtful approach to this integration cannot be overstated. Educational institutions, industry partners, and regulatory bodies must collaborate to create an educational ecosystem that embraces the possibilities offered by AI while upholding the ethical and creative standards that define the architectural profession.

References

1. Carpo, M.: The Second Digital Turn: Design Beyond Intelligence. MIT Press, UK (2017)
2. Leach, N.: Architecture in the Age of Artificial Intelligence: An Introduction to AI for Architects. Bloomsbury Academic. Bloomsburry Visual Arts. UK (2022)
3. Picon, A.: Digital technology and architecture: towards a symmetrical approach. Technol. Archit.+ Des. **6**(1), 10–14 (2022). https://doi.org/10.1080/24751448.2022.2040297
4. Fallon, K.: Early computer graphics developments in the architecture, engineering, and construction industry. In: IEEE Annals of the History of Computing, vol. 20, no. 2, pp. 20–29 (1998). https://doi.org/10.1109/85.667293
5. Clark, K.: Gothic Revival: An Essay in the History of Taste. John Murray Pubs Ltd. UK (1995)
6. Frampton, K.: Modern Architecture: A Critical History. Thames & Hudson, UK (1988)
7. Frampton, K.: Studies in Tectonic Culture. MIT Press, USA (1995)
8. Lynn, G.: Animate Form. Princeton Architectural Press, USA (1999)
9. Picon, A.: Digital technology and architecture: towards a symmetrical approach. Technol. Archit.+ Des. **6**(1), 10–14 (2022). https://doi.org/10.1080/24751448.2022.2040297
10. Lewicka, A.: Architectural work output - space presentation forms. Czasopismo Techniczne **2015**, 91–95 (2016). https://doi.org/10.4467/2353737XCT.15.287.4690
11. Heseltine, P.: BIM and the Art of Architecture. Wiley & Sons (2003)
12. [Video] (2023) We tried to compete with AI... [AI vs. ARCHITECT]. Last accessed 23.10.05 from https://www.youtube.com/watch?v=N709ZrxoIP0&t=563s&ab_channel=DamiLee
13. Midjourney (n.d.) Prototyping and visualization tool. Retrieved from https://www.midjourney.com/home/?callbackUrl=%2Fapp%2F
14. OpenAI (n.d.) DALL-E: Image generation from text. Retrieved from https://openai.com/blog/dall-e/
15. Enscape (n.d.) Real-time rendering & virtual reality. Retrieved from https://www.enscapeitalia.it/
16. Autodesk Forma (n.d.) AI-driven platform for spatial analysis. Retrieved from https://www.autodesk.it/
17. UrbanFootprint (n.d.) Data-driven urban planning. Retrieved from https://urbanfootprint.com/
18. Autodesk (n.d.) Dreamcatcher: Generative design tool. Retrieved from https://www.autodesk.com/solutions/generative-design
19. Rhino (n.d.) Grasshopper: Algorithmic modeling. Retrieved from https://www.grasshopper3d.com/

20. TestFit (n.d.) Building configuration software. Retrieved from https://testfit.io/
21. Autodesk (n.d.) Revit: BIM software. Retrieved from https://www.autodesk.com/products/revit/overview
22. DESTINI (n.d.) Estimator: AI-driven construction cost estimator. Retrieved from https://www.beck-technology.com/destini-estimator/
23. UpCodes (n.d.) AI: Proactive compliance checks. Retrieved from https://up.codes/ai
24. Procore (n.d.) Construction management software. Retrieved from https://www.procore.com/
25. PlanGrid (n.d.) Construction productivity software. Retrieved from https://www.plangrid.com/
26. SAM (n.d.) Semi-automated mason. Retrieved from https://www.construction-robotics.com/sam100/
27. Hadrian X. (n.d.). Robotic bricklaying. Retrieved from https://www.fbr.com.au/view/hadrian-x
28. SiteAware (n.d.) Digital replication and analysis of buildings. Retrieved from https://www.siteaware.com/
29. Canvas (n.d.) Building evaluations and renovations. Retrieved from https://www.canvas.build/
30. El Moussaoui, M.: The ocular-centric obsession of contemporary societies. Civil Eng. Archit. 8(6), 1290–1295 (2020). https://doi.org/10.13189/CEA.2020.080613
31. Leach, N.: Architecture in the Age of Artificial Intelligence: An Introduction to AI for Architects, pp. 15–26. Bloomsbury Visual Arts. UK (2022)
32. Unwin, S.: Analysing Architecture. Routledge (2009)
33. Kolarevic, B.: Architecture in the Digital Age: Design and Manufacturing (2010)
34. Groat, L., Wang, D.: Architectural Research Methods. Wiley, USA (2013)
35. Eastman, C., Teicholz, P., Sacks, R., Liston, K.: BIM Handbook: A Guide to Building Information Modeling for Owners, Managers, Designers, Engineers, and Contractors. John Wiley & Sons (2008)

Teaching Architecture in the Age of Fragility

Camillo Frattari[⊠]

Polytechnic of Milan, Milano, Italy
camillo.frattari@polimi.it

Abstract. The origin of academic architectural education can be found in the inauguration of the Académie d'Architecture in 1671 in France based on the idea of knowledge as the transmission of a 'style'. This idea continues to characterize contemporary architectural education but in a different way, through the pursuit of a personal aesthetic expression or in the satisfaction of real estate and construction market requirements as a cynic acceptance of reality. To conceive architectural education only as serving the market is to lose the meaning of architecture itself as a way to design, and improve, humanity. The contemporary age is characterized by the notion of fragility that represents the many uncertainties of our time related to different issues involved in politics, economy, energy, ecology, and demography. In the field of architecture and urban design, the main causes of fragility come from the phenomena of planetary urbanization and climate change that are visible in disaster, migration, periphery, inequality, diversity, and planetary crisis. While the 'Academié' model of education is founded on the knowledge of historical buildings in their aesthetical prerogatives, an architectural education focused on the contemporary issues of fragility needs to be based on the knowledge of projects that faced these challenges, to define a conceptual framework of a new agenda with new theories, technics, and references. Teaching architecture in the age of fragility means promoting the critical role of architects inside, and for, human society through projects of adaptation, hospitality, community, process, coexistence, and imaginary.

Keywords: Architectural education · Fragility · Adaptation · Hospitality · Community · Process · Coexistence · Imaginary

1 Introduction

The origin of academic architectural education can be found in the inauguration of the Académie d'Architecture on the 3rd of December 1671 in France. This was the first institution whose entire focus was on the study of architecture and the specific training of architectural students [1].

The first director of the Académie, the architect Nicolas-François Blondel, outlined in a lecture the dual role of this new institution. The first role was to study the history of architecture, survey historic buildings, and define the most correct form of Classicism. The second role was to teach the found knowledge to 'students'. Initially, the program of the Académie comprised weekly public lectures given by Blondel, but in the following

© The Author(s) 2025
M. Barosio et al. (Eds.): EAAE AC 2023, SSDI 47, pp. 94–101, 2025.
https://doi.org/10.1007/978-3-031-71959-2_12

decades was devised a teaching structure that resembled what is practiced in many current Western architectural schools.

Aesthetics as a dominant value remains in the sudden 'revolutionary architecture' of eighteenth-century France as the work of Etienne-Louis Boulée, Claude-Nicolas Ledoux, and Jean-Jacques Lequeu that express in a 'visionary' way the ideals of the time [2].

On the opposite, the Bauhaus school founded by Walter Gropius in 1919 in Weimar focused on artistic self-expression and handcraft techniques that negate classical style. It proposed a training model that reflected the needs of a 'New Architecture' of 'standardization' and 'rationalization' for mass production [3].

The idea of knowledge as the transmission of a 'style' or a 'technique', continues to characterize contemporary architectural education but in a different way, through the pursuit of a personal aesthetic expression or in the satisfaction of real estate and construction market requirements as a cynic acceptance of reality. The outcome is an architectural production that emphasizes the expression of the architect as a style, or on the other hand his disappearing in the mannerism of the mass production.

In 1971, Charles Jencks figured out in a diagram titled 'Evolutionary Tree to the Year 2000' [4], six coherent traditions that tend to self-organize around underlying structures: logical, idealist, self-conscious, intuitive, activist, unself-conscious. These deep structures act like 'attractors' not only because of personal preferences but as a result of typecasting and the ways in which the market requires architects to possess a distinct style and level of expertise.

The diagram was intended to be a prediction of architecture until the year 2000, so after 30 years he made a revised version of the diagram titled 'The Century is Over. Evolutionary Tree of Twentieth-Century Architecture' to summarize what happened. First, the diagram showed that almost 80 percent is not made by architects, or at least is the outcome of wider processes that are unselfconscious from an artistic point of view: 'building regulations, governmental acts, the vernacular, planning laws, mass housing, the mallification of the suburbs, and inventions in the technical/industrial sphere'. The second observation is that the ecological imperative, started as a polemical movement, has been adopted by each of them in different ways so 'green architecture' comes from everywhere like a label on different products. The last one is the existence of a 'reactionary modernism' favored by corporate forces of production and patronage, looking for an impersonal, abstract, semi-classical sobriety [5].

To conceive architectural education only as serving the market is to lose the meaning of architecture itself as a way to design, and improve, humanity. This reduces education to the explanation of a manual for a new kind of working class, it becomes merely instruction.

2 Fragility

The contemporary age is characterized by the notion of fragility that represents the many uncertainties of our time related to different issues involved in politics, economy, energy, ecology, and demography [6].

Fragility is the quality of an object or system to be easily 'broken' even by a weak force, and it can be easily referred to people, cities, territories, and ecosystems to explain the instability of contemporary living conditions. In the field of architecture and urban design, the main causes of fragility come from the phenomena of planetary urbanization and climate change that are visible in disaster, migration, periphery, inequality, diversity, and planetary crisis. While the 'Academié' model of education is founded on the knowledge of historical buildings in their aesthetical prerogatives, an Architectural education focused on the contemporary issues of fragility needs to be based on the knowledge of projects that faced these challenges, to define a conceptual framework of a new teaching agenda, with new theories, technics, and references.

The following case studies illustrate several experiences that could be seen as a response to the issues of fragility and offer a wide range of conceptual tools for architectural education.

2.1   Disaster > Adaptation

In the past 50 years, the number of natural disasters has increased and affected territories all over the world that need to be restored from traumatic events or enhance their resilience. In response to the catastrophe brought on by the precariousness of Italian land, Marco Navarra (NOWA), developed the project 'Loco Grande: variante al progetto del canale fugatore' [7] in Giampilieri after the 2009 Messina flood.

Since the restoration projects involved every river flow from upstream to downstream, they required the modulation of instruments fitted to varied situations, ranging from the landscape to the urban context, following the flood. The Messina civil engineers' first project for the hydraulic canal in the "Vallone Puntale" area was based on several strict decisions that provided few options for reducing the impact of the work on the land. The options were limited to burying the reinforced concrete box and covering the side walls with stone from the surrounding area.

The alternative put forth by NOWA with the motto 'Riparare Fiumare', drastically reexamined the purely technical solution to the issue and proposed a sophisticated urban project that was mindful of the locations and the community relations entwined with it, a point at which the previous hypothesis would have irreversibly destroyed. Certain project tools, including the section and topographic models, needed to be improved due to the unique nature of the interventions along the rivers (also known as 'fiumare').

The capability of rethinking space in a nomadic key is made possible by the compulsive repetition and superimposition exercise. Emphasis is placed not only on the archaeological fragments that arise like permanencies following a traumatic occurrence but also on the traces that are erased and moved along the route. The result is a project that combines infrastructure, landscape, public space, and belonging to the notion of adaptation.

This case highlights the importance of considering the land as a fundamental part of the design project through the comprehension of geology and morphology using the architectonic tool of sections to develop an integrated design with landscape and infrastructure, in symbiosis with natural events.

2.2 Migration > Hospitality

Natural disasters, hazards, wars, and poverty that arise in many countries push millions of people to migrate. From refugee camps to local hosting centers, the role of architects seems to be merely technical support in emergency management focused on providing shelters and survival for a short time response. Looking at the persistence and the complexity of these phenomena emerges the need for design strategies that accept migration as a context in which to operate and not only as a transitional condition.

Studio ABVM of Bonaventura Visconti di Modrone in collaboration with Leo Bettini Oberkalmsteiner tackled the issue of migration with their 'Maidan tent' project [8]. The disembarked, the repatriated, and the relocated live in the unseen cities that are the refugee camps in Greece along the Macedonian border. Humanitarian emergencies are now occurring in tent cities that were first built as temporary solutions.

The 200 square meter Maidan tent, which can accommodate 100 people, is the venue for social and sharing events including movie screenings, birthday celebrations, World Cup TV watching, and Ramadan celebrations. It will soon house a market for fruits and vegetables.

The circular Maidan tent structure is separated into eight pieces, each with two concentric zones. In addition to providing some privacy, it fills the social and interactive role of a public square, which is becoming less and less common in the age of globalization. Steel and aluminum are used in its 16-m diameter construction to provide a light and strong structure. The fabric canopy resists fire, wind, and water.

Gender inequalities were present in the requests. Men wanted a location to play ping-pong, play cards, smoke shisha, drink tea, and sell cigarettes. Women requested that the Maidan be divided into two sections, with doors, enclosed places, and spots for the kids to play.

In these 'ephemeral cities' that appear as new kinds of urbanization, the case of the Maidan tent expands the notion of hospitality through a new type of public space for a peculiar context. Light structures and flexible spaces are used to give form to a shared functional program according to different cultures and religious beliefs, and accepting migration as a new living condition that needs space and architectural responses.

2.3 Periphery > Community

The growth of urbanization that involves cities in the last decades, has increased the size of the periphery as a critical but predominant area. In that context, the density of population and informal settlements compounded by social and crime problems is proportioned to the absence of public services and facilities and demanding urban regeneration to avoid gentrification and top-down strategies.

The works realized by 'El Equipo Mazzanti' are emblematic interventions for the issue of the periphery as the case of the Parque Biblioteca España [9], which is situated on a hillside that has been impacted by violence since the 1980s due to the Medellin drug trafficking network. The government's social master plan program includes it in its efforts to provide everyone with equitable access to social and economic possibilities. A facility with an auditorium, training room, administrative room, and library on a separate volume was requested by the program. The plan was to divide the program into three sections: the auditorium, the rooms, and the library. These sections would then be connected by a bottom platform that would allow for flexibility and autonomy, hence increasing people's engagement since each section would function independently. The project is divided into two main structures: the building, which is made of rocks, and the platform, which integrates the cover and turns it into a square that explores the valley. In this way, the building gains more authority as a gathering spot, multiplies its connections, and allows it to grow as a point of reference.

More than just a building, it suggests creating an 'operative geography' that is part of the valley, akin to a mechanism for organizing the program and the zone. It does this by displaying the unknown directions of the erratic mountain contours, not in the sense of a metaphor but rather as an organization of the form in the location, a folded building sliced like the mountains. The project is a statement about architecture that belongs to the community and represents the context also in its morphology and appearance.

2.4 Inequality > Process

Urban regeneration of areas occupied by informal settlements usually causes phenomena of gentrification that push away residents toward the next periphery. Due to the economic values of housing and public facilities, they are often not affordable to all, and so invalidates the possibility to improve urban living conditions through architectural interventions.

The housing of Quinta Monroy [10] realized by Alejandro Aravena and Elemental faced the issue of inequality. Quinta Monroy was the last informal settlement in the heart of Iquique, a city located 1,500 km north of Santiago in the Chilean desert, at least at the time the project was started. Due to the poor living conditions, the government decided to replace the settlement by providing new housing units for the families residing there.

The project's first crucial choice was to remain on the same land, which was three times more expensive than the area typically designated for social housing. This was done to prevent evicting the current residents and moving them to the periphery, where land is less expensive but may cause marginalization and negatively impact property value growth. In addition, the 7,500 USD subsidy per family permitted the construction of a maximum of 36 square meters, which is half the area of a typical middle-class home. The project employs a typology that effectively utilizes the available space, allows for controlled house extensions to prevent crowding, and encourages self-build procedures in order to address these concerns. Families receive half of a nice home furnished with first-rate services, along with technical assistance so they can complete the additions on their own.

While architects usually claim to have total control over the formal outcome of the architectural project as an accomplished work of art in itself, Elemental focused on the process as the object of architectural design to develop a system open to future modification.

2.5   Diversity > Coexistence

Diversity of faith and culture is often the cause of social conflicts, especially in the era of globalization and melting pot society in which interactions between people from different backgrounds are more frequent.

The Abrahamic Family House [11] realized by David Adjaje Associates approaches the issue of diversity of culture and belief in a symbolic place. The project consists of three religious buildings situated atop a secular visitor pavilion: a church, a synagogue, and a mosque.

The house will foster the ideals of peaceful coexistence and acceptance among various beliefs, nationalities, and cultures by acting as a community for interreligious dialogue and exchange. Visitors will be able to take part in religious services, hear readings from the Holy Scriptures, and participate in sacred ceremonies within each house of worship. The fourth area, which is independent of any one religion, will act as a center for all good-willed people to unite as one. Additionally, the community will provide event-based and instructional programming.

The form is derived from the three faiths, and it is made by carefully defining what is similar versus what is different through the lens of these revelations. The next area of discovery is the shared ground, the public area between the three buildings, where the differences meet. The garden, which is situated between the three chambers and the three faiths, is used as a potent metaphor—a secure haven where civility, community, and connection coexist. The podium breaks down barriers to inclusion, enables one to engage with each space without feeling excluded, and promotes the celebration of this shared history and identity.

The case of the Abrahamic family house reveals how architecture through formal dialogue between different languages and identities, reflecting cultural and religious values, can build a space of shared feeling that unifies the diversity of beliefs, and thus people, realizing coexistence.

2.6   Planet > Imaginary

Global warming, the primary cause of climate change, is a process that started with the Industrial Revolution and increased by the accumulation of habits and production activities consolidated throughout the years. The planetary challenge to invert climate change requires an approach that could not be so immediate and solved in the short term. It needs to reinvent the way we conceive the whole society and thus architecture.

Four Dioramas [12] by Nemestudio is a speculative design that investigates the crisis of the Planet in terms of climate and resources. The project constitutes the physical exhibition of the Pavilion of Turkey at the 2021 Venice Architecture Biennale titled 'Architecture as Measure', which consists of four dioramas and a large table. The project queries the politics and implications of seemingly insignificant aspects of architectural construction and compares them with their planetary dimensions, such as resource extraction geographies, material supply chains, maintenance, and care in Turkey, in an effort to spark a renewed sense of planetary imagination.

Four dioramas each depict a portion of a narrative about the solidarity and survival of multiple species. Some of the current environmental controversies are reflected in the story through the archaeology of an imaginary future Turkey. An abandoned marble quarry depicted in the Diorama of Quarry is the result of centuries of resource extraction. We observe the extractive construction activities of ancient times and have a view towards the outdoor museum. Diorama of Logistics is a sizable warehouse that facilitates a significant amount of cargo during a multispecies migration to a new location. Diorama of Maintenance and Care is a repair site in the New Land where ongoing maintenance and care are provided for both the constructed structures and the endangered non-human species. A reconstruction site for Earth's future occupants is the Diorama of Formwork. In the distance, one can make out massive Carbon Monuments and a variety of statues representing the more-than-human imagination of the Ancient Lands of Anatolia. All of these depict the new mythologies of the 'New Land' and thus the future.

While architectural education typically involves transmitting established values, techniques, and practices based on the works of past masters or buildings, whether recent or remote, the work of Nemestudio exhibited in Four Dioramas, restores importance to the role of imaginary as a way to shape the present towards a different future.

3   Conclusions

Architectural education based on the classical academic approach of the transmission of a style is affected by the dominant market rules. The issues of fragility that characterize the contemporary age such as disaster, migration, periphery, inequality, diversity, and planet, require a different educational system based on social topics, instead of purely aesthetics, and referred to works that spread the field of architecture in contamination with external instances, maybe unexplored yet.

The case studies previously addressed are used to define notions that help to define an educational agenda: the emergency caused by a Disaster is faced by a project of 'Adaptation'; the phenomenon of Migration needs to embrace the notion of 'Hospitality'; the conflicts of the 'Periphery' are solved by spaces for 'Community'; the scarcity of resources that produce inequality can be overcome by the 'Process'; the 'Diversity' of beliefs is reflected in a peacefully 'Coexistence'; the crisis of the 'Planet' reveals the need of a new 'Imaginary' to design an alternative future.

Teaching architecture in the age of fragility means promoting the critical role of architects inside, and for, human society through projects of adaptation, hospitality, community, process, coexistence, and imaginary. Adaptation refers both to the use of architectural tools to understand the consistency of context in the whole and to design architectures as adaptive organisms. Hospitality expands the notion of public space to the possibility of giving dignity to unstable living conditions in a neglected place. Community regards the feeling of belonging to an architecture made possible by people's participation and the sensitive appearance of the construction. Process as the object of architectural design to develop a system open to modification. Coexistence refers to the potential of architectural language and space to create a dialogue between cultures and people. Imaginary is a way to subvert the present scenario of habits and socioeconomic structures to prefigure a truly sustainable future.

References

1. Griffin, A.: The Rise of Academic Architectural Education: The origins and enduring influence of the Académie d'Architecture, 1st edn. Routledge, Oxon (2019)
2. Kaufmann, E.: Three revolutionary architects, Boullée, Ledoux, and Lequeu. Trans. Am. Philos. Soc. **42**(3), 431–564 (1952)
3. Gropius, W.: The New Architecture and the Bauhaus. Mit Press, Cambridge MA (1965)
4. Jenks, C.: Architecture 2000: predictions and methods. Studio Vista, London (1971)
5. Jenks, C.: Jencks' theory of evolution: an overview of twentieth-century architecture. Archit. Rev. **208**(1241), 76–79 (2000)
6. Chiffi, D., Curci, F.: Fragility: concept and related notions. Territorio **91**, 55–59 (2019)
7. Navarra, M.: Architettura geologica: Traiettorie circolari nell'antropocene. Techne **11**(22), 35–41 (2021)
8. Maidan Tent. https://www.maidantent.org/, last accessed 2023/10/28
9. Parque Biblioteca España, El Equipo Mazzanti. https://www.elequipomazzanti.com/es/proyecto/parque-biblioteca-espana, last accessed 2023/10/28
10. Elemental, Incremental housing projects, Quinta Monroy. https://www.elementalchile.cl/en, last accessed 2023/10/28
11. Abrahamic Family House, Adjaye Associates. https://www.adjaye.com/work/the-abrahamic-family-house, last accessed 2023/10/28
12. Four Dioramas, Nemestudio. http://nemestudio.com/projects/venicebiennale, last accessed 2023/10/28

Risk, Trust and Big Beautiful Mistakes*: Keys to Innovation in Architectural Education

Johanna S. Gullberg[1] and Gro Rødne[2]([⊠])

[1] Henning Larsen Architects, Trondheim, Norway
[2] Faculty of Architecture and Design, NTNU, Trondheim, Norway
gro.rodne@ntnu.no

Abstract. By presenting examples from Making is Thinking and other educational initiatives in the architecture programme at the Norwegian University of Science and Technology, NTNU, this paper points to why risk-taking is especially needed in architectural education today, as global crises demand new and critical ways of working with architecture. From a Scandinavian horizon, the paper problematizes the paradox that educational systems instead of promoting education as the uncertain and troublesome business it is, tend to please students in their longing for predictability. The paper describes risky learning spaces where learners start by doing, preferably in teams and off campus, thereby counteracting predictability and opening up for unexpected mishaps to occur – mishaps with innovative potential. The paper shows that such risky learning spaces introduce collective, emotional and embodied dimensions in architectural education. Moreover, the paper contributes with experience-based knowledge regarding how trust – a precondition for risk-taking and a notion underexplored in research on higher education – may be established in educational settings. Building on the idea of a fruitful tension between risk and trust in educational spaces, the paper ends with a list of gutsy proposals for a pedagogy of mistakes. *"Big beautiful mistakes" is an expression used with inspiration from architect and educator Sami Rintala.

Keywords: Architectural education · Design-build · Material learning spaces · Risk · Trust · Pedagogy of mistakes

1 Introduction: The Research Question

In this paper, it is claimed that risk-taking is essential in architectural education. Yes, according to the authors' experiences as educators at the Norwegian University of Science and Technology (NTNU), risk-taking is in fact a requirement for the reshaping of the role of the architect that is so urgently needed if architecture is to be societally relevant now, in our times of crises. The paper builds upon previous publications by the authors, see especially Johanna S. Gullberg's doctoral thesis [1] and an article by Gro Rødne and Leif Martin Hokstad [2]. The paper is meant to provide educators, in particular those who teach in architecture and other aesthetic fields in higher education, with instructions for how to implement risk in curricula. It is therefore guided by the following research

© The Author(s) 2025
M. Barosio et al. (Eds.): EAAE AC 2023, SSDI 47, pp. 102–111, 2025.
https://doi.org/10.1007/978-3-031-71959-2_13

question: *How can learning spaces in architectural education be set up to encourage risk-taking, and so that the effects of risk-taking can be processed and appreciated?* The paper is based on the idea that architects may contribute to educational research by developing perspectives on material and spatial dimensions of learning spaces.

2   Risk: Why It Is Needed, and Why It Tends to be Avoided

In architectural education, risk-taking is needed because it makes learners develop abilities at critical thought and action which enable them to take part in forming both the curriculum and the professional field of architecture. These abilities are essential, because the only thing certain is that the future is uncertain. According to the World Economic Forum 2023 [3], resilience, flexibility and agility are among the top five skills growing in importance by 2027, with creative thinking on top. Architects need to feel at home in complexity and manage a multitude of parameters, often completely contradictory, in order to achieve holistic solutions that create value for societies in a rapidly changing world.

It is therefore problematic, as educational researchers show, that universities reduce uncertainty. Gert Biesta [4] and Ray Land [5] are both worried that standardization of higher education leads to that learning becomes associated with smooth success while we lose sight of what it actually is: a practice that is slow and full of risks. The idea of the student as consumer can, says Ronald Barnett, be seen as a logical consequence of that universities exist in a market-oriented society [6]. But also, universities themselves, Barnett continues, inhibits free thought, for instance by pushing neighbouring fields to compete, and by rigging assessment systems where the survival of courses depend on benevolent feedback from students. Barnett thinks that these external and internal circumstances contribute to that many educators avoid putting students in "challenging situations", and, in turn, to that students get fewer opportunities to practice their ability at criticality [6].

Architectural educators should embrace challenging situations instead of avoiding them. There is a common understanding, exposed for instance at the EAAE conference in Torino 2023, that a paradigm shift must come within the field of architecture. However, the myth of the individual master architect still lives. It does so although architectural practices do much more than serve the market of the building industry [7], and in spite of repeated calls for that the spatial settings and pedagogical guidelines of architectural education must be changed to reflect the fact that architecture is a collective practice full of compromises [8]. Changes to architectural education are more urgently needed than ever. As Beatriz Colomina and her colleagues say in the introduction to the book *Radical Pedagogies*, we live in a moment of "global crises, ecological catastrophe, and rapidly increasing inequities [when] the challenge to inherited disciplinary hierarchies can, and must, happen in the spaces of education" [9].

When established curricula for design, history and theory courses as well as learning spaces and professional habits are to be questioned, both learners and educators will be presented with risks. It is well known that although individuals and systems may be aiming for change, resistance to risky experiences of learning, will occur. For instance, in addition to facing their own habits, educators will most likely need to help learners tackle "design fixation" phenomena where learners are stuck with an idea or design darling where good ideas or previous successes with similar challenges may in fact block better solutions [10, 11]. Figures 1 and 2 are based on experiences made at NTNU and may function as somewhat strange reminders of why it is important to conquer resistance to learning and move towards student active or even student-driven learning, where learners pose questions rather than gather information, look at problems creatively and flexibly from several perspectives, communicate with clarity, and cope with challenges [12].

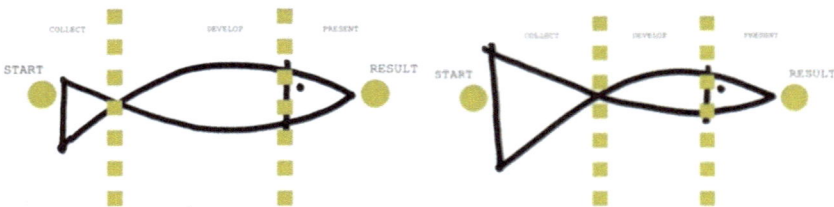

Fig. 1. The linear design process – where learners collect information, develop a design and then present it – seen as the body of a fish, with the fish to the right showing that students tend to feel comfortable with hoarding information and thereby shrink the design process, Trondheim, 2023 (Illustration by G. Rødne, after B. O. Braaten).

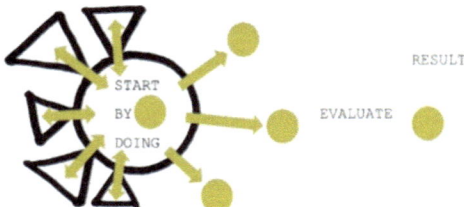

Fig. 2. This paper proposes that the design process should look more like this pulsing creature: start by doing, and let analyses, problem definitions and designing happen simultaneously, Trondheim, 2023 (Illustration by G. Rødne).

In the pulsing model, Fig. 2, problems will be found and approaches to those problems tested out. This will involve the appreciation of mistakes, or rather outcomes that would be regarded as unexpected or strange by conventional protocols, but which – if they are taken seriously – can lead to innovation.

3   How to Set Up Risky Learning Spaces

However, learners in the Scandinavian context, having been brought up in rich well-fare nations, are usually inexperienced in meeting challenges and resistance. Their reactions to trouble may be characterized by stress rather than creativity, not least because national basic education have taught them linear and causal logics. For the time being, top grades from such basic education is the only entrance ticket to architecture studies at NTNU. One aim with this article, is simply to point out that mismatches in admission requirements at NTNU and other architecture schools should be mapped and adjusted.

Mismatches between linear logics and the need for appreciating experimentation and mistakes become evident in the very first course in the architecture programme at NTNU. It is a month-long design-build course where students complete an entire design process, from idea to full-scale structures, on sites in the city of Trondheim (Fig. 6). Other pedagogical approaches at NTNU which present learners with risk, are the learning perspective *Making is Thinking* (MT) [13], implemented in various formats across the curriculum, and the Master courses *Design in Context* (DC) and *Experimental Practice* (EP). The three approaches all work both on and off campus, and they share aims for student-driven learning processes where hands-on making is a gateway to awareness of oneself as a learner and as an actor in real-life contexts dealing with societal challenges like gentrification and segregation. Not least may hands-on making and a varied use of bricolage techniques be keys to awareness of sustainability logics, teaching students to take care of what there is rather than make perfect plans for a distant future. DC is organized as an architectural office where student teams are exposed to risk by being given the freedom and responsibility to manage real projects from the first client dialogue to completion. Embodied experiences have here shown to be connected to transformed perspectives. As Sami Rintala, one of the tutors, says, the design-build workshop format lets learners who are not used to making start "a physical change that allows mental change to take place" [14]. EP and MT both acknowledge artistic and crafts-based dimensions of architecture. They set up non-linear design processes with hands-on experiments for learners to make practice-based investigations of and responses to wicked problems in a world in radical change.

Based on previous research findings [2] and on what is going on in the mentioned pedagogical approaches, the outlines of three features of risky learning spaces appear, a) Start by doing, b) Work in teams, and c) Work off campus. In order to concretize paths towards the needed systemic transformations of architectural curricula, a, b and c will be further described in the following. In addition, there is a basic prerequisite to keep in mind. Educators must dare to take the risk to d) Leave the students alone. Because, without lived experiences of failures and problem-solving, learners will never become active and critical architects.

3.1 Start by Doing

As mentioned, architecture students at NTNU are thrown into an intense design-build process, a crash course in the risks and possibilities of starting by doing. Yet, this experience does not prevent them from getting stuck in linear design processes at other points during their studies. A basic tactic of MT is to always start by doing exercises, for instance blind drawing or collective model-making, that loosen up tensions and potentially nudge students to trespass any fear of mistakes (Fig. 3). Bricolage techniques enable learners to work with elements considered as garbage or found objects like sand and sticks. To transform an empty space into a shared and tangible learning space by filling it with visual and physical tests is a key to risk-taking. In this process, it is essential to pay attention – for instance to the possibilities of found sand and sticks. It is about saying "yes to the mess" [15], improvising with the richness of unknown resources of people and places rather than sticking to comfortably planned intentions. In an educational setting, such improvisation demands a scheme that is robust and yet flexible.

Fig. 3. Workshop Cirka Teater and N. E. Holtan, MT, Trondheim, 2016 (Photos by Cirka Teater and NTNU).

3.2   Work in Teams – With Peers, Educators, Professionals From Other Fields

To work with others involves idea-generation both through friction and playful interaction, where "who did what" becomes uninteresting [2]. As mentioned above, university structures should support exchanges between disciplines although this may lead to unpredictable outcomes. At NTNU, first-year tutors invite art and music students to perform in the students' full-scale structures. Both EP and MT fuel architectural investigations by enacting artistic perspectives. For instance, workshops on textile art and weaving are held by EP, and MT has long-term collaborations with the theatre company Cirka Teater [16]. In the meeting with the theatre, it has shown to be especially challenging for architects to relax in working with embodied and emotional dimensions of material space [1]. The theatre company's methods and creative bravery do however help learners and educators to step out of their comfort zones. Who can hold on to habitual positions, when the "clients" are The Stick Man and The Oyster Lady, or the site is on a bunker wall?

3.3   Work Off Campus – in Real-World Settings, with Citizens

Working in a real-world setting, with actual stakeholders and complex problem-solving, is important to understand that no design process is free from obstacles. This is a basic idea of all DC projects, teaching learners to handle logistics and communication, but also letting them feel that they can steer the world, project by project, towards sustainability. In EP, links between theory and live projects are enhanced, for instance when permaculture principles guided design interventions at a farm outside Trondheim. In collaboration with Cirka Teater and other actors within the cultural field, MT has set up public events and exhibitions in Trondheim. Another MT example of working off campus is from 2021, when a temporary intervention was set up to make locals care for a neglected area in Larvik (Fig. 4). In spite of negotiations regarding pandemic and other security issues, MT engaged a diverse group of citizens in creating an architectural interpretation of a baroque garden. The intervention sparked the municipality to rebuild the garden and renovate the buildings surrounding the site. The learning outcomes from working off campus involve social and embodied dimensions which are hard to mimic in a traditional studio setting. Moreover, as in this case, students may get a lived experience of that temporary interventions can make a real difference.

Fig. 4. MT workshop in Larvik, 2021 (Photos by Larvik municipality and G. Rødne).

4 Trust: A Precondition for Risk-Taking

As earlier mentioned, this paper stems from a Scandinavian perspective and thus a generally non-hierarchical learning tradition, where teamwork is common and barriers between educators and learners are low. These are great outsets for learning spaces characterized by mutual trust, where educators may dare to engage learners in forming the curriculum, and learners may dare to share mistakes and doubts, thereby becoming able to enter transformative states of learning. As shown, however, architectural educators still have a way to go in designing structures and spaces that allow for sharing experiences. In the following, examples of exercises for building trust in design courses will therefore be briefly introduced, as well as thoughts on how forms of assessment, reflection and feedback may be arranged throughout processes and in relation to those exercises. Because, as educational theorist David Carless says, "without trust, students may be unwilling to involve themselves fully in learning activities which may reveal their vulnerabilities" [17]. Trust, he continues, is built on openness, reliability, honesty, benevolence and competence – and to build learning spaces where risk-taking and mistakes are acknowledged, educators must therefore develop not only professional knowledge but also interpersonal, communicative skills [17].

5 How to Establish Trust

It is crucial that students feel at home in complex and uncertain situations, and that they learn to recognise mistakes as a necessity. In fact, failures are essential for all forms of creative work, and liberating laughter serves as a creative trigger [18]. To work in teams is risky but also a key to trust. To create a collective drawing, for instance, is a fun exercise with the serious intention to conquer obstacles of self-censorship and move towards collective pride. DC encourages learners to take on roles in teams and thereby build their individual competences. As Nina Haarsaker, architectural educator at NTNU, puts it, teamwork can be like being a band – instead of all "playing the guitar", we should trust each other in developing different means to enrich our joint work. While it is urgent, as mentioned above, that students are left alone, it is also important that educators level with the students. As Rintala says, "learners will trust educators who earn their position by working with the group, making discoveries together and thereby showing that investigations continue also after university exams" [14]. While authority in the traditional master–apprentice model is given through the educator's predefined position, the authority of the educator who participates on equal terms may be described as co-created with the students.

5.1 Breed Trust Through Hands-on Collaboration

Before dealing with the risks of real-world settings, learners may need to build up confidence and trust in themselves and others. Within the frame of MT and the research centre TransArk [19], two material settings for such preparation, the FormLab and the Sandbox, have been developed, as well as a cogenerative model for learning through reflection. The FormLab challenges the conventional idea of the design studio by being an accessible learning lab where students and staff may test hands-on techniques, and thereby learn to acknowledge failures as a necessary and productive condition for creativity. The interactive digital and analogue Sandbox is placed in the FormLab and aims to reduce the fear of making mistakes. The distance between idea and hand is reduced as three-dimensional sketches are easily made and remade, while a camera and a 3d scanner document all phases of the sketch process. As they work together around the sandbox, drawing in and placing objects in the kinetic sand, learners may relax in sharing ideas by responding to what their peers are doing. In other words, trust is bred through hands-on collaboration.

5.2 Develop Relevant Forms for Reflection and Feedback

There are several reasons for architectural educators to learn more about structures for reflection, assessment, and feedback. If assessment – instead of being delivered as a final mark – is dialogic and continuous throughout a course, it may contribute to an atmosphere of trust [17]. Moreover, by training themselves in making and communicating judgements regarding both their process and the quality of their work, learners prepare for an unpredictable future [20]. That the relevance of learning to reflect on processual dimensions appears to increase when designers use artificial intelligence as a tool for decision making, is a hypothesis the authors wish to explore in the future. MT and EP consciously work with how forms for assessment and reflection, such as process books and peer to peer conversations, can support the experimentation sought for. Gullberg's doctoral thesis [1] shows how Morten Levin's cogenerative model for action research may be developed within architectural education (Fig. 5). The model was used to make a case study in the Master course collaboration between Making is Thinking and Cirka Teater in 2016. As mentioned above, embodied exercises presented learners with high risk in the collaboration with the theatre company. What was also found, however, was that participation in the sequence of learning arenas for reflection set up by the researcher to constitute a cogenerative dialogue, made learners engage more decisively in the risky exercises of the course. In other words, the introduction of a safe structure such as the cogenerative model promoted risk-taking.

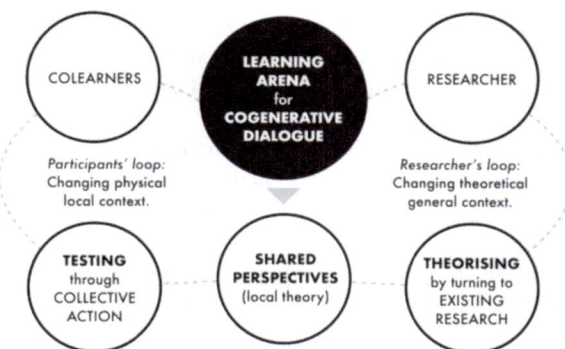

Fig. 5. The cogenerative model for action research, 2021 (Diagram by J. S. Gullberg [1], based on diagrams by M. Elden and M. Levin [21], and M. Levin [22].)

Fig. 6. Full scale building by 1st year students in Trondheim, 2023 (Photos by A. Gilberg and Z. Živanović, NTNU.)

6   Conclusion: Proposals for a Pedagogy of Mistakes

This paper concludes with a list of proposals to educators. Proposals for a pedagogy of mistakes, perhaps, or for embracing the fruitful and contradictory tensions between risk and trust. The list is a sort of to-do list for responding to the research question addressed: *How can learning spaces in architectural education be set up to encourage risk-taking, and so that the effects of risk-taking can be processed and appreciated?*

1: **Evaluate and adjust admission procedures**, so that they match what is going on in architectural education. This includes recognising the value of hands-on skills and creative risk-taking, as well as seeing architectural education in the context of national educational systems.

2: **Leave the students** to struggle on their own and listen to where they want to go. But also, level with learners by working alongside them in real-life situations.

3: **Focus on process** rather than final results. Set up models for collective reflection and dialogic assessment. What if a process focus may lead to critical ways of working with artificial intelligence in design processes?

4: **Work off campus** and leave space for the unexpected in course designs. Rebuild the architecture school, both its organisation and its physical premises. Let the FormLab and other in-school premises for experimentation go wild and spread out. Do practice-based research to evaluate and develop learning spaces.

5: **Connect demands on sustainability** to the aesthetic and material core of the practice of architecture. Use aesthetic methodologies and techniques, for instance the bricolage approach, to enhance creativity while also implementing issues of sustainability and circularity in design processes.

6: **Promote risk** by establishing theory to support pedagogical undertakings, for instance by combining architectural, aesthetic and educational theory.

7: **Have fun**. Creative play, idea generation and implicit mistakes may ignite liberating laughter levelling everybody in shared experiences of making big, beautiful results, be they mistakes or successes.

References

1. Gullberg, J.S.: Cogenerating Spaces of Learning: The Aesthetic Experience of Materiality and Its Transformative Potential within Architectural Education. Doctoral thesis, NTNU. NTNU Open access Homepage (2021). https://ntnuopen.ntnu.no/ntnu-xmlui/handle/11250/2740718, last accessed 2024/02/06
2. Rødne, G., Hokstad, L.M.: 'Making is Thinking': from design fixation to provocative competence. ARENA J. Archit. Res. (AJAR) **7**(1) (2022). https://doi.org/10.5334/ajar.294
3. World Economic Forum Homepage. https://www.weforum.org/agenda/2023/05/future-of-jobs-2023-skills/, last accessed 2024/02/06
4. Biesta, G.J.J.: Beautiful Risk of Education, 1st edn. Routledge, Oxfordshire (2013)
5. Land, R.: Toil and trouble: threshold concepts as a pedagogy of uncertainty. In: Land, R., Meyer, J.H.F., Flanagan, M.T. (eds.) Threshold Concepts in Practice, pp. 11–24. Sense Publishers, Rotterdam (2016)
6. Barnett, R.: The thoughtful university: a feasible Utopia. Beijing Int. Rev. Educ. **1**(1), 54–72 (2019)
7. Harriss, H., Hyde, R., Marcaccio, R. (eds.): Architects after Architecture. Routledge, New York and Oxon (2021)
8. Salama, A.: Spatial Design Education: New Directions for Pedagogy in Architecture and Beyond. Routledge, New York and Oxon (2015)
9. Colomina, B., Galán, I.G., Kotsioris, E., Meister, A.-M.: Introduction. In: Colomina, B., Galán, I.G., Kotsioris, E., Meister, A.-M. (eds.) Radical Pedagogies, pp. 11–20. The MIT Press, Cambridge and London (2022)
10. Bilalic, M., McLeod, P., Gobet, F.: Why good thoughts block better ones: the mechanism of the pernicious Einstellung (set) effect. Cognition **108**(3), 652–661 (2008)
11. Crilly, N.: Fixation and creativity in concept development: the attitudes and practices of expert designers. Des. Stud. **38**, 54–91 (2015)
12. Kallick, B., Zmuda, A.: Orchestrating the move to student-driven learning. Educ. Leadersh. **74**(6), 53–57 (2017)
13. Making is Thinking Homepage. https://makingisthinking.net/, last accessed 2024/02/06
14. Rintala, S.: Email sent to Rødne G 2023/10/04 (2023)
15. Barrett, F.J.: Yes to the Mess: Surprising Leadership Lessons from Jazz. Harvard Business Review Press, Brighton (2012)

16. Rødne, G., Gullberg, J.: Hvorfor er det åpenbart å samarbeide med et teaterkompani i arkitek-tutdanningen? – Seks svar. In: Sæther, S.A. (ed.) Cirka 40 år med teater – En historiefortelling og en teaterantologi, pp. 224–231. Cirka Teater, Trondheim (2023)

17. Carless, D.: Trust and its role in facilitating dialogic feedback. In: Boud, D., Molloy, E. (eds.) Feedback in Higher and Professional Education: Understanding It and Doing It Well, pp. 90–103. Routledge, Oxon and New York (2012)

18. Carlsen, A., Landsverk Hagen, A., Clegg, S., Gjersvik, R.: Liberating laughter: how playful energy and humor opens up people, situations and ideas. In: Carlsen, A., Clegg, S., Gjersvik, R. (eds.) Idea Work, pp. 182–197. Cappelen Damm AS, Oslo (2013)

19. NTNU TransArk Homepage. https://www.ntnu.edu/transark, last accessed 2024/02/06

20. Tai, J., et al.: Developing evaluative judgement: enabling students to make decisions about the quality of work. High. Educ. **76**, 467–481 (2018)

21. Elden, M., Levin, M.: Cogenerative learning: bringing participation into action research. In: Foote Whyte, W. (ed.) Participatory Action Research, pp. 127–142. SAGE Publications, Thousand Oaks (1991)

22. Levin, M.: Co-generative learning. In: Coghlan, D., Brydon-Miller, M. (eds.) The SAGE Encyclopedia of Action Research, pp. 109–112. SAGE Publications, London (2014)

Transformations of Public Spaces - Sustainable and Ethical Approach to Architectural and Urban Design Using Mixed Cultural Background

Patrycja Haupt[1]([⊠]), Mariusz Twardowski[1], Luca Maria Francesco Fabris[2], Riccardo M. Banzarotti[2], Andres Ros Campos[3], J. L. Gisbert[3], and Pedro Verdejo Gimeno[3]

[1] Cracow University of Technology, Krakow, Poland
phaupt@pk.edu.pl
[2] Politecnico Di Milano, Milano, Italy
[3] Universidad CEU Cardenal Herrera, Moncada, Comunitat Valenciana, Spain

Abstract. Transformations of public spaces are an integral part of city life. These processes although addressing site specific problems are universal for the urban centres across European countries struggling with deurbanization, social exclusion, massive tourism, climate change and pollution. The study is based upon the outcomes of the series of international workshop showing the approach to public space in different location. The aim of research is to show what problems addressed during the design workshop had a universal dimension and which ones were more site specific, addressing the needs of the local community. The basic research question asked was also how to transform public space in order to create active, reach, inclusive and climate neutral environment for the inhabitants. What are the factors in the public space creation that would result in active public space? The research methodology was based on a multi-criteria comparative analysis of activation factors in selected public spaces and evaluation of students' proposals ac-cording to those factors. These transformations were often driven by specific objectives, such as creating iconic landmarks, improving public safety, or generating economic activity on the other hand some of the proposals were concentrating on local community needs introduced by the meetings with local stakeholders. A hybrid approach that combines the strengths of both approaches provided by the workshop methodology was proved effective didactic method.

Keywords: Public space transformation · sustainable public space · ethical approach to architectural and urban design · quality of public space

1 Introduction

Vivid and active public space is considered to be the essence of the city. It serves as an essential venue for social interaction, community engagement, and cultural expression [1]. Due to contemporary goals to make the cities more compact and in that way sustainable there is a need to rediscover new spaces that were abandoned, used for different

© The Author(s) 2025
M. Barosio et al. (Eds.): EAAE AC 2023, SSDI 47, pp. 112–121, 2025.
https://doi.org/10.1007/978-3-031-71959-2_14

purposes, damaged [2, 3]. Transformations of these spaces have become increasingly important as urban centers face complex challenges struggling with the problems of depopulation, social exclusion or the detrimental impacts of mass tourism. Contemporary public space has also become a tool to deal with climate change, and environmental pollution. Sustainable and user sensitive approach to architectural and urban design is crucial in addressing these issues [4–6].

The presented research focuses on the teaching methodology that was an essential outcome of the Erasmus + project entitled"The Activation of the Public Spaces of the City Centres through Ethical and Sustainable Design Based on the Local Communities Participation / Response / Proaction, that was conducted by Cracow University of Technology as a Leader in partnership with Politecnico di Milano, Universidad Cardenal Herrera CEU Valencia and Eurokreator s.c. from October 2020 till July 2023. A series of international workshops were completed as a part of the project that aimed to implement a sustainable and ethical approach to public space design, taking into account the mixed cultural backgrounds and diverse needs of urban communities. The methodology encouraged students to explore both top-down and bottom-up approaches to public space transformations [7–9].

2 Methodology

The project's central objectives revolve around the transformation and activation of public spaces, with a primary focus on instilling sustainable and ethical principles in public space design. The course methodology is structured to teaching, data collection and result assessment methodology (Fig. 1).

Fig. 1. Project objectives, by Patrycja Haupt.

Teaching techniques encompass interdisciplinary learning, international workshops, and cultural immersion, fostering an appreciation for the significance of cultural context in public space design. Additionally, stakeholder engagement ensures the consideration of site-specific requirements, while cross-cultural collaboration promotes the celebration of diversity. The data collection methodology involves thorough site surveys, stakeholder interviews, data analysis, and comparative assessments, providing a solid foundation for well-informed proposals. To assess results, a methodology is implemented, which includes criteria-based evaluations, peer and expert reviews, community feedback, and long-term monitoring, all of which collectively guarantee the effectiveness and ethical alignment of student proposals. This comprehensive approach equips students with the knowledge and skills needed to address the multifaceted challenges associated with public space transformation, all while promoting cultural sensitivity and sustainability in the field of urban design.

3   Workshop Framework

The methodology in the framework of the project was based on various teaching methods. There were three main stages introduced in the curriculum of each of the partners' courses. The first stage, pre-workshop was based on online activities and consisted of the introduction in a form of webinars for all of the partner university students about the deep context of the site and broad approach to the urban characteristics of the city. The second stage was the main interaction between the students of various cultural background and involved on-site workshop, working in international groups. The period of the semester after the workshop was devoted mostly to individual work on the development of the project. The final stage was an online presentation of the final designs in presence of all the students and tutors involved in the workshop, also the local stakeholders involved in the design site area. (Fig. 2).

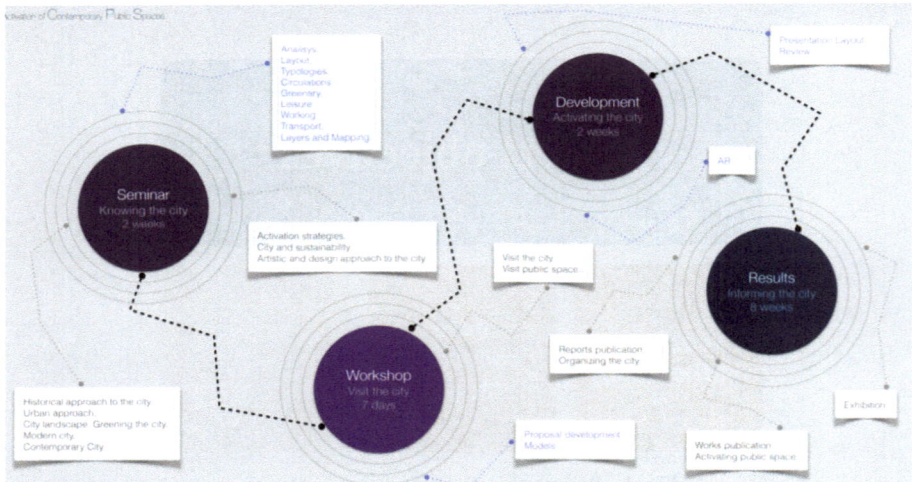

Fig. 2. The graph of the teaching methodology, by A. Ros Campos.

The teaching methodology revolves around a structured framework of international workshops. These workshops take place in various urban locations chosen by the partner university of the project, with distinct cultural and socio-economic backgrounds, enabling students to gain exposure to a wide range of public space challenges and opportunities. Each project partner's choice of location within the city revealed a range of distinct public space challenges including historical significance, cultural heritage, lack of pedestrian connections, conflicts in the use of space, deprivation of any activities and abandonment (Figs. 3, 4, 5).

Fig. 3. Milan, project site visit 2021, by Piotr Broniewicz.   **Fig. 4.** Krakow project site visit 2022, by Piotr Broniewicz.   **Fig. 5.** Valencia project site visit 2022, by Piotr Broniewicz.

The first workshop took place in Milan. The Italian partner have chosen a site of Piazza Tirana that allowed the students to approach the problems of public space in different scales from urban scale of a district of town through the residential complex surrounding the public space ending with the public space design within the perimeter of the buildings. The site that presents a unique context of the pedestrian connection by the over ground passage above the railway. Here, the emphasis as put on revitalizing an urban square in a vibrant, cosmopolitan city. In order to perform a complex and wise design the student had to understand the urban context of the city. Milan is a global fashion and business hub, and the public space has to reflect those qualities in the city's urban fabric. Transforming this space requires an understanding of Milan's contemporary urban dynamics. On the other hand it is also essential to think of the city through the prism of its historical significance. Therefore the proposals for its transformation should preserve and celebrate this heritage while addressing modern needs. Interaction with local stakeholders, including business owners, residents, and cultural institutions, is crucial to understanding the square's role in the daily life of Milan's residents. The activation factors for that site, apart for the friendly space for users might require solutions such as enhancing pedestrian access, creating event spaces for fashion shows or cultural festivals, and providing outdoor seating for cafes and restaurants.

In Krakow the chosen site was an area adjacent to Nowy Kleparz, on one hand a historic marketplace in Krakow, on the other a part of Twierdza Kraków fortification including Fort Luneta directly on the site. This design challenge was fascinating site for public space transformation being deeply rooted in history and offering unique opportunities for change. The students design tasks included preservation of cultural heritage while on the other hand adapting architecture and creating public space or adapting it for modern uses. The strong local community in this area was also an issue that had to be taken into consideration in terms of local community expectations from both residents and business perspective while respecting the site's cultural background. Balancing these elements was crucial, as well as respecting the local identity of a space where local vendors offer goods as in former and contemporary marketplace. Creating spaces for art exhibitions and performances together with recreational areas for the residents was one of the site priorities.

Valencia, a beautiful coastal city in Spain, has offered a unique public space connecting the gardens of Turia and the embankment of the Mediterranean sea site has been chosen as a workshop location due to its distinct characteristics and challenges. This location offers a diverse set of opportunities for public space transformation, blending natural elements with urban life. Valencia's background that is known for its rich cultural heritage, and this mixed cultural background becomes an inspiration for the design process solutions. The clue elements to consider were the respect and celebration of the city's traditions and cultural identity. In order to understand the site-specific needs, students had to engage in interactions with the local community. Data collected from the site surveys highlight specific activation factors including promoting outdoor activities, supporting local artisans and cultural events, and enhancing the connection between the gardens and the sea.

In each of these locations, the teaching methodology focused on blending global best practices with site-specific requirements, enabling students to appreciate the cultural and historical backgrounds while addressing contemporary urban challenges. The data collection and result assessment methodologies are tailored to each site, ensuring that the proposals effectively address the unique needs of these diverse public spaces.

4   Workshop Outcomes

4.1   Multi-criteria Comparative Analysis for Data Collection

The aim of the analysis was to identify the key factors contributing to the activation and sustainability of public spaces. In order to achieve that goal a tool of the multi-factor chart was used. The information of the site according to this tool was divided into 4 key areas such as physical arrangement of space, cultural context of space ethical and environmental issues. (Figs. 6, 7).

The result of this study which was a group activity was to evaluate the effectiveness of different public space transformations. Special attention was drawn to the features such as accessibility, safety, social inclusivity, environmental impact, and economic vitality. The analysis were performed in the first stage of the design process.

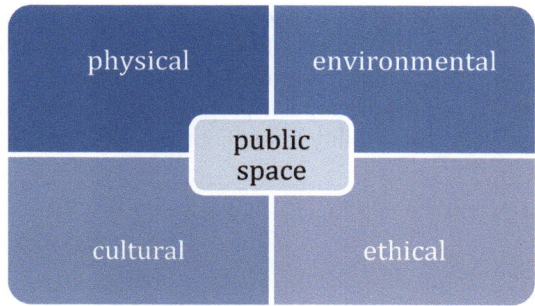

Fig.6. Fields of analysis, by Patrycja Haupt.

			Kryteria analizy (Opis i ilustracje) / Criteria of analysis (Description and pictures)
ASPEKTY FIZYCZNE PRZESTRZENI / PHYSICAL ASPECTS OF SPACE	**ARCHITEKTURA /** ARCHITECTURE	KOMPOZYCJA (el. kompozycji – osie, dominanty, akcenty, konkretność wnętrza urbanistycznego, krawędzie, otwarcia, zamknięcia widokowe) / COMPOSITION (elements of composition - axes, dominants, accents, specificity of the urban interior, edges, openings, view closures)	*[map illustration]* Dominanty: Kościół św. Józefa. Akcenty: Fontanna, pozostałości torów po dawnej linii tramwajowej. Kompozycja: Plac ma kształt trójkąta, otoczony trzema ulicami. Osie kompozycyjne: Trzy główne Krawędzie: Trzy ulice Otwarcia i zamknięcia: Plac jest zwieńczony budynkiem kościoła, a dookoła otoczony i zamknięty pierzeją kamienic. Przebicia w postaci ulic i bram. / Dominants: Church of St. Joseph. Accents: The fountain, the remains of the tracks on the old tram line. Composition: The square is triangular in shape, surrounded by three streets. Compositional axes: Three main Edges: Three streets

Fig.7. Analytical chart (physical part showing compositional analysis, by B. Mierczak, D. Meres, K. Łukasik).

4.2 Student Proposals and Methods of Assessment

During the workshops, students were tasked with developing proposals for public space transformations. These designs were required to address the specific needs of the chosen location and to consider the identified activation factors. The proposals were expected to aim to create public spaces that are active, inclusive, and environmentally sustainable.

Based on students proposal several elements play a pivotal role in shaping public spaces into vibrant and active places (Fig. 8). These components encompass aspects like providing easy access, ensuring safety, promoting inclusivity, engaging the local community in the design process, offering functional amenities, recognizing the cultural context, accommodating events and gatherings, prioritizing environmental sustainability, encouraging economic activity, organizing regular programming, integrating art, interactive features, and emphasizing sustainability throughout the design and management.

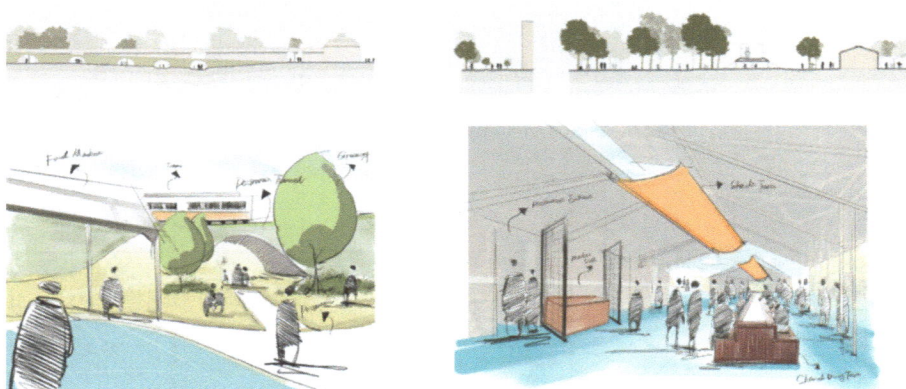

Fig. 8. Bridge connection in Kleparz, Kraków, design 2020, by Team06 Solomiia Vezhanska, Mu Lin, Zuzanna Matuszna, Marwan Afifi.

To evaluate the proposals developed in the workshops, a comprehensive methodology was employed. It included assessing each proposal during the defense based on predefined criteria linked to activation factors, engaging in peer review for collaborative improvement, obtaining expert feedback from professionals and mentors, seeking input from the local community and stakeholders, to ensure the proposals continue to meet their intended objectives and adapt to changing needs.

The transformations in these workshops are guided by a blend of top-down and bottom-up objectives. Top-down approaches focus on creating iconic landmarks, enhancing public safety, and generating economic activity. On the other hand, bottom-up approaches emphasize local community engagement, social inclusivity, and environmental sustainability.

4.3 Teaching Methodology Effectiveness

The teaching methodology demonstrated in these international workshops has proven to be an effective didactic tool for nurturing sustainable and ethical approaches to public space design. The combination of top-down and bottom-up approaches, along with the diverse cultural backgrounds of the locations, encourages students to think critically and creatively about the transformation of public spaces [10–12]. The effectiveness of teaching methods can be assessed by the distribution of addressing different groups of public space problems. The aim of the project was to sensitize students to sustainable, ethical and cultural approach to public space design using the basis of their knowledge on spatial solutions [13, 14]. To assess whether the series of introductory workshops all of the design solutions presented for three sites were evaluated to establish which groups of problems identified in the initial query were addressed in the final design (Fig. 9).

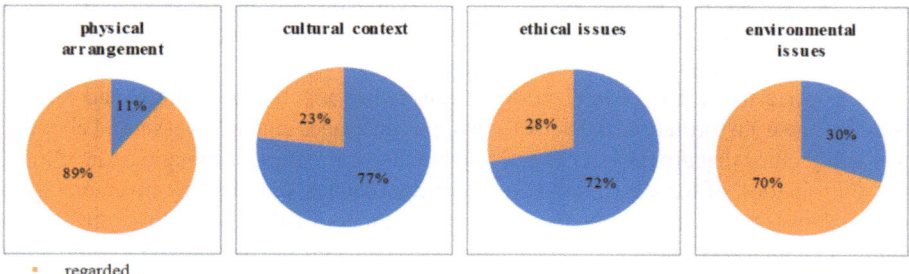

regarded
disregarded

Fig. 9. The result of the assessment of the workshop projects – addressed issues, by Patrycja Haupt.

The comparison results revealed some interesting insights. Notably, 89% of the proposals successfully tackled physical problems, indicating a strong focus on addressing practical and tangible issues in public spaces. Surprisingly, 28% of the proposals addressed ethical concerns, surpassing the 23% that focused on cultural context [15]. This unexpected finding highlights the recognition of the importance of ethical considerations in public space design, possibly reflecting a growing awareness of the need for responsible and ethical urban development.

Additionally, 70% of the proposals placed a significant emphasis on sustainability, underlining the growing significance of environmentally conscious design practices in modern urban planning. This demonstrates a proactive approach in integrating sustainable elements to create more eco-friendly and resilient public spaces [16–18].

In summary, while the prominence of physical problem-solving was anticipated, the greater attention given to ethical aspects over cultural considerations underscores the evolving priorities in public space design. Moreover, the substantial focus on sustainability reflects the commitment to creating more environmentally responsible and resilient urban environments.

5   Conclusion

Transformations of public spaces play a vital role in addressing the contemporary challenges faced by urban centers across European countries. A teaching methodology that offers an approach based on four groups of criteria influencing the activeness of public space offers students a comprehensive understanding of the complex nature of public space design. By utilizing a multi-criteria comparative analysis, students are able to identify key activation factors and develop proposals that create active, inclusive, and environmentally sustainable public spaces. This methodology equips future architects and urban planners with the tools necessary to tackle the evolving needs of our cities in an ethical and sustainable manner.

The students' proposals, as evidenced by their emphasis on practical problem-solving and the prioritization of ethical considerations over cultural ones, reflect the evolving landscape of public space design. The substantial focus on sustainability underscores the commitment to creating environmentally responsible and resilient urban environments. These insights not only provide a valuable roadmap for future public space transformations but also emphasize the need for inclusivity, sustainability, and ethical responsibility in shaping vibrant and dynamic urban spaces.

In a world marked by urbanization, social inclusion, climate change, and cultural diversity, the transformation and activation of public spaces emerge as essential elements in creating cities that are not only functional but also harmonious, engaging, and ethical. The combination of top-down and bottom-up approaches, enriched by diverse cultural perspectives, equips the architects and urban planners of the future to address these complex challenges effectively. Through thoughtful design and inclusive planning, urban environments can be shaped that inspire and serve their inhabitants while respecting their cultural heritage and the environment.

References

1. Gehl, J.: Cities for People. Covelo, London Island Press, Washington (2010)
2. Bambó, R., Nicolás, P., Medina, C., Ezquerra, I., García-Pérez, S., Monclús, J.: Quality of public space and sustainable development goals: analysis of nine urban projects in Spanish cities. Front. Archit. Res. **12**(3), 477–495 (2023)
3. Kim, S., Kwon, H.: Urban sustainability through public architecture. Sustainability **10**(4), 1249 (2018)
4. Dudzic-Gyurkovich, K.: The search of shade. Contemporary projects of shadowing elements in public spaces of european cities. Housing Environ. **20**, 40–49 (2017)
5. Klimowicz, J.: Selected examples of the use of greenery and water as a natural space elements improving local climates conditions. Housing Environ. **24**, 26–33 (2018)
6. Haupt, P.: Naturalne elementy kompozycji w kształtowaniu współczesnej przestrzeni miejskiej. Relacje budynku z otoczeniem. Wydawnictwo Politechniki Krakowskiej, Kraków (2015) (in Polish)
7. Kazanecka-Olejnik, L.: Użytkownicy i struktury aktywnych transformacji przestrzeni publicznych w mieście. Politechnika Wrocławska, Wrocław (2022) (in Polish)
8. Strydom, W., Puren, K.: A participatory approach to public space design as informative for place-making. Civil Eng. Urban Plan. Archit. **4**, 33–40 (2013)
9. Foth, M.: Participation, Co-Creation, and Public Space. J. Public Space **2**(4), 21–36 (2017)
10. Depani, A.: City making doesn't have to be top down vs bottom up. Archit. Rev. **06**, 2022 (2022)
11. Carmona, M.: Principles for public space design, planning to do better. Urban Des. Int. **24**, 47–59 (2019)
12. Nguyen, T., Han, H., Sahito, N.: Role of urban public space and the surrounding environment in promoting sustainable development from the lens of social media. Sustainability **11**(21), 5967 (2019)
13. Njungea, E., Asilsoy, B.: A study about public participation in the universal design of public spaces. Mimarlık Fakültesi Dergisi – J. Faculty Archit. Cilt **2**(2), 37–47 (2020)
14. Li, J., Dang, A., Song, Y.: Defining the ideal public space: a perspective from the publicness. J. Urban Manage. **11**(4), 479–487 (2022)

15. Lamri, J., et al.: The 21st Century Skills: How Soft Skills Can Make the Difference in the Digital Era. The Next Society (2019)
16. Gil-Mastalerczyk, J.: Architectural education in the formation of the built environment with sustainable features. World Trans. Engng. Technol. Educ. **18**(2), 146–151 (2020)
17. Widera, B.: Education of architecture students in the light of the European Green Deal. World Trans. Engng. Technol. Educ. **19**(1), 79–84 (2021)
18. Jabłońska, J., Ceylan, S.: Sustainable architecture in education. World Trans. Engng. Technol. Educ. **19**(1), 96–101 (2021)

Building a Community Through a Design Build Studio Program

Arda Inceoglu[✉]

MEF University, Istanbul, Turkey
inceoglua@mef.edu.tr

Abstract. This paper aims to provide a comprehensive and critical assessment of the outcomes stemming from a Design-Build program, a pedagogical approach widely adopted by educational institutions worldwide. These programs are instrumental in equipping students with vital practical skills, often unattainable within the confines of a conventional studio environment. While the objectives of this program align with those of similar initiatives in various educational institutions, an examination reveals an unexpected and substantial outcome. Beyond its primary goals, the Design-Build program has played an integral role in instilling a culture of collaboration and camaraderie within the school, thereby significantly contributing to the overall success of its architectural education. All stages of the program consist of collaborative processes, instilling from an early age the importance of working together by helping each other than individual competition.

Keywords: Design Build Studio · Architectural education · public benefit · learning by doing · community · collaboration

1 Introduction

Design-Build Projects represent specialized pedagogical instruments within architectural education programs. They provide students with invaluable opportunities for hands-on, full-scale construction experiences, fostering the application of their creative abilities for the greater public good, and emphasizing principles of collaboration [1]. Worldwide, numerous academic institutions incorporate Design-Build Studios into their curricula, each employing various methodologies, and utilizing a range of tools [2].

Design-build live projects provide students with hands-on experience, enhancing their understanding of construction techniques, project management, and client communication [3]. Students develop a range of skills, including problem-solving, teamwork [4], and technical skills, which are critical for their future professional practice [5].

Design-build live projects offer significant educational benefits by providing students with practical, hands-on experience. They also positively impact communities by addressing local needs and fostering engagement [7]. However, challenges such as resource constraints, time management, and quality control must be addressed to maximize the effectiveness of these projects [8]. Best practices include strong partnerships,

M. Barosio et al. (Eds.): EAAE AC 2023, SSDI 47, pp. 122–131, 2025.
https://doi.org/10.1007/978-3-031-71959-2_15

comprehensive planning, and mentorship. Design-build live projects are a valuable component of architectural education, offering dual benefits of practical student training and community improvement. Future research should focus on long-term outcomes of these projects for both students and communities, and the development of frameworks to address common challenges.

In the curriculum of the Architecture program of MEF University, there exists a mandatory Design-Build Studio component for all enrolled students [9]. The inception of the DBS program was driven by a set of pedagogical and academic objectives. Over the course of time, we have observed unanticipated outcomes of this program that significantly influence our educational framework within the institution. These outcomes are the topic of the main discussion in this paper.

2   Structure of the Design-Build Program

When establishing the Design-Build Program the primary consideration revolved around its overarching objectives: What valuable experiences would the students derive from their participation in this program? This pivotal question naturally influenced every facet of program development. Through a comprehensive survey of analogous initiatives worldwide, we identified four principal goals: Practicing Architecture, Experiential Learning in Construction and Materials, Leveraging Architecture for the Public Good, Collaborative design. These goals are congruent with the objectives of numerous programs worldwide, as demonstrated in Table 1.

Table 1. List of stated goals of Design Build Programs

Goals of global Design-Build programs [1]	Goals of MEF Design build program
Hands-on experience	Experience of building their own design
Simulation of professional experience	Practice profession early on
Experimentation with materials, techniques	Experience with materials
Investigation of craft	Learn simple construction techniques
Collective design and building	Teamwork in design and construction
Community service through architecture	Using skills for public benefit

From the beginning the program emphasized collaboration as the most important goal. Collaboration occurs on different levels. First of all, all design is collaborative. Students work in teams on the design. Instructors make sure that everybody is involved in the decision making, which is the most challenging aspect of the design process. Also, all construction is collaborative. Instructors, assistant students and first year students all work in all aspects of the construction together as a team. In some projects, students are involved with the stake holders (a community, students from a primary school) and develop the program in collaboration.

The second question concerning the structural configuration of the Design-Build Program pertains to the scope and scale of its projects. As posited by Corser and Gore

[10], there are two predominant approaches within the realm of design-build programs. In the first approach, projects assume a public and large-scale character.

However, as Corser and Gore contend, these endeavors tend to be less conducive to the exploration of novel concepts. The exigencies of erecting intricate structures within condensed time frames and constrained resource parameters limit the scope for experimentation. Conversely, the second approach entails projects primarily geared toward the experimentation with 'materials and processes.' Nevertheless, they are impeded in their capacity to address broader public concerns. Commonly, such projects manifest as temporary installations.

However, a third approach emerges as a synthesis of the salient elements present in the previous two paradigms. In this instance, projects maintain their focus on public welfare, embodying a deliberate design to furnish essential functionality to a community. These endeavors assume a notably reduced scale, affording students the latitude to engage in unbridled experimentation concerning materials and construction techniques.

In our Design-Build program we have adopted the latter approach as our guiding principle. Our aspiration was to ensure that the projects, though on a relatively modest scale, would make a meaningful contribution to the local community. In addition to this, we aimed to infuse an element of 'extraordinary' design into the projects. This objective served not only pedagogical purposes but also carried cultural significance. In the physical landscape of Turkey, there is a notable dearth of well-designed public projects. Even in their small-scale manifestation, our projects assumed the role of exemplars, offering the public a glimpse of the transformative potential of architecture. Certainly, the definition of 'extraordinary' differs for students, people in the community, architects. What we try to achieve is to make sure that our designs are different from what the users have seen in their environment, involve elements of surprise, include well-designed details and also serving the initial purpose.

Another pivotal question that demanded deliberation in the formulation of our projects pertained to participant selection. Across the globe, many design-build initiatives enlist volunteers, involve students at advanced stages of their education, or are integrated into graduate-level programs. From its inception, our program was conceived with the intention of inclusivity, extending participation to all our students. An easier approach would be work with students who have performed the best in their first year, the ones who are the most interested and hard working or the ones with best construction skills. We deliberately have not chosen this path, in the projects we include students who have failed their courses, did perform badly in the studio, are not hard working at all. While this inclusivity is a commendable notion, it introduces evident logistical challenges. Given our institution's annual intake of approximately 120 students, this commitment necessitates the execution of eight to ten projects on an annual basis.

The final inquiry regarding the program's structure concerns the timing of these projects within the students' educational journey. The optimal approach would be to involve students from various levels of experience, thereby ensuring each team benefits from a mix of students with both experience in design and proficiency in digital technologies. However, the Design-Build program was initially slated for the summer of our very first year. This decision was borne out of pragmatic considerations rather than pedagogical ones. Given that the school was in its nascent stage, the primary aim was to initiate the program at the earliest possible juncture.

It is worth noting that conducting a design-build program solely with first-year students is atypical, given their relatively limited knowledge levels. Nonetheless, their enthusiasm to participate in the realization of real projects for genuine beneficiaries is consistently high. Had we been an established institution integrating a new design-build program into its existing curriculum, we would have likely explored alternatives, such as delaying its introduction to later academic years. This non-negotiable initial placement has had implications for various facets of the program and its impact on our academic institution.

2.1 Typical Process of a Design-Build Project

Since its inception in 2015, we have successfully completed sixty projects. These annual projects traditionally commence immediately after the conclusion of the Spring term. In recent years, our program involves a cohort of 120 students. The entire process, from design inception to construction completion, typically extends over a duration of four weeks, with the actual construction phase taking around one week. Project organization, stakeholder engagement, and sponsorship procurement are facilitated by the faculty before this four-week period. The design requirements are typically delineated during the Spring term through preliminary discussions with clients/users.

As the Spring term draws to a close, project proposals are presented to first-year students, inviting them to make selections. Subsequently, considering the students' preferences and considering factors such as group size and gender balance, the composition of project teams is finalized. Furthermore, we announce student assistantships and appoint one or two assistants for each project team. The selection process of assistants is merit-based, hinging on an evaluation of applicants' performance in the Design-Build projects they participated before.

After the semester's conclusion, project teams convene to receive the design brief. These interactions with clients/users transpire on-site or through online conferences when working in remote locations. The design process exhibits flexibility in its execution across various groups. However, as a foundational principle, we endeavor to involve all team members in collective decision-making. Typically, individual students or smaller subsets within the group generate distinct design proposals, which are subsequently scrutinized collectively. Instructors then facilitate a session where the most meritorious aspects of these proposals are amalgamated, culminating in the development of a final scheme that embodies the collaborative spirit of the team.

Following the finalization of the scheme, the design undergoes further development, typically by breaking it down into its components, with smaller groups assigned to the refinement of each part. During this phase, there is a focus on materials, details, and construction. Given that first-year students possess limited knowledge in these domains, instructors do offer guidance. At the culmination of this phase, 1:1 scale prototypes of joints are fabricated to ascertain the precision of the proposed details. The entire design process, from its inception to conclusion, typically encompasses a span of approximately two weeks.

The construction phase of a project typically spans one week. In most instances, adaptation to unforeseen site conditions is executed. Common challenges encountered on the construction site include material shortages, extreme climatic conditions, prolonged construction timelines, interpersonal tensions arising within the team due to demanding work and living conditions. Additionally, the progression of construction in many cases hinges on local support for materials and equipment, which may experience delays, consequently impeding the construction process.

Upon the completion of construction, we conclude our involvement in the project. Final phase of the project is the post-production process. The student assistants gather all material developed during the design phase, photographs and films taken during construction, make new drawings and renderings if necessary. In most cases the project is submitted to the annual selection of architectural projects in Turkey [11] (Fig. 1).

Fig. 1. Example of before and after a project.

3 Evaluation of MEF University Design-Build Program

Since its establishment in 2015, our Design-Build Program has garnered immense popularity among our student body. Despite the program's physically demanding nature, most of our students express a strong desire to participate in it repeatedly. This is possible since older students are afforded an opportunity to participate in the program by assuming roles as assistants.

It is noteworthy that, in several instances, the students who excel in the design studio may differ from those who excel during construction. This phenomenon has constituted one of the important outcomes of our Design-Build program. Several students, who encountered challenges during their initial year of design studios for various reasons, including difficulties in comprehending abstract concepts, difficulties in creating precise and immaculate designs, or a lack of enthusiasm for architecture, find themselves actively engaged in the physical act of construction. The tangible and concrete nature of the construction site often kindles an interest and passion for both building and, subsequently, design among these students. This newfound enthusiasm significantly contributes to their academic success when they return to school in the ensuing fall semester.

The program has played a pivotal role in expeditiously elevating our institution's profile within professional circles. Owing to the recognition garnered through the exhibition of several of our projects in the annual selection of architectural projects in Turkey, prominent architects have had the opportunity to acquaint themselves with our school's distinctive pedagogical approach. Over the span of six years, from 2016 to 2022, a total of nine of our Design-Build projects have been chosen for inclusion in these exhibitions.

Incorporating DBS projects into their portfolios has notably facilitated our students in securing internship opportunities at architectural firms internationally. An examination of the internship listings on our website [6] reveals a substantial proportion of our students engaging in internships at architectural offices spanning the globe. It's imperative to underscore that this pattern sets our students apart from their counterparts in Turkish architectural schools, where such a global reach is typically less common.

Annually, we administer a questionnaire to solicit feedback from our graduates, encompassing various dimensions of their educational experience. Among the survey items is an open-ended query pertaining to the Design-Build Studio Program. We have collated and summarized responses to this question over the course of four years, presenting the findings in Table 2. Notably, the response rate to our annual graduate questionnaire has consistently been substantial, with 45% of all our graduates participating and offering insights. The question reads like this: '*How do you evaluate the 'Design and Build' program that MEF University has been running since its first year?*' All the answers to this question are analyzed and summarized into relevant categories here for ease of comparison. For instance, this is an answer, translated into English: '*Most importantly, working with different students has made it much easier for me to embrace other people's ideas or put forward my own in a work environment.*' This answer is categorized into '*Collaboration skills*' category. Due to the open-ended nature of the questions, some answers can contain more than one statement, such as this: '*I believe it is the most productive program at the faculty. It not only instills confidence in students and teaches practical skills during their educational years but also leads to the emergence of projects that capture the attention of offices.*' In this case, answers would fall into three different categories: '*Confidence*', '*Construction skills*', '*Looks good in the portfolio and job applications.*'

After employing this categorization method on all responses, it becomes evident that a significant majority (74%) of the answers align with the articulated objectives of the program, as presented in Table 2. This alignment is in line with expectations, as our program goals are congruent with the objectives commonly shared by Design-Build programs globally. (Table 1). A surprising outcome is the absence of responses that could be classified as '*Satisfaction of using skills for common good, public benefit*' category. One plausible interpretation for this phenomenon is that the pursuit of public benefit is so intrinsic to the program that it is considered an implicit and fundamental aspect, obviating the need for explicit mention.

Table 2. Graduate questionnaire results (cumulative results of 2019, 2020, 2021, 2022 open ended question about evaluation of the design build studio). Comparison of questionnaire results with expected goals.

Goals of MEF University Design build program	Graduate questionnaire results	N
Practice profession early on	Learned a lot/Excellent beginning	72
Experience of building their own design	Practicing architecture/Design vs reality	27
Teamwork in design and construction	Collaboration skills	19
Learn simple construction techniques	Construction skills	16
Experience with materials	Experience with materials / Learning about details	16
Using skills for public benefit	-	-
		%74
	Responsibility / Confidence / Initiative	14
	Passion for architecture/Motivation/Fun	14
	MEF spirit / best experience at MEF	7
	Big difference with other schools	7
	Looks good in the portfolio and job applications	6
	Design as communication	4
		%26

Nonetheless, a noteworthy portion (26%) of responses deviate from the anticipated answers, as delineated in Table 2 and marked in orange. These responses can be categorized into two overarching domains. The first pertains to the acquisition of soft skills relevant to personal and professional development, *(Responsibility/Confidence/Initiative)* and motivation (*Passion for architecture/Motivation/Fun*). These represent invaluable attributes for any educational institution to instill in its students. Notably, the cultivation of a healthy confidence, the encouragement of using initiative, and the fostering of a sense of responsibility are not exclusive to architecture education but are intrinsic to the broader spectrum of educational objectives. These essential skills are the building blocks through which students evolve into designers (or intellectuals) characterized by a distinct degree of individuality, transcending the role of mere technical practitioners. Motivation and a genuine affection for architecture, qualities that may not universally manifest at the outset of one's educational journey, constitute essential prerequisites for effective learning. It is imperative to underscore that instilling a fervent passion for architecture is among the most pivotal objectives in the initial year of architectural education [13]. These elements serve as the driving forces that underpin the entire learning process.

The second category pertains directly to distinctive facets of the program. Given that no other educational institution in Turkey offers a comparable program, our students perceive it as a unique opportunity to distinguish themselves. This distinctiveness further reinforces the first category by fueling heightened motivation among the participants. It is also worth noting that a substantial number of students underscore the program's role in encapsulating the 'school spirit,' signifying its defining influence on the character and identity of our educational program.

These responses are indicative of the students' profound sense of belonging and attachment to the school. Evidently, because of their participation in the DBS program, students perceive the acquisition of skills and values that are deemed indispensable to their educational journey. Furthermore, it appears that the program equips them with a distinct competitive advantage, setting them apart from other architecture graduates and expediting the initiation of their professional careers. To augment this list, one can include another category of responses, 'Collaboration skills,' underscoring the crucial role of collaboration within any cohesive community.

In consideration of these multifaceted factors, it becomes evident that a primary and unexpected contribution of our Design-Build program is the cultivation of a 'School spirit' among our students, fostering a strong sense of community. In architectural education, as in other domains of higher education, the cultivation of individuality in students is undeniably imperative. However, concurrently establishing an environment characterized by collaboration rather than competition, where all individuals experience a shared passion, mutual support, and collective motivation, holds intrinsic value. It is within this conducive atmosphere that most students can attain success, transcending the conventional archetype of the extroverted, ambitious, and self-motivated individuals, and ensuring that all students have the opportunity to thrive. Certainly, this is not to say that there are no students who have individual priorities, see themselves competing with their peers, complain about their grades being lower than their friends. Of course there are these students as well. However, beyond the results of the questionaries, our studio environment is another proof of the collaborative sprit in the school. All our studios are located in a single space, all student working next to each other. Even if they wanted to, they cannot hide their work from anybody. The organization of space is another aspect where collaboration is promoted.

4 Conclusion

In summary, our Design-Build Program has exceeded our initial expectations and made substantial contributions to the overall success of our institution. While our initial objectives primarily centered around the practical aspects of education, such as material knowledge, construction experience, and problem-solving skills, we have discovered that the program's most significant impact extends beyond the realm of tangible skills.

The most profound contribution of the program lies in its role in fostering a sense of community within our school. It has cultivated a strong peer learning environment, shifted our institution from a competitive to a collaborative ethos, and cultivated a culture of diligence and commitment. Furthermore, it has ignited a passion for architecture in a substantial majority of our students, all made possible through the unifying role of the institution as a thriving community.

It is intriguing to note that this outcome has, in part, arisen from what was initially perceived as a challenge. Operating a Design-Build Program primarily with first-year students, despite their limited experience overall has, paradoxically, proven to be an asset. This approach has allowed us to engage all students right at the commencement of their educational journey, fostering a sense of community and shared purpose.

The key takeaway from this experience is not merely that every institution of higher education must institute a Design-Build Program. Rather, it underscores the significance of developing unique programs tailored to the particular ambitions and characteristics of the institution. Such programs can serve as a potent tool in cultivating a sense of community and, in turn, facilitate the attainment of the institution's pedagogical objectives with remarkable efficiency.

References

1. Hinson, D.: Design as research: learning from doing in the design-build studio. J. Archit. Educ. **61**(1), 23–26 (2007)
2. Design for the common good network map. https://www.designforthecommongood.net/about/map, last accessed 2024/05/28
3. Anderson, J., Priest, C.: Learning by doing: the educational value of design-build projects. J. Archit. Educ. **70**(2), 230–243 (2016)
4. Bell, B.: Good Deeds, Good Design: Community Service through Architecture. Princeton Architectural Press, Princeton (2004)
5. Bozikovic, A.: Design-build for social impact. Archit. Rev. **244**(2), 45–52 (2018)
6. Chiles, P., Holder, A.: Live Projects as critical pedagogy. J. Educ. Built Environ. **5**(1), 5–24 (2010)
7. Dean, A., Hursley, T.: Rural Studio: Samuel Mockbee and an Architecture of Decency. Princeton Architectural Press, Princeton (2002)
8. Smith, K.A.: Design-build education: integrating project delivery for improved learning outcomes. Archit. Des. **81**(2), 76–83 (2011)
9. MEF University FADA Design-Build Program website. https://www.fada.mef.edu.tr/Portfolio, last accessed 2024/05/28
10. Corser, R., Gore, N.: Insurgent architecture: an alternative approach to design-build. J. Archit. Educ. **62**(4), 32–39 (2009)

11. Arkitera Annual Architectural Selection 2022. https://www.arkitera.com/haber/turkiye-mim arlik-yilligi-2022ye-secilen-projeler-belli-oldu/, last accessed 2023/10/12
12. MEF University FADA Website showing global internships. https://www.fada.mef.edu.tr/int ernships, last accessed 2023/10/23
13. Snell, K.: Towards a New Paradigm in Architectural Education, vol. 2. Faculty Publications and Scholarship (2014). https://source.sheridancollege.ca/fast_arch_publ/2

Affirmation of a Discipline: Ephemeral Tectonics of Architecture Lesson

Sinsa Justic[✉]

Faculty of Architecture, University of Zagreb, Architectural Design, Zagreb, Croatia
author@uni-hannover.de

Abstract. Architecture is identified by its presence, its perceived reality. Architecture study is research in essential, defining values of architecture. Within multitude of discipline's constitutive elements, the objectification of architecture's perceived state is still one of its vital and specific tasks. No matter technology, economy or social normative imperatives, architecture is aestheticized, reflective research. A meaningful creative act no matter of aesthetic code or material circumstances. Architecture's immaterial, yet identifying difference is processed by exact conceptual decisions. Study of architecture must sustain a course where a working process is identified as a concept of work itself, where ephemeral emerges out of exact, conceptual execution. Ephemeral is a perceived state, it requires a sensitive person. In time of study, the search of ephemerality through tectonics of architecture lesson is sensitizing pedagogical tool. It is a foundation for architecture's cultural and human role. In times of intertwining disciplines, architecture must keep tools for protecting its disciplinary identity and intangible, defining differences. A course unit where precise and consistent production leads to authentic if ephemeral presence is valuable tool for affirmation of Architecture's immaterial, unmeasurable identity.

Keywords: Ephemeral · Tectonic · Presence · Exactitude · Affirmation · Discipline

1 Introduction

The course units presented in this paper are offered to students of architecture during the third and fourth semester of the bachelor study. Being part of a relatively small course, they are worth 1.5 ECTS, or less than one percent of bachelor study total credit. Units are aimed at enhancing perceptive sensitivity and at developing processual discipline and resource consciousness. Units explore the exactitude in precise use of simplest of resources to produce artefacts whose presence is perceived to be more than the sum of their parts. The experience of a finalized product, an impression of its correctness and a memorable image is hoped for as a reward for a student's work.

Study develops, inspires, and affirms architectural discipline by research into its specific tools, limits, and essences. Architecture study continuously exposes causality between exact processes and ephemeral outcomes as valuable architecture lessons about constituencies of architectural presence and its defining feature: a real perceived quality.

© The Author(s) 2025
M. Barosio et al. (Eds.): EAAE AC 2023, SSDI 47, pp. 132–141, 2025.
https://doi.org/10.1007/978-3-031-71959-2_16

Architecture develops its own reality, its own state of being and the specific intensity of its presence we perceive.

In his book For an Architecture of Reality, Michael Benedikt wrote about intensity of our perceptions in a way which corresponds beautifully with specific role of architecture lessons as presented in this article [1]. While reading his description of our experience, in times when the world is perceived afresh, with our perceptions freed from being sentimental, one thinks about students of architecture at the beginning of their study. At the dawn of a new sustainable paradigm, could their perceptions be set to neutral and undesiring, and their inspiration encouraged by the simple correspondence of appearance and reality or by the perceived rightness of things as they reveal their origin.

2   The Meaning

The purpose of lessons presented here is to reveal processual origin of things, their perceived value, their presence. The presented processes are not descriptive. For the purpose of discourse, and as a reference, [2] we call them "meaningless". *Meaningless work* is liberated from direct purpose. While seemingly "useless" or "unprofitable", it can be encouraging when the result produced outgrows our initial expectations. It rewards a student's commitment and determination to proceed with the process. It deepens our perception, makes us resource conscious, and reminds us of sensations of the ordinary, often hidden by a routine.

Explaining the idea, Walter de Maria wrote that Meaningless work exists between mediums, in transmutable form, and against interpretation. He says Meaningless work makes us aware of the experience, its context and reality, enabling new ways of seeing and liberation from superfluous and superficial categorizations and explanations [2].

Our experience of the real is independent of references; the presence of a work is produced every time anew. To produce it, one must invest time and creative force, some courage and anticipation of the result, into intensive production. This is exactly where seemingly meaningless comes to its full purpose: to enable meaning to grow out of process and to shape whole of measurable and immeasurable, accurateness of a work and the ephemerality of its perceived state.

The concept of Meaningless work is presented to students as an ironic, conceptually consistent tool against superficial creativity. It encourages appreciation of work whose perceived ephemerality levitates in the presence of the material object as a certain clarity and self-evidence of its origin.

3   Lessons

3.1   Anamorphosis: Production of Ephemeral Presence

Students of architecture perceive the world of architecture afresh.

As an attempt to produce real and resource conscious, ephemeral presence students were asked to produce an anamorphosis, a direct relation with viewer. Each student produces an unfinished object. Three sides of a cube were produced in firm paper, cardboard or foamboard. Several sets of surfaces and lines are marked on them in different techniques including collage, cutouts, linear or shaded drawing. Seen from the right position in space, these points, lines, or surfaces would seemingly complete a cube in multitude of positions. An ephemeral presence occurred at viewpoints in the *expanded field* [3] of the object, as seen in Fig. 1.

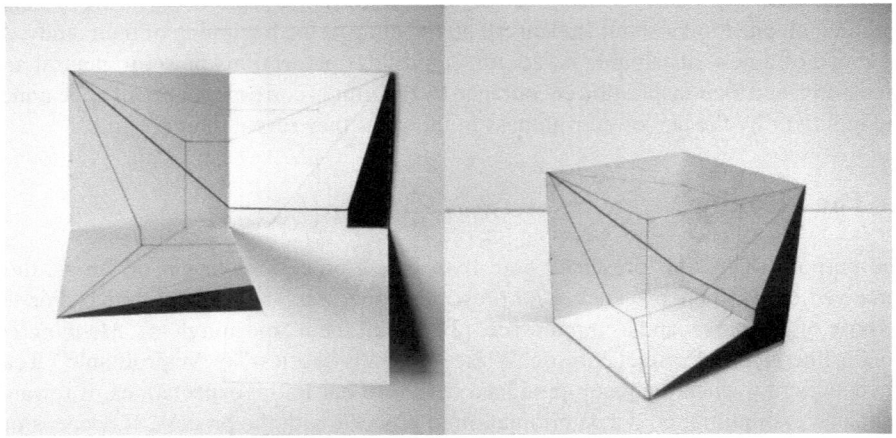

Fig. 1. The origin of the object; graphite pencil, acrylic paper 400g/m^2.

Drawing is transcended beyond being descriptive. Rudolf Arnheim proves that the visual combination of lines is controlled by the law of simplicity. He observed that when a combination of lines produces a simpler figure than the mere sum of separate lines would, it is then seen as one integrated whole [4]. When seen from a correct position, thinner lines which form a simple shape seemingly come forward, in front of the more intensive executed elements. An airy impression of transparent spaciousness is produced and could have been modulated by technique and intensity in production of these elements.

Searching for their position in space would bring them into ephemeral state between a moment before and moment after the emergence of a cube. A small object produced transitions of one's horizon plane, transforming the line as visual element into a "transitive verb" [5].

The surprising impression of transparency was recorded by photography or video sequence. To proceed with the technically exact process, a cloud of cube-emerging viewpoints was then drawn accurately in plans of a studio room and could have been visualized in several ways: by overlapped photographs of a student searching for viewpoints, in relation to other objects arranged in its field or marked by light and recorded with long exposure camera action, as seen in Fig. 2.

An important lesson to students of architecture is communicated here. Ephemeral as immaterial, perceived quality of object can emerge out of precise use of simple resources. In his Six memos for the new millennium, Italo Calvino, for example, describes *Exactitude* primarily as a well-defined and well calculated plan of the work. For him *Exactitude* evokes a clear visual image and arises by precise use of a language [6]. He argues that exactitude is precondition for vague or open. Students are invited to understand it as a metaphor or some advice for an understanding of some fundamentals in architecture discipline's value system.

Fig. 2. Origin of the drawing: visualization of the "cloud" of viewpoints, Martinovic Mia, student, 4th semester.

A more contemporary description of ephemeral quality occurring through the exactitude-indeterminacy duality is found in Jesse Reiser and Nanako Umemoto's Atlas of novel tectonics. Drawing upon Edmund Huserl's concept of protogeometry, a science dealing with vagabond or nomadic morphological essences, a particular tectonic quality is described as "anexact-yet-rigorous." [7].

According to it, and of importance to this discussion, an ephemeral presence is produced when exact geometry is superimposed to matter in real space. A potential in sophistication of production processes which can control and alter projections of geometries into matter is recognized, producing new tectonic quality and perceived presence of an object. To process further a lesson about potentially an-exact-yet-rigorous origin of the ephemeral, a more material course unit is offered to students.

3.2   A Cast: Atmosphere of an Ordinary Room

Unit translates a real space of students own or well-known and daily used room into an artefact, and space of an artefact into imaginary space whose presence is in distant resonance with existing everyday space it begun with. A real and imaginary, exactness of the geometry and the resistance of the material interfere.

A precisely executed geometrical shape is cast in gypsum through the production process whose logic is as evident as cause-and-effect. Unit also emphasizes importance of a key, Beuysian, processual balance between chaotic and thoughtful as productive of real plasticity. According to it, a real shape is processed between two opposing principles, as Beuys would call it, a chaotic willingness and a thoughtful shaping which meet and are in mutual conflict [8].

Within a relatively small cube, 10cm in size, a robust artefact and its atmosphere are produced in several phases: at the beginning, the student thinks in reverse mode to produce a core element, a precise Styrofoam shape of a room's negative space to which secondary elements are added. Their architectural role is dual – while they hold a core element in position within mold, they will also become light and viewing openings to the inner space of the cast artefact – a cast of a room. The Styrofoam negative is then photographed in two ways: held by hand which shows its real scale, and against neutral background to reveal its shape and plasticity, to provoke thoughts about its formal qualities and, more importantly, to trigger curiosity about its soon-to-be cast inner space. The process steps from the negative to the produced artefact are shown in Fig. 3.

Fig. 3. Styrofoam negative, 10cm mold, a 10cm cube with cast interior of a room inside.

Styrofoam negative is positioned into mold and gypsum is poured. This is an irreversible point in a production process, an ephemeral moment; the end of Styrofoam construct's life and a beginning of a memory of it. It is exactly at this point when a wet and muddy process of materially limited accuracy is confronted with its tectonic opposite: a precise, CAD axonometric drawing of a student's own room being processed. To avoid any trace of sentimentality, a spatially provocative drawing is required; an axonometry is constructed in a way that all walls, floor, ceiling, and furniture is visible from two equal and simultaneously active viewpoints. El Lissitzky's Proun Room [9] and Joseph Albers's Structural Constellations [10] are used as illustrative references. To remain complementary with the general materialness of unit and the real, physical size of an artefact, this CAD drawing wanted to give an impression of scaleless space of continuously active spatial transitions.

Finally, a cast is being extracted from mold and after drying and maturing for couple of days it is being sanded, repaired minimally, and photographed. A structured photography process included light to plasticity modulations. It showed the atmosphere in cast inner space of the artefact, provoking ephemeral presence balanced between its resonance with existing space of an ordinary room and an impression of scale open to imagination. See Fig. 4.

3.3 *Concetto Spaziale*: Imaginary Within Real

As opposite to the substantial materiality of previous unit, a complete resource free unit is offered to students to provoke and intensify observation within the space of their

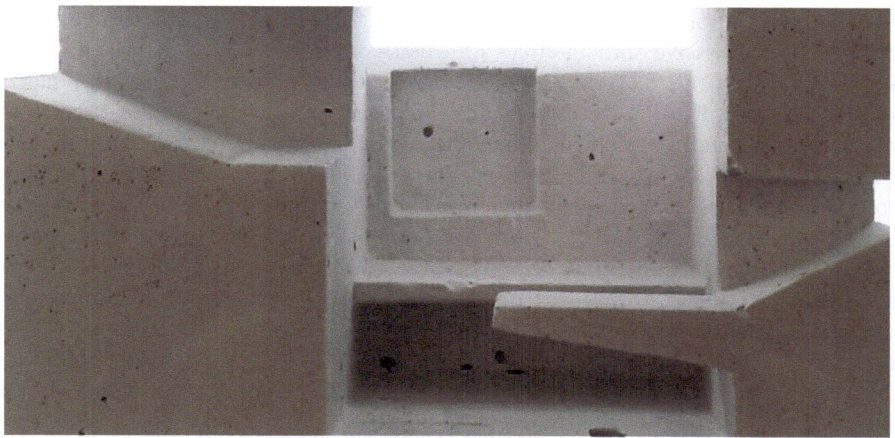

Fig. 4. Atmosphere of an ordinary room, student: Bobetko Rebecca., 4th semester.

everyday living. The unit explored an inflexion point between the stare and seeing, and a shift between observed and finally perceived. Students were asked to look and see the potential spaces in between, around or in objects of everyday use in their surroundings. Interspaces and inner spaces are photographed, modulated by simple photographic tools, and presented as atmospheres on a different architectural scale, suggested by human figure mounted in a photograph.

Lucio Fontana's paradigmatic work title was used as a reference for imaginary cut through the real by power of student's imagination. There is no suggestion of action or use, just stillness, atmosphere, and scale. Simple drawings were made, suggesting possible architectural structures behind produced atmospheres, as shown in Fig. 5.

Fig. 5. Origin of an atmosphere, student: Matušić Nikolina., 4th semester.

3.4   A Field

When understood as transient, impermanent, or open-to-chance quality we perceive within presence of exactly executed work, ephemeral becomes an almost ethical value of its own production process. In his seminal *Field Conditions* [11], Stan Allen's interpretation of progressive shift in ethics of artistic productions is of great pedagogical importance to students of architecture. He described transfer of object conceptualization to structured distribution and designed interference of local rules. A predictably ephemeral, designedly coincidental outcome is enabled when repeated local rule or its exact value is open to material quality in spatial field. Anexact-yet-rigorous quality is produced when local rule meets the material world. A moire effect occurs as vibrant yet unsteady presence upon overlayed grids and produces constant change when its geometry meets spatial shift. A precise design of production process is capable of controlling and conceptualizing indeterminacy of its outcome. If architecture is understood as a production of presence, then processual exploration of conditions for "openness" of its perceived state is a valuable pedagogic tool.

Under a work title *A Field*, a group of units is offered to students of architecture to thermalize and interpret formative potential within a distribution of a local rule. Units are characterized by iterative processualism and serial work, as they produce a pattern, or a work in progress. Students experiment with repetition and distribution of a local rule to gain insight in production of ephemeral presence by exact processes and basic resources.

A Field as a condition for student group work. In this case, distribution of local rules is interpreted as distribution of instruction for work. When instruction meets the individual labor of each student, it becomes the exact frame for an-exact outcome. Each student makes one unit according to the same instruction. In a final work, an individual sensitivity of a draftsmen is regulated by the given instruction - a local rule.

Example: An ordinary blue ink pen is given to each student. Each one of them produced a simple unit according to instruction: a sketchbook sheet of paper is cut along marked lines to form a square fold measuring 20x20 cm. Before cutting, an intensive, almost tactile coating of blue pen is layered down for two hours over right, inner side of a fold. Paper is cut, fold and fixed into a marked position on a white board as one of twenty-five elements forming a square meter, to show individual differences, a small deflection within an exactness of produced whole and to bring a new meaning to each student's uniquely executed work.

An ordinary blue pen is given an unconventional role. As a result, almost as a discovery, beauty of one square meter of-almost-Yves Klein Blue hue emerged, its hapticity saturated through twenty-five individually different hand/body pressures, as visible in Fig. 6.

Field as a condition for individual student work. Given for an individual student production, the unit called for exploration of repetition or seriality.

Student work is a documentation of research in which beginning, and outcome are connected by logical order of decisions while idea is being developed and confirmed by systematically conducted production process. Student conceptualizes individually, produces work, and writes down a concept, a sequence of conceptual decisions in the form of a script detailed enough to enable reproduction without the author being present. Instruction is a tool of exactitude, a declaration of concept. As a pedagogical tool it is aimed at objectification and translation of ephemeral presence into purely technical instruction. Student works are in range from purified variations of CAD graphics to

Fig. 6. Origin of one square meter of-almost-YKB.

processual little actions. Exactitude of their limits could have been set in range from quantity of material to duration of action. The resulting work is not referential, it does not have a narrative. It temporarily denounces its immediate purpose being an attempt in research or production of authentic self-evidence.

Fig. 7. Origin of a Blur, Pavelić Iva, 4th semester.

Figures 7 and 8 show an example:

1. Six reels of adhesive tape, length 66m in total and 48mm wide is pasted onto a room window, in layers, following defined pattern, one reel to cover three layers for window size 60 × 12 cm;
2. After the end of each reel, the window was photographed, the photographs record the emergence of dimly, opaque, and translucent membrane made of transparent material;
3. As the final step, this, now firm material is to be peeled from the window, cut into equal pieces, and rearranged to form a rectangular prism.

Fig. 8. Origin of a Blur, Pavelić Iva, 4th semester.

4 Conclusion

Architecture study teaching process is technical, efficient, and exact, where architecture is measurable. Its learning outcomes are regulated to achieve needed qualifications. Today, discipline's own search for sustainability and participative societal role are normative. But this ecological and societally aware intentions do not guarantee architecture its presence, a unique value to complement and improve spaces and landscapes of our everyday existence. Architecture is a creative practice; it starts genuine creative processes and a search for atmospheres that produce its defining values. Its own tools need constant reaffirmation and protection from decay. This is especially relevant in time of study which itself is introductory, preparatory, and formative phase of practicing architecture.

Study ensures a time for thinking about architecture, its defining frames, and ephemeral margins. It goes beyond architecture. The search for the ephemeral is part of architecture's essential sensitivity. It enables the occurrence of its identifying difference. This difference is described as minimal, almost negligible, something that objectively does not exist [12]. Architect's ability to construct this difference is architecture discipline's vital interest. To produce it, one must search for it. One of architecture's pedagogy goals is to trigger and encourage this search, to motivate students by showing that occurrence of this difference is achievable by conceptual thinking and processual approach to even the simplest of resources. Architecture lesson of appropriate tectonic shows student that meaning emerges with consequent production within conceptually framed conditions.

In time of study, the search of ephemerality through tectonics of architecture lesson wants to develop a special vigilance and sensibility. Just as Yves Klein wrote decades ago, man will inhabit space by force of his sensitivity [13]. For students of Architecture, architectural space arises impregnated with sensitivity of an architect.

Along with its professional goals, the study of architecture could be a beautiful way of student's personal development. If architecture persists on irreplaceability of an architect, the study of architecture should encourage its students into diverse research into origins and meanings of architecture's presences.

Study is a foundation for architecture's cultural and human role. In times of intertwining disciplines, architecture must keep tools for protecting its disciplinary identity and intangible, defining differences. That's why, in addition to all its lessons of exactitude, architecture pedagogy still must provoke idealistic research into satisfying and meaningful presences as primary and most beautiful mode of societal usefulness. And a course unit where intensive production of an ephemeral idea leads into meaningful presence is an important tool for the affirmation of the Architecture's immaterial, unmeasurable identity.

References

1. Benedikt, M.: For An Architecture of Reality, 1st edn. Lumen Books, New York (1987)
2. De Maria, W.: Meaningless work. In: McFadden, J. (ed.) Reaktion Books. London (2016)
3. Krauss, R.E.: The originality of the Avant-Garde and Other Modernist Myths. The MIT Press, Cambridge, Massachusetts, London, England (1986)
4. Arnheim, R.: Art and Visual Perception, A Psychology of the Creative Eye. University of California Press, Berkeley, Los Angeles, London (1974)
5. Serra, R., "Shift", Arts Magazine (1973) in Krauss, R.E.: The originality of the Avant-Garde and Other Modernist Myths. The MIT Press, Cambridge, Massachusetts, London, England (1986)
6. Calvino, I.: Six Memos for the Next Millennium. Harvard University Press, Cambridge, Massachusetts (1988)
7. Reiser, J., Umemoto, N.: Atlas of Novel Tectonics. Princeton Architectural Press, New York (2006)
8. Denegri, J.: Dosier Beuys, DAF, Zagreb (2003)
9. Riedijk M (2009) De Tekening, The Drawing, 010 Publishers, Rotterdam
10. Saletnik, J.: Joseph Albers, Late Modernism, and Pedagogic Form. The university of Chicago Press, Chicago and London (2022)
11. Allen, S.: Practice: Architecture Technique+Representation. Routledge, New York (2009)
12. Čeferin, P.: What (in the World) is Architecture Doing, Oris, no.83., Zagreb (2013)
13. Klein, Y.: Truth Becomes Reality, Ed. Morley S (2010) Documents of Contemporary Art; The Sublime. Whitechapel Gallery, London, and The MIT Press. Cambridge, Massachusetts (1961)

Stumbling as a Praxis of Design Practice. A Pedagogical Experiment in "Theory and Criticism of Architectural Action"

Caterina Quaglio[✉] and Edoardo Bruno

Politecnico di Torino, Turin, Italy
caterina.quaglio@polito.it

Abstract. The Thesis Seminar "Theory and Critique of Architectural Action" is a multidisciplinary educational experience designed to explore the methodological aspects associated with the development of a research project in architectural design, with a focus on Master's Degree Graduation. The Seminar identifies six key methodological issues in research related to project practices and approaches them from a multidisciplinary perspective, subjecting each component to critical examination. Through lectures and Seminar meetings, each discipline challenges these issues based on its scientific status, providing specialized disciplinary insights. In doing so, the Seminar equips students with the necessary theoretical and practical tools to test the application of a robust methodological structure to address various challenges arising from the real world in their research activities. The structure of the Seminar is not intended to establish a specific sequence among the research activities object of inquiry, nor does it seek to offer exhaustive solutions to every issue raised. Instead, the Seminar's aim is to provide students with critical and methodological instruments within a collegial space for open discussion and debate. The pedagogical objective is to enable students to critically evaluate and systematize interdisciplinary knowledge and skills acquired during their previous academic journey, allowing them to position original research programs within the broader framework of design practice.

Keywords: architectural design · design processes · design pedagogy · tactical design · learning by doing

1 Introduction: Question and Positioning

1.1 The Thesis Seminar as a Platform for Reflecting on Architectural Research

After approximately five years of education characterized by an alternation of disciplinary courses and design laboratories, students are confronted with the challenge of exploring architectural design as the subject of their thesis research. Often, this challenge is resolved, by producing an architectural project as a means of addressing a selected issue – as made evident by the majority of design thesis outputs. This approach, which leans towards practical application, bears little resemblance to scientific research.

© The Author(s) 2025
M. Barosio et al. (Eds.): EAAE AC 2023, SSDI 47, pp. 142–151, 2025.
https://doi.org/10.1007/978-3-031-71959-2_17

The latter demands the establishment of how a working hypothesis can lead to original, methodologically rigorous results capable of making an impact within the relevant scientific community.

In light of these considerations, this contribution draws from the experience and outcomes of the inaugural edition of the interdisciplinary thesis Seminar, "Theory and Criticism of Architectural Action", to explore and test, within the pedagogical realm, the intricacies and tools of scientific research in architecture. This Seminar, held at the Politecnico di Torino for the first time in 2023, assembled a group of approximately twenty-two students dedicated to developing research projects within the domain of architectural design with the aim (i) to foster a dialogue between individual experiences and expert contributions and, by doing that, (ii) to emphasize shared methodological and theoretical issues.

The Seminar was conceived with a clear premise: the elaboration of an architectural thesis does not merely entail the development of an architectural design. This is not to suggest that architectural design is excluded from this process; rather, it is contextualized as both an object of study and an investigative tool, rather than as the main expected outcome.

1.2 The Thesis Seminar as a Platform for Reflecting on Architectural Research

The adoption of this standpoint is underpinned by the intention to grasp the "salient character of reality" (Ferraris, 2015: p.55, translated by the authors) which defines architectural design, investigating its mechanisms and effects in order to "distinguish dreams from reality and science from magic" (Ferraris, 2015: p. 30, translated by the authors). In this sense, architectural design is placed in its intrinsic capacity – a technique intended more as a *poiesis* in the sense of fabricating, than a *praxis*, pictured in the middle of a direct and synchronic action – to be understood as a research laboratory, opening up a series of issues that are inherent, though not entirely overlapping, with proper scientific research.

The laboratory, in fact, implies the sequencing of a series of actions that, when tested in a "controlled" environment, allow the designers-researchers to gain experience, thereby enhancing their skills and/or knowledge. Input hypotheses are validated and/or falsified, necessitating a series of subsequent cycles (Latour & Woolgar, 1979). The outputs obtained from the laboratory experience may or may not include an architectural design, but it is essential in any case to define a criterion of generalizability that transcends a specific context; otherwise, the scientific research status of the work performed is at risk of being compromised.

Starting from the premise of considering the work of thesis students as a research laboratory, the Seminar addressed a series of methodological issues. These issues were specifically addressed within the scope of the course's educational project, but they have general scientific significance for architectural studies. (i) What are the physical and temporal boundaries that research/design activities compel us to navigate through a meticulous process of selection and interpretation, which results in continuous compressions and dilations throughout the work? (ii) What sources are called into play, defining heterogeneous archives whose taxonomy and degrees of priority are tied to operations that have little to do with the project as the invention of a tabula rasa? (iii) What are the operations that the laboratory carries out in the field, where the emergence of disputes and deviations question the initial hypotheses and the role of architectural design itself as an activity that needs to reinvent itself in each situation? (iv) Under what conditions is the transfer of practices possible between distinct contexts, and within what limits?

In addressing these issues consciously and rigorously, the practice of design resembles a profession in which experience is accumulated (Schön, 1983), through a process of progressive professionalization that, while retaining certain defining elements, allows for adaptation to a socio-technical context in constant flux (Hughes, 1963).

1.3   The Project as an Object of Study? from Architecture as Individual Expression to Architecture as Technique

The laboratory, therefore, encourages us to move beyond the conception that architectural design must be entirely confined to the dichotomy constituted, on one side, by the interpretative capacity of the subject (expression) and, on the other, by the operability of the object (measures). The relationship established between the designer and its design, a subject dear to the construction of narratives and historiographies (Olmo, 2023, p: 164), assumes a different role within the research laboratory. Here, it is investigated beyond individual experience, constructing general definitions based on the effects it carries.

According to this interpretation, the designer-researcher – who, in the Seminar, is a student working on a research thesis – goes beyond the role of the "initiate" into design. Rather than an individual who hones its art of designing through emulation, training and repetition (Barioglio et al, 2023), research is characterized by the transmission and overlap of those remnants and adjustments inherent in the process of knowledge innovation. This represents a fundamental point for the sciences of design, in a perspective aimed at moving beyond the notion that university classrooms should be transformed into places of indoctrination and the "transmission of initiatory knowledge" (Armando & Durbiano, 2023: p. 139, translated by the authors).

The position we intend to raise is that it is more appropriate to consider the theory of architectural design as a scientific discipline that can be described and transmitted as a sequence of traceable actions, a know-how in which dexterity – where cunning, or "*metis*", is not a morally reprehensible act (Jullien, 2015) – and the ability to interpret and predict data based on experience are an integral part of the research and action process.

2   "Stumbling" as a Praxis of Architectural Design Practice

2.1   Changing Perspective: Conducting Research (on Design) Through Design

Research on design necessitates a shift from "free enunciations" to cycles of methodologically conscious laboratory activities – involving return, repetition, and alteration – which develop in direct relation to a real process/context, thereby producing traceable effects within a controversy. The ability to "measure" the capacity to overcome certain obstacles or changes in trajectory resulting from external conditions – once again, invoking actions typical of scientific laboratory activities – allows us to question the extent of the project's power of action concerning a given contingency.

Hence, there is a need for experimentation with pedagogical approaches that place methods and tools at the center for measuring the adjustments in design action (Federighi & Bruno, 2022). In the pedagogical experience of the thesis Seminar, it is precisely the recurrent "practice of stumbling" that becomes the focal point of research activities on design action, bringing this aspect to the forefront and making the project simultaneously the object and method of scientific investigation.

2.2 A Laboratory of Tactics

Moving away from the short-circuit between ideation and realization (Armando & Durbiano, 2023: p. 119) as a linear expression of architectural design – a "misplaced trigger from the outset" (De Carlo, 1968: p. 138, translated by the author) – opens up the possibility of circumscribing design practice to a technique, which by no means implies an exit from the realm of knowledge and, therefore, from its possibility of transmission in the field of design pedagogy. In fact, by adapting to a know-how model, architectural design becomes, in the realm of research, both an object of investigation and an operational tool through which generalizable concepts can be deduced from empirical experiences of engaging with reality.

Architects' education is thus intertwined with the ongoing interaction between an operational direction – namely, acquiring the competencies and functional tools for legitimizing design in social exchanges – and a critical direction – giving tactical value to one's own laboratory. For both of these domains, direct engagement with the current context as an arena for experimentation is not only inevitable but also desirable. This outlines a specific mode of conducting research in the field of architectural design: the malleability of projects, the discrepancies between initial promises and final results, take on speculative significance as traces of a process aimed at measuring the effects of design action in its interrelation with the socio-technical context in which it operates (Armando & Durbiano, 2023: p. 169).

3 The Seminar "Theory and Criticism of Architectural Action"

3.1 The Structure: 6 Key Methodological Issues

To delve deeper, both theoretically and practically, into the specificities that characterize a research path in architecture, we have identified six key methodological issues as the backbone of the Seminar program: "Hypothesis", "Scale and time", "Maps and events", "The empirical research", "The fieldwork" and "The staging".

The identification of these specific nodes is based on a series of criteria aligned with the research objectives:

– they are central to current scientific debate, encompassing a multitude of conflicting positions and interpretations;
– they transcend specific objects of study, being therefore suitable for interdisciplinary investigation;
– they have highly operational implications, lending themselves to "in-action" experimentation.

Within the Seminar, these six methodological nodes were not addressed as a linear sequence, but rather as a network of interconnected issues. With the aim of outlining each area of investigation in a shared yet open-to-interdisciplinary-contributions manner, we have first identified the operational horizon of each node and made explicit some key questions and challenges.

"The hypothesis": it frames the primary research question arising from contextual conflicts, disputes, and issues. This operation puts the basis for forward steps, defining or anticipating a tentative sequence of future actions.

"Scale and time": it highlights the specific point of observation in the research. This includes addressing blurred spatial and geometrical boundaries, historical periodization, and future scenarios that constitute the scientific laboratory for research operations.

"Maps and events": it involves detecting, narrating, and tracking the selected dispute, connecting stakeholders, documents, and places. The mapping operation aims to identify the sphere of influence of each "actant" involved in the process (Latour, 2005) and enhance predictive capabilities of design action.

"The empirical research": it includes determining research sources, operations, and their sequence. The heterogeneous nature of sources can be tackled by developing archives and taxonomies to support research activities.

"The fieldwork": it comprises all operations carried out directly on-site, involving techniques to test sources and archives, collect multiple points of observation, and interpret the data gathered for recalibrating the primary research question.

"The staging": it pertains to developing a selected narrative, a plot, or design operations capable of staging the elaboration of the thesis.

Each of these methodological operations, as described, was explored at the same time through various types of disciplinary contributions and direct research experiences conducted by thesis students on their respective case studies. Moreover, a common ground for the examination of each "node" was progressively built up via the construction of collective reflection on research practice in architecture, particularly a working approach characterized by a partial and contingent understanding of the process/context in which it operates.

3.2   The Pedagogical Approach: Interdisciplinary Learning-By-Doing

The educational objectives and pedagogical approach of the thesis Seminar aim to reflect an interdisciplinary learning-by-doing methodology. More specifically, each methodological issue was addressed in the Seminar through two parallel approaches. On one hand, students were required to engage in guided practical exercises related to their own work, stimulating reflection in practice (Schön, 1983). On the other hand, specialized didactic contributions further developed the theoretical and methodological conceptualization of the six nodes. Three main types of contributions were in particular provided throughout the Seminar:

1) "Thematic lectures": these lectures are intended to build a broader theoretical framework on specific topics.
2) "Operational lectures": these lectures focus on specific analytical tools and research methods.

3) "Interdisciplinary seminars": these sessions encourage cross dialogues on each topic, fostering broader discussions and insights.

Thematic and operational lectures provided the theoretical underpinning of the Seminar. They were delivered by invited experts from a range of disciplines, including anthropology, philosophy, sociology, and history. Ideally, the list of disciplines and contributions involved – while limited in the Seminar by the constraints of a semester-long teaching experience – can be extended to encompass all perspectives that facilitate dialogue on a common ground of interest. In parallel, the interdisciplinary seminars progressively addressed the specificity of design practice and architectural action compared to other practices.

The chronological organization of the seminar was designed to enable a continuous intertwining of different types of contributions and learning exercises. Each methodological issue was developed over a two-week module, commencing with the launch of a practical exercise. This approach allows students to start working and thinking by doing on a topic while progressively expanding their knowledge and interpretative tools. The outputs produced were collectively discussed at the end of each module but remained subject to continuous updates and modifications throughout the whole Seminar, integrating and complementing contributions from other nodes.

For example, in the first module, dedicated to the research hypothesis, students were asked to frame their main research question and develop an action plan to address it. Nevertheless, the request to explicitly articulate the research hypothesis and the actions aimed at investigating it, accompanied the entire development of the work, devoting specific attention on any deviations and modifications in progress. The second and third modules challenged students to delimitate and map the object of study in all its dimensions – chronological, spatial, social, etc. The fourth and fifth exercises paid specific attention to the technicalities of research work, including the selection, production, and interpretation of direct and indirect sources (Fig. 1). Finally, the last module, pertaining to the "staging" of the work done, implied to identify the narrative strategy most aligned with the research work's objectives.

The Seminar's outputs reflect a shift not only in perspectives but also in the objects observed within the research work. The focus moves towards the design process rather than its outcome (the built architecture), emphasizing the mobilization of multidisciplinary competencies in action and how they intersect with and modify the ongoing process. Spatial, technical, or conceptual architectural drawings are often developed alongside or in relation to diagrams that aim to incorporate time and actions in both analytical and design operations (Fig. 2).

All the students worked on real controversies or processes, with the aim to consider the effects of their design choices, if not measurable, at least plausibly predictable. Furthermore, design projects, conceived as technical objects, become negotiation tools within the process rather than photographs of a pre-determined product. In doing so, the Seminar aspires to reproduce, within the limits of an educational experience, the actual implications of a multi-actor design process, making it a specific object of research and experimentation for future professionals.

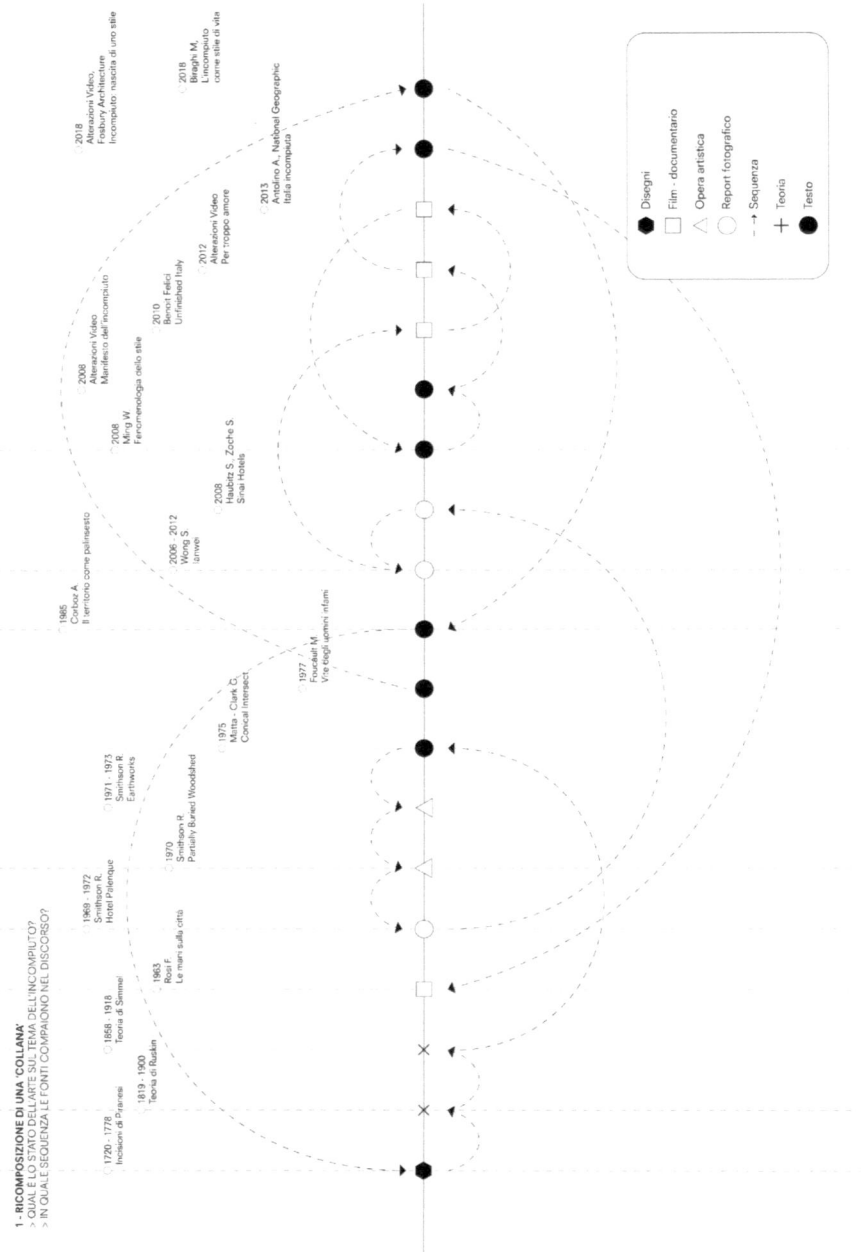

Fig. 1. A preliminary taxonomy of the sources considered in the development of exercise number 5. It is necessary to underline both the different types of materials considered as well as their looping relationships along the research, where they are activated according to a precise tactic (diagram by Antonio Nicoletti).

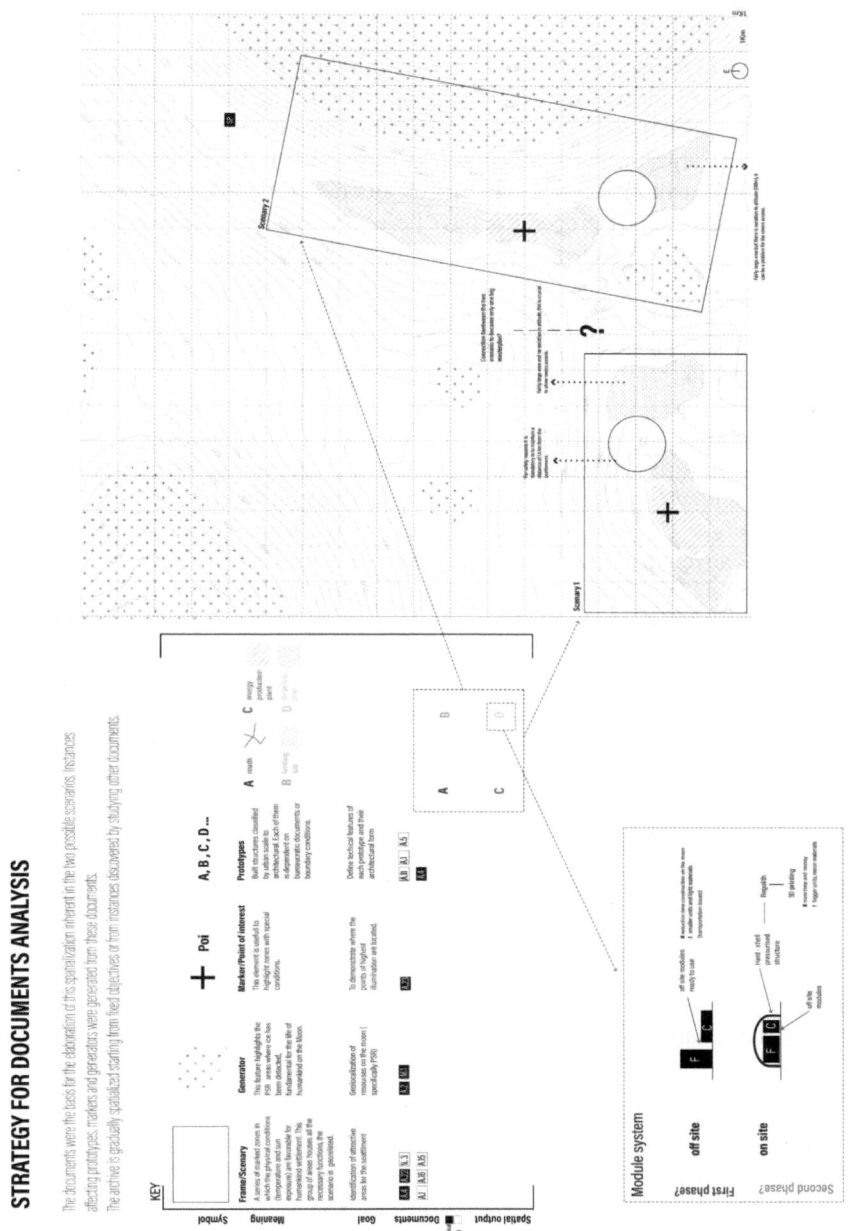

Fig. 2. Spatialization of different design scenarios based on documents collected along the research activity for exercise number 6. Their recombination is affecting preliminary traces which needs to confront subsequent contingencies (diagram by Mammino Mariapia).

4 Final Remarks

The experience of the Seminar allowed us to highlight some final remarks and questions from that can be of interest for a common discussion. The first question relates to the relationship between theory and practice: How to transmit and an operative knowledge through a set of didactic tools which are un-coded and infinitely expandable? In this sense, the Seminar framed a possible answer in erasing boundaries between thoughts and actions, while looping procedures have been promoted as pedagogical experiments. As a matter of fact, un-coded tools were turned into working hypothesis within the laboratory – instead of ontological horizon. In doing so, the progressive expansion of each student's research toolkit allowed for a deep understanding of an iterative technique, which surpass – or at least re-configure – the conception of architectural design as a weak scientific approach.

The second question addresses transdisciplinarity, questioning the potentials for effective integration among disciplines. The students had several inputs from experts and researchers from different fields, helping them to configure architectural design as an open-source field where different ontologies can cooperate in moving towards an-action oriented laboratory. Testing through practical research operations their observations significantly blurred the boundaries between structured scientific fields.

Finally, the main pedagogical question underlying the Seminar is if research project can be developed – and thought – as an exercise in tactics. The atypical outputs produced by the students as outcomes of the Seminar – alternative design solutions, maps of controversies that can boost or refuse specific actions, taxonomies of datasets and sources, etc. –, suggest an interest in further developing an experimental pedagogy on architectural design focused on the juxtaposition of iterative sequences of micro-decisions as a way to unveil spaces for innovation and, more generally, further dissemination among the scientific community of an alternative approach to design practice.

References

Armando, A., Durbiano, D.: Critica Della Ragione Progettuale. Saggi 935. Il Mulino, Bologna (2023)

Barioglio, C., Campobenedetto, D., Quaglio, C.: Transformative design teaching. Challenging the didactic assumptions of polytechnic schools through the lens of the professional role of architects. In: Flynn, P., O'Connor, M., Price, Dunn, M. (eds.) Rethinking the Crit, pp. 121–134. Routledge, London (2023)

De Carlo, G.: La piramide Rovesciata. Architettura Oltre il'68. Quodlibet, Macerata (1968)

Federighi, V., Bruno, E.: The Detroit Great Game: Explorations Around Architectural Design and Its Agency. AADR – Art Architecture Design Research, Berlin (2022)

Ferraris, M.: Manifesto del Nuovo Realismo. Laterza, Bari (2015)

Hughes, E.C.: Professions. Daedalus **92**(4), 655–668 (1963)

Jullien, F.: De l'être au Vivre: Lexique Euro-Chinois de la Pensée. Gallimard, Paris (2015)

Latour, B.: Reassembling the Social: an Introduction to Actor-Network-Theory. Oxford University Press, Oxford, New York (2005)

Latour, B., Woolgar, S.: Laboratory Life: The Social Construction of Scientific Facts. Sage Publications, Beverly Hills (1979)

Olmo, C.: Storia Contro Storie: Elogio del Fatto Architettonico. Donzelli Editore, Rome (2023)
Schön, D.A.: The Reflective Practitioner: How Professionals Think in Action. Basic Books, New York (1983)

A Cosmopolitan Architectural Education

Massimo Santanicchia[⊠]

Iceland University of the Arts, Reykjavík, Iceland
massimo@lhi.is

Abstract. This paper is the result of forty dialogues conducted during my PhD research, which took place between 2018 and 2020, among students and educators' of the Nordic Baltic Academy of Architecture. These conversations were initiated by three questions: What skills should students have after studying architecture? How should these skills be taught? How can architectural education be of special importance to our society? The answers to these questions were analysed and interpreted by following the Grounded Theory approach. What emerged from these dialogues is the shared conviction to use architectural education as a complex project to advance the knowledge, traits, attitudes, values, and behaviours necessary to respond to global challenges whilst creating conditions for students and their educators to locally engage as active citizens. This combination of global awareness and local activism is at the base of formulating the Theory of Cosmopolitan Citizenship in Architectural Education whose purpose is to help students and educators cultivating a language, activating a pedagogy, and developing a scholarship capable of advancing new political agencies to codesign healthier, safer, and a fairer world in a changing social, ecological, and political environment.

Keywords: Architectural Education · Cosmopolitan Citizenship · Nordic Baltic Academy of Architecture

1 Introduction: The Nordic-Baltic Academy of Architecture

I am head of the department in architecture at Iceland University of the Arts (IUA) and – together with my cohort – in charge of designing the educational experience for our students. Since 2016, I have been attending meetings of the Nordic Baltic Academy of Architecture (NBAA), an organisation of educators from nineteen schools of architecture from the Nordic and Baltic countries. NBAA was established in 1993, with the aims to share knowledge, promoting common views, concerns, and interests in the broad field of architectural education and research [1].

In Autumn 2018 as part of my PhD research on architectural education in the Nordic-Baltic area and as one of the educators actively involved in the NBAA, I started conducting interviews with students and educators of the network, thinking together about the responsibility, value, and meaning of architectural education. The fundamental results of these conversations are the essence of my PhD titled: "Becoming cosmopolitan citizen-architects: A Reflection on Architectural Education in a Nordic-Baltic Perspective"

M. Barosio et al. (Eds.): EAAE AC 2023, SSDI 47, pp. 152–160, 2025.
https://doi.org/10.1007/978-3-031-71959-2_18

which was defended at the University of Iceland in November 2022 [2]. In this article the essence of my PhD research is presented in a concise and focused way to address specifically the need to develop a language, pedagogy, and scholarship in architectural education as a project to advance the knowledge, traits, attitudes, values, and behaviours necessary to respond to global challenges whilst creating conditions for students and their educators to locally engage as active citizens (Table 1).

Table 1. The Nordic Baltic Academy of Architecture

Country	Name of the School
Denmark	AArch: Aarhus School of Architecture KADK: Royal Danish Academy of Fine Arts
Estonia	EKA: Estonia Academy of the Arts TalTech: Tallinn University of Technology
Finland	A: Aalto University TUNI: Tampere University of Applied Sciences O: University of Oulu
Iceland	IUA: Iceland University of the Arts
Latvia	RTU: Riga Technical University
Lithuania	VDA: Vilnius Academy of the Arts VGTU: Vilnius Gediminas Technical University KTU: Kaunas University of Technology
Norway	BAS: Bergen School of Architecture NTNU: Norwegian University of Science and Technology AHO: Oslo School of Architecture and Design
Sweden	C: Chalmers School of Architecture and Design KTH: Royal Institute of Technology UMU: Umea School of Architecture LTH: Lund School of Architecture

2   Research Questions and Methods

A total of 40 semi-structured interviews were conducted between 5 November 2018 and 26 March 2020 among twenty-nine educators and fourteen students from the NBAA network. All interviews were initiated by three research questions:

- What skills should students have after studying architecture?
- How should these skills be taught?
- How can architectural education be of special importance to our society?

Each question addresses architectural education from a different viewpoint to reveal dispositions, skills, pedagogies, and multiple societal agencies associated with the practice of architectural education. "Skill" was explained to interviewees not only as an ability to do something (an expertise), but rather as the combination of knowledge, attitudes, values, and behaviours considered vital to becoming an architect. Through these questions I could conduct an in-depth exploration of an area in which the interviewees have a substantial personal experience that is architectural education.

All interviews were recorded, for a total of 32 h and 13 min, they were then transcribed verbatim, and promptly emailed to participants, who were invited to make comments or amendments. All interviews were then analysed following the Grounded Theory (GT), a rigorous method for collecting and analysing qualitative data to construct theories grounded in the data itself [3]. Each interview was first singularly analysed, and its essential key topics were highlighted. The analysis of each interview was then compared to the other interviews to find out recurrent topics, shared cultural themes, and essential common features. By doing so a theoretical direction emerged, one that positions the global challenges (climate crises and social injustice) at the core of architecture education and indicates that each school of architecture has the duty to educate future practitioners for active societal agency to contribute to their solution.

Through an iterative relational process of analysis of the recorded interviews, their comparison, and interpretation based both on my personal experience and the use of pertinent literature from the field of world citizenship education, I could construct a communal perspective grounded in its Nordic–Baltic context, that is a theory that I call of Cosmopolitan Citizenship Architectural Education (CCAE), whose purpose is to advance the societal scope and meaning of architectural education.

3   A Cosmopolitan Architectural Education

3.1   The Findings from the Nordic-Baltic Voices

What emerged from the forty Nordic–Baltic dialogues is the acknowledgment that the grand challenges—the climate crisis and social inequality—are the most important issues that need to be faced for the continuation of life on our planet and their addressing ought to become the foundation of any design process and the purpose of architectural education. Nordic-Baltic respondents underline the importance of conceiving architectural education as an explorative and formative process to help students finding both their interests and their societal agency, by acquiring the knowledge, traits, attitudes, values, and behaviours necessary to collaborate in bettering the world starting from their own communities. The Nordic-Baltic voices view the process of learning as one devoted helping students developing critical skills (the capacity to question everything), social awareness (the ability to understand what you see), self-reflection (understanding the impact of your design choices), imagination (being able to conceive of and represent what is not there yet), and action (the ability to pursue your ideas beyond the school's limits). Fostering social agency through the design process demands of schools to create the learning conditions for students and their educators to be exposed to diversity of thoughts, to different ways of knowing and doing, and to create conditions for collaboration with otherness, that is to create a learning environment that supports students developing their political roles.

The strong commitment of the Nordic-Baltic voices to operate both as active citizens and agents of positive global change led me to investigate the fields of citizenship education, cosmopolitanism, and global citizenship education. The reviewing of this literature served as the theoretical framework through which analyse and interpret the Nordic–Baltic voices and led me to build a theory which I call Cosmopolitan Citizenship Architectural Education (CCAE). This theory is grounded in the NBAA's context, and yet it reflects my multicultural and multidisciplinary interests and the historical context – a time of challenges to biodiversity, human health, and well-being. As such CCAE theory has "a direction, an orientation, a purpose" [3] that of helping students and educators cultivating a language, activating a pedagogy, and nurturing a scholarship capable of educating a new generation of architects committed to respond to the grand challenges and to bring about positive societal change. To these societal agents, I have given the name of cosmopolitan citizen-architects [4–10].

3.2 Theoretical Framework for Cosmopolitan Citizenship in Architectural Education

Cosmopolitan: from the Greek kosmopolitēs, 'citizen of the world.'

Citizenship: a juridical status and a civic and political agency that positions everyone in terms of rights and responsibilities into a larger societal context [11].

The concept of cosmopolitan citizenship is based on the understanding that we all inhabit the same living system. Cosmopolitan citizenship does not mean homogenization of ideas, nor the obliteration of cultural differences but it celebrates the diversity that exists in the world by also recognizing its common traits [12, 13]. Cosmopolitan citizenship is a societal project of redefining who we are and how we relate to each other, as diverse and equal beings who live in a common and shared world.

The United Nations Educational, Scientific and Cultural Organization (UNESCO) explains cosmopolitan citizenship as a project of education which requires the acquisition of the knowledge, skills, attitudes, values, and behaviours necessary to become active promoters of more peaceful, tolerant, inclusive, secure, and sustainable societies [14]. Cosmopolitan citizenship education requires both an acute awareness of the state of the world – its problems, injustices, and possibilities – and the intention to engage for solutions, to care for and with Others. This type of education emphasizes political, economic, social, and cultural interdependency and interconnectedness that exist between the local and the global, and the shared responsibilities that each individual carries as a distinct yet equal citizen of a shared and common world [15].

Educating for cosmopolitan citizenship requires constant interactions between different people to develop social awareness and new perspectives. It involves attaining knowledge, skills, attitudes, and behaviours necessary to understand that all Earthlings are part of the same ecological and social system, and to envision a common future wherein no one is excluded and to actively engage as agents of care for life on Earth. Cosmopolitan citizenship is indissolubly linked to solidarity, empathy, emancipation, freedom, and the pursuit of global justice; as such, it is practice oriented because it requires critical civic engagement with real cases [16].

Even though the term "cosmopolitan citizenship education" was never mentioned by any of the respondents it is my understanding that the concept of cosmopolitan citizenship captures the shared thoughts on architectural education that have emerged from the Nordic–Baltic dialogues. With this theoretical framework presented I know focus in explaining what a language, pedagogy, and scholarship of cosmopolitan citizenship pertain.

3.3 The Language of Cosmopolitan Citizenship in Architectural Education

The language of Cosmopolitan Citizenship Architectural Education helps students acquiring a broader vocabulary of concepts and ideas to redefine what architecture is and can do. This language speaks of architectures as plural, heterogeneous, situated, and collaborative processes which are always in relation to places, communities, and people, and yet such processes are also profoundly influenced by global forces. A CCAE language is world-related and place-based, and it explains architectures as the social and ecological relations involved in their practice—a practice that transcends the design of buildings to include processes of thinking, theorising, and writing that relate humans and their environment [17, 18]. This is a practice that is holistic and receptive of arts and humanities, science and technology, and new social, technological, and ecological challenges [19–21]. It is one that can be used in multiple ways: as a critical process of enquiry [22]; as a vehicle for raising social awareness [23, 24]; as a tool for imagining and advancing agendas of social justice [23, 25]; and as a collaborative project aimed at living together harmoniously [26].

The language of CCAE validates students' different voices, interests, and different ways of practicing architecture. It explains creativity as a collaborative project based on thinking together. It invites students and their teachers to consider school's time not as a rehearsal for future practice, but as a time invested to challenge the status quo, a time for action, and a time to forge the conditions for civic engagement between academia and the world outside the school. CCAE language encourages the creation of a caring learning environment to empower students in developing their own architectural practice as well as their societal agency to contribute making a positive difference in the world. The language of CCAE is further influenced by the work of international architectural commentators who celebrate the value of architectural education beyond building design, who expand architecture's agency by making the field of research more receptive to diverse voices and conditions [2, 27].

3.4 The Pedagogy of Cosmopolitan Citizenship in Architectural Education

The pedagogy for CCAE aims to redesign power relations in the design studio—the very core of Nordic–Baltic architectural education, by making it more receptive and inclusive of different ways of being, thinking, and making architectures; more collaborative among students, their educators, and their community; and more concerned with exploring the design process to advance social and ecological justice. A pedagogy for CCAE is committed to educating critical thinkers capable of addressing issues of societal relevance and making them at the base of the design process. The pedagogy for CCAE aims to form self-reflective, collaborative, and socially aware beings equipped with the social skills to engage in dialogue with diverse people (experts and non-experts), to cooperate and collaborate (with everyone), and to form future practitioners capable of bringing diverse forms of knowledge and experience together in their communities.

A pedagogy for CCAE is committed to honouring the two fundamental purposes of architectural education, that is to educate ethical professionals and world citizens [20]. Such pedagogy aims to equip future architects with the skills, attitudes, traits, and behaviours necessary to move away from current destructive practices and towards the environmental, social, and economic justice necessary to protect life on our planet. Cosmopolitan citizenship architectural pedagogy invites each school in the world to define both its local mission and its global relationship contributing therefore to redefining how we live together [2].

3.5 The Scholarship of Cosmopolitan Citizenship in Architectural Education

Cosmopolitan citizenship education aims to advance the very idea of scholarship. More than thirty years ago, Ernest Boyer, in his influential report Scholarship Reconsidered: Priorities of the Professoriate, invited universities "to clarify campus missions and relate the work of the academy more directly to the realities of contemporary life" [28]. Many universities today are working on expanding the meaning and scope of scholarship, strengthening their societal relevance and public engagement. Ronald Barnett for instance advocates for the "ecological university" as one "that takes seriously the world's interconnectedness and the university's interconnectedness with the world" [29].

The scholarship of CCAE recognises that fairer knowledge is constructed when diverse perspectives and standpoints are included in the research process, and when knowledge production reflects critically on the nature of academic research by asking what purposes it serves and whom does it support and discriminate. The scholarship of CCAE allows social and ecological events to further shape, bias, and influence the nature and scope of academic research, and it celebrates its unique local bonds, and it acknowledges that every place is never dissociated nor dematerialised from the global context [30]. The scholarship of CCAE recognises academic researchers' responsibility to disseminate their outcomes via open and clear platforms freely accessible to a larger audience. It recognises the importance for academics to be part of the most pressing moral, political, and cultural questions and hence collaborate across disciplines with other academics and practitioners to advance just and fair solutions. The scholarship of CCAE promotes forms of scholarly activism by creating a learning environment conducive for students and teachers to transgress university boundaries for civic engagement and the pursuit of social and ecological justice. The scholarship of cosmopolitan citizenship aims to create a new era: the Cosmopolicene, a collaborative, inclusive, and caring age where the interconnectedness among us all is valued and protected, and where development translates in social and ecological justice [2].

4 Thinking from the North

Each school of architecture visited in this research represents a microcosm devoted to the production, discussion, and dissemination of architectural thinking [22]. Each school is a place where "the ethos of a profession is born" [31]; where attitudes are shaped and carried into professional life; where a legacy is passed down from one generation to the next; where architects' possible societal roles are imagined and then enacted.

My intention with this research is to think together with my Nordic–Baltic colleagues and students on architecture's social and political responsibilities and obligations. During this dialogical process it emerged vividly that the essence of architectural education is based on forming civic minded, engaged professionals who can use their acquired skills in multiple ways for the betterment of their community.

I cannot, nor do I wish to, claim that any of these findings belong exclusively to the NBAA network, nor to architectural education only. I can only claim that these forty interviews have a common ground that of connecting architectural education to cosmopolitan citizenship education. It is this indissoluble link between society and architectural education, this societal sense of responsibility that is, for me, the key to understanding architectural education in the Nordic–Baltic region [2].

5 Conclusions

This paper exposes the main findings presented in my PhD "Becoming cosmopolitan citizen-architects: A Reflection on Architectural Education in a Nordic-Baltic Perspective". It continues therefore supporting the idea that architectural education is a field of study not only receptive to the notion of cosmopolitan citizenship, but one that helps to advance it. As such, architectural education is of paramount importance for both educating not only future designers of buildings but for preparing students for active cosmopolitan citizenship. It is of paramount importance to educate future architects who can contribute to shape new perspectives and new stories of what architectures are and can do, architects who can enact new societal agencies necessary to face and respond to the present grand challenges and those yet to come. Educating cosmopolitan citizen architects means supporting the long tradition that envisions the essential role of an architect as a public servant devoted to the protection of the common good. It recognises that each architect has a political agency, and as such an architect is asked to be socially relevant and to use the practice of architecture in multiple ways as a critical process of enquiry, as a vehicle for raising social awareness, as a tool for collective imagination, and as a collaborative project aimed at caring for and repairing the common good, besides the undisputed role of architects as buildings' designers.

Educating cosmopolitan citizens means becoming inquisitive-knowledgeable-self-reflective-critical-empathic-collaborative-caring beings, who are instigative of hope and have the courage to act in the now for the pursuit of a better world. Becoming cosmopolitan citizen architects means learning to include the Other, future generations, and unrepresented voices in the design process to achieve social and ecological justice, it means learning to make design decisions grounded in their social and environmental context and equally influenced by the understanding of their local and global implications. Becoming cosmopolitan citizen architects means challenging the myth of the star architect as a solitary genus, towards forming architects who are more prone to dialogue and collaboration, who understand the value embedded in celebrating the social and ecological relations present in each design process. This is about understanding that the ongoing environmental crisis needs to constitute the premise and scope of scholarly investigation; be part of educational discourse, form our individual and collective planetary consciousness and unite us as we move towards solutions. Becoming cosmopolitan citizen architects is connected to lived experiences, it is a process and an ongoing activity that aims to complete us as humans in our ever-changing realities and connecting us with the world [2].

References

1. NBAA [Nordic Baltic Academy of Architecture] (2023). https://www.nbaainfo.org/purpose/. Accessed 05 Jan 2024
2. Santanicchia, M.: Becoming cosmopolitan citizen-architects: A Reflection on Architectural Education in a Nordic-Baltic Perspective, PhD, University of Iceland, School of Humanities (UI), School of Education (UI), Faculty of Icelandic and Comparative Cultural Studies (UI), Faculty of Education and Diversity (UI) (2022)
3. Charmaz, K.: Constructing Grounded Theory, p. 3. Sage, Thousand Oaks (2014)
4. Santanicchia, M.: Systems thinking and systems feeling in architectural education. In: Lorentsen, E., Torp, A. (eds.) Formation: Architectural Education in a Nordic Perspective, pp. 258–275. Architectural Publisher B, Copenhagen (2018)
5. Santanicchia, M.: Becoming citizen architects: a case study of a design studio in Reykjavik. Build. Mater. (22), 116–136 (2019)
6. Santanicchia, M.: Becoming cosmopolitan citizens architects, a reflection on architectural education across the Nordic Baltic academy of architecture NBBA: a student's perspective. In: Roth-Čerina, M., Cavallo, R. (eds.) The Hidden School Papers: EAAE Annual Conference Zagreb 2019, pp. 312–335. European Association for Architectural Education, Delft (2020a)
7. Santanicchia, M.: Design Education for World Citizenship, pp. 171–179. DIID, Rome (2020b)
8. Santanicchia, M.: architectural education for cosmopolitan citizenship: five stories, two questions, and one directive. MD J. (10), 64–73 (2020c)
9. Santanicchia, M.: Architectural Education for a New Beginning, pp. 467–470. The Architect (special issue: Reboot) (2021)
10. Santanicchia, M.: Becoming cosmopolitan citizen architects: an educator's reflections on architectural education across the Nordic Baltic academy of architecture. Nordic J. Arch. Res. **34**(1) (2022)
11. Kymlicka, W., Norman, W.: Return of the citizen: a survey of recent work on citizenship theory. Ethics **104**(2), 352–381 (1994)
12. Appiah, K.A.: Cosmopolitanism: Ethics in a World of Strangeness. Penguin Books, London (2006)
13. Brown, G., Held, D. (eds.): The Cosmopolitanism Reader, p. 6. Polity Press, Cambridge (2010)
14. UNESCO (2015) Global Citizenship Education: Topics and Learning Objectives. https://unesdoc.unesco.org/ark:/48223/pf0000232993. Accessed 05 Jan 2024
15. Osler, A., Starkey, H.: Changing Citizenship: Democracy and Inclusion in Education. Open University Press, New York (2005)
16. Harvey, D.: Cosmopolitanism and the Geographies of Freedom. Colombia University Press, New York (2009)
17. Tharp, B.M., Tharp, S.: Discursive Design: Critical, Speculative, and Alternative Things. MIT Press, Cambridge (2019)
18. Harriss, H., Hyde, R., Marcaccio, R. (eds.): Architects After Architecture: Alternative Pathways for Practice. Routledge, London (2021)
19. Rendell, J., Hill, J., Fraser, M., Dorrian, M. (eds.): Critical Architecture. Routledge, London (2007)

20. International Union of Architects (UIA): UIA and Architectural Education: Reflections and Recommendations (2011)
21. UNESCO-UIA: UNESCO-UIA Charter for Architectural Education (2017). https://www.uia architectes.org/webApi/uploads/ressourcefile/178/charter2017en.pdf. Accessed 05 Jan 2024
22. Ockman, J. (ed.): Architecture School: Three Centuries of Educating Architects in North America. MIT Press, Cambridge (2012)
23. Cruz, T.: Rethinking uneven growth. In: Gadanho, P. (ed.) Uneven Growth: Tactical Urbanisms for Expanding Megacities, pp. 48–55. Museum of Modern Art, New York (2014)
24. Yaneva, A.: Five Ways to Make Architecture Political: An Introduction to the Politics of Design Practice. Bloomsbury Academics, London (2017)
25. Hyde, R.: Future Practice: Conversations from the Edge of Architecture. Routledge, London (2012)
26. Fitz, A., Krasny, E.: Critical Care: Architecture and Urbanism for a Broken Planet. MIT Press, Cambridge (2019)
27. Frichot, H., Sandin, G., Schwalm, B. (eds.): Architecture in Effect: After effects: Theories and Methodologies in Architectural Research, vol. 2. Actar, New York (2018)
28. Boyer, E.L.: Scholarship Reconsidered: Priorities of the Professoriate, p. 13. Carnegie Foundation for the Advancement of Teaching, San Francisco (1990)
29. Barnett, R.: The coming of the ecological university. Oxford Rev. Educ. **37**(4), 439–455 (2011)
30. Plumwood, V.: Shadow places and the politics of dwelling. Aust. Humanit. Rev.Humanit. Rev. **44**, 139–150 (2008)
31. Cuff, D.: Architecture: The Story of Practice, p. 43. MIT Press, Cambridge (1991)

Bridging Methods and Disciplines: An Architectural Pedagogy for Rural Areas Community

Stefano Sartorio[✉]

Dipartimento di Architettura E Studi Urbani, Politecnico di Milano, Milan, Italy
stefano.sartorio@polimi.it

Abstract. Architectural pedagogy in rural areas necessitates a thoughtful consideration of both methodological and disciplinary approaches within the educational process. Is it enough to define the way of teaching Architecture as a method or as a discipline? In a constantly changing world, facing multiple fragile conditions (e.g., environmental crisis, social disparities, demographic changes), the way of teaching, and learning, architecture shall consider over-coming this narrow boundary. An effective balance between method and discipline is needed, where architectural pedagogy should be responsive to the variations of fragile territories and communities. This capability enables students and rural communities involved to develop more sensitivity to context, recognizing which design solutions are best tailored to their realities. Inner areas, for instance, which suffer from constant depopulation and lack of connection, though rich in historical/natural heritage, are entangled in a complex multiplicity of design scenarios. Especially in Italy, whose territory is 60% made of inner areas [1], this condition should be approached by universities for the architects of the future. Moreover, architectural pedagogy in rural areas should carefully consider the interplay between method and discipline as a participatory; a community-oriented method allows the co-creation of contextually relevant solutions within the educational process of teaching architecture. The contribution proposes a view on this challenge, in which architectural education has a profound role in shaping a desirable future, starting from the ongoing experiences developed in some inner areas case studies, such as the one of Vione (Valle Camonica, Lombardy) and of Morino (Valle Roveto, Abruzzo).

Keywords: Cross-Disciplinarity · thick boundaries · Italian Inner Areas · Rural Challenge · Participatory Education

1 Introduction

1.1 Rural Areas: The Perks of Architecture and Pedagogy

It is crucial to discuss the possible impact of architectural education in the frame of a global shifting condition under multiple layers of topics (environmental crises, social inequalities, population changes…). So, what could it be its role in shaping the architects

© The Author(s) 2025
M. Barosio et al. (Eds.): EAAE AC 2023, SSDI 47, pp. 161–170, 2025.
https://doi.org/10.1007/978-3-031-71959-2_19

of the future? Nowadays, there can be a shift in teaching architecture from defining a specific knowledge inside the discipline or a series of methods that can be generalised and exploited for practice, as promoted in the call of this conference.

The following contribution addresses the topic in two stages: in the first part (*ch.2 Hypotheses*) reasoning around a broader and inclusive definition of our discipline and methods of teaching architecture. The visualization of the problem "discipline" vs "method" (the proposed couple of antinomian concepts) was the starting point for developing the *thesis (ch.3)*. In the second part, *case studies (ch.4)*, examples provide a double-faced analysis to overcome the strict separation between the antinomian concepts.

As the world faces multiple conditions of change, the topic is restricted to the intricate condition of inner areas. In Italy this term describes lagging regions; remote rural communities living with few or non-sufficient parameters of formal education services (at least a high school - liceo, ITS, or CFP), health structures (full operative hospitals at least in DEA di I Livello), and public connections systems (railways stations of Silver level) or that requires more than 27,7 min to reach a polarity of those services [2]. This inquiry transcends the realm of only architectural education and extends to a broader spectrum of addressing contemporary global challenges, though can and should be approached by an architectural pedagogy for a better framing of their situation.

Indeed, the method employed in architectural pedagogy for rural areas should be participatory, experiential, and community oriented. Traditional didactic approaches may not be suitable in fragile settings, such as rural ones, where local communities possess rich and various, knowledge and experiences that may not collide with the disciplinary knowledge of architecture. Students and rural communities must be engaged directly with a set of tools from various disciplinary and methodological nuances. This hands-on approach facilitates a co-creation process, ensuring that architectural interventions are contextually appropriate, meeting community's actual requirements and student's formative objectives.

2 Hypotheses

2.1 Logical Assumptions/Thick Boundaries

Understanding the interplay between method and discipline is then crucial for effectively educating future architects who can address the unique challenges of rural contexts.

Method: Via Latin, from Greek *methodos* [μέθοδος]' pursuit of knowledge', from *meta* [μετ'] (expressing development) + *hodos* [ὁδός] 'way'. 1: A particular procedure for accomplishing or approaching something, especially a systematic or established one.

Discipline: via Old French, from Latin *disciplīna*' instruction, knowledge'. 1: the practice of training people to obey rules or a code of behaviour, using punishment to correct disobedience. 2: a branch of knowledge, typically one studied in higher education.

Starting from the above meaning and the etymological definition of the terms [3] while method refers to the instructional strategies, techniques, and tools utilised to deliver knowledge and skills, discipline pertains to the core principles; it frames the shape of thinking and practicing of one subject, sometimes also the philosophy beneath a pragmatical subject. As per their definitions, the two are entangled; the author wonders whether practicing and teaching architecture is closer to the former or the latter.

Supposing a scenario where a *scientific status* serves as the arbiter, firmly delineating the boundaries of a discipline such *teaching architecture*, it would demarcate the domains within the discipline and those that exist outside it, prescribing the methods of teaching and learning that are permissible. The whole definition of *teaching architecture*, with its subjects, topics would be contained in this perimeter. And so, per the same logic, it would appear the same perimeter in every *teaching a subject*. However, when we step outside the realm of disciplinary boundaries and peer in from an external point of view, it would appear that specific knowledge and teaching methodologies operate in isolation (Fig. 1). These seemingly disparate entities may, at first glance, lack the interconnections. However, there must be something in between the isolated knowledge of the disciplines. They are not isolated monads but rather components of a rich tapestry of interdisciplinary interactions.

Indeed, addressing the definition of "Thick Boundaries" [4]: *In nature, we see many systems with powerful, thick boundaries. The thick boundaries evolve as a result of the need for functional separations and transitions between different systems. They occur essentially because wherever two very different phenomena interact, there is also a 'zone of interaction' which is a thing in itself, as important as the things which it separates.*

As Christopher Alexander presents in his exploration of 15 fundamental properties, these thick boundaries take on a life of their own, resembling zones of interaction. This zone of interaction takes presence itself. Likewise, in Aristoteles' *Third Man Argument* (τρίτος ἄνθρωπος), these zones of interaction shall be something that stands in between each discipline. We can look at them in a different way: they evolve to accommodate the functional separations and transitions between distinct systems into becoming something that is an interplay between the systems themselves.

So, what could this zone of interaction be? In the imagined scenario in which every discipline could stand by itself, separate from the other, the author identifies the link in the tools and methods that are shared between disciplines (Fig. 2). Furthermore, the same concepts could be applied to the specialisations within the same discipline (e.g., Architecture and Landscape Architecture). They would serve as a bridge between disciplines, a role that assumes even greater significance when we focus on rural areas. These areas, characterized by their intricate blend of challenges, demand innovative solutions, which may be born out of the interdisciplinary.

2.2   Architecture as a Discipline/Architecture as a Method: Precedent Notable Bridges

Of course, connections between different realms of knowledge (disciplines) are not a novelty, and disciplines themselves can also become tools displayed in the learning of other disciplines. Indeed, for the proposed thesis, it is essential to stress the benefits of bridging methods in disciplines.

This is the case of John Snow's *On the Mode of Communication of Cholera (1855)*, in which the use of architectural knowledge, through diagrams and urban maps, demonstrated how to prevent the spread and fight back cholera, in Soho in mid-XIX century. John Snow employed a dot map to illustrate the concentration of cholera cases around

Fig. 1. *Monades*: it would appear that specific knowledge and teaching methodologies operate in isolation, Milan 2023 (original drawing by the author, property of Stefano Sartorio).

water pumps, effectively showing the intersection of disease spread with a geographical reference in the urban space. Additionally, he harnessed statistical data to highlight the correlation between the quality of the water source and cholera outbreaks. His groundbreaking work revealed that homes served by specific pumps faced a staggering fourteen times higher cholera incidence compared to others. During that era, the medical discipline possessed limited knowledge regarding the disease and its transmission. It was only through the innovative use of an architectural tool, enabling the spatial visualization of the issue, that Snow discerned the intricate connection between water distribution and cholera clusters. This interdisciplinary bridge, linking the realms of medicine and urban planning, led to a brand-new discovery. John Snow's study, thanks to the interpolation of tools from another discipline, thus stands as a pivotal moment in the annals of public health history.

Speaking of bridging tools between disciplines, Pagano's work for the VI Milanese Triennale can be described as an interesting example to dwell on. In his research, the architect sees photography as an instrument of truth, capable of rendering reality as it is, objectively. The aim was to promote "the true autochthonous tradition of Italian architecture: clear, logical, linear, morally and also formally close to contemporary taste" [5]. Although the architectural objective is stated in the exhibition catalogue, there is also another research undergoing the use of photography: an ethnographic survey on Italian regional traditions [6]. Indeed, the architect continued with his work of cataloguing the rural structures he encountered in his wanderings around Italy, always depicting the evolutionary scenario of those artifacts used by the rural population. In all his panels,

Fig. 2. *Thick boundaries*: In the imagined scenario in which every discipline could stand by itself, the link in the tools and methods that are shared between disciplines, Milan 2023 (original drawing by the author, property of Stefano Sartorio).

he showed a sequence picturing not only the architectural space but also registered and testimonies the instruments used by farmers, rural families, and shepherds. In this way, from Bilò's point of view, photography is a tool that transcends the mere role of documenting a spatial composition; it is indeed a method of recording human relations to the physical space of living and working; an undiscussable ethnographic portrait of rural populations.

Similarly, another case of bridging tools between disciplines is the innovative contribution of interviews (from the socio-psychological disciplines) to urbanistic research in *the image of the city* [7]. This example perfectly shows the interplay of interviews and drawing to discover a psychological fact: people's perception of the space. Lynch's research led to find out of people's orientation (mind maps) and the drawing interviews were bridging the two disciplines: architecture and psychology.

Further examples can be found if we consider that a discipline itself can become a method of inquiry (a tool) for another discipline. Architecture as a discipline can be considered a third educator for pedagogical purposes, becoming a proper tool for the discipline of pedagogy; and *vice-versa* pedagogical knowledge can lead and modify spatial decision in the discipline of architectural composition [8].

3 Thesis

3.1 Application on the Field

As per the example given in the precedent paragraphs, and following the syllogism of *ch.2 Hypotheses*, it is plausible whether the definition of a scientific status is to remain within the dominium of "Architectural education" but using tools and methods from a different discipline. More than that, overcoming the strict separation of knowledges, framed in disciplinary dominiums and disciplinary methods can lead to various positive impacts. Methods can be seen as bridges between disciplines (Fig. 3). Indeed, teaching (and learning) architecture by merging with other disciplines may lead to new solutions to the global shifting conditions we are living in.

To conclude, there is no need to revolutionise one discipline to evolve the way of teaching/learning it. Especially in the field of architecture, which involves intricate interactions among the physical environment, the technological one, and the structuring of cultural and social aspects within the profession. We can look at what stands at the margins of the discipline (Fig. 4), exchanging tools and methods between knowledges, and having benefits to the core of the discipline that deals with current global emergencies.

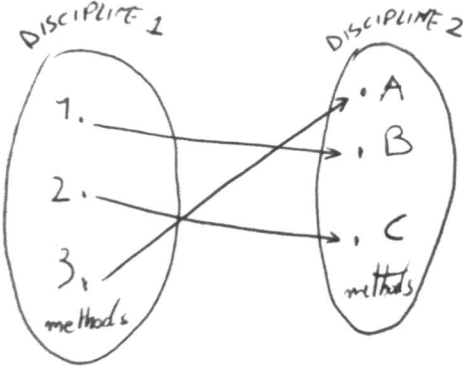

Looking at the margins of teaching in architecture,
to deal with current global emergencies has benefits to the core of the discipline.

Fig. 3. *A bijective funcion*: Methods can be seen as bridges between disciplines, Milan 2023 (original drawing by the author, property of Stefano Sartorio).

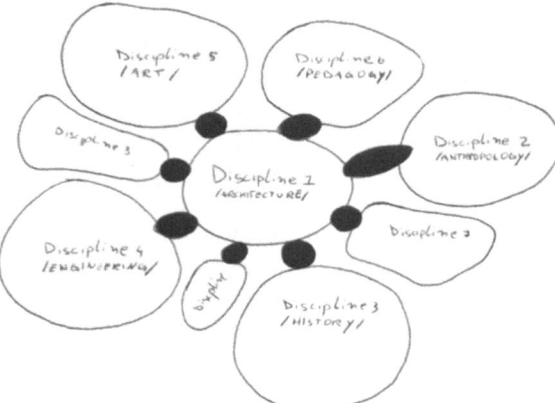

Fig. 4. *What stands at the margins of the discipline*: exchanging tools and methods between knowledges, and having benefits to the core of the discipline, Milan 2023 (original drawing by the author, property of Stefano Sartorio).

4 Case Studies

Considering the condition of Italian inner areas, they are recently become of central interest in various disciplines. Design studios and researchers of various universities had described and analysed those territories. Amongst the many, two specular examples of architectural pedagogy, related to the lagging rural areas are presented.

The first one *(Ch. 4.1)* stems from the acquired experience as Teaching Assistant during Architectural Design Studio 3 at Politecnico di Milano and held by prof. Emilia Corradi and prof. Alisia Tognon between 2020–2023. The didactic activity takes place thanks to the project "Digital Twins Laboratories" funded by the AUIC school, Politecnico di Milano.

In this former case, teaching architecture involved the use of other disciplines methods and tools, meeting one community's need. The second one *(Ch. 4.2)* is an ongoing experience, in which one rural community hosts students from various disciplines. In this case, both the rural community and the students passed through an architectural pedagogy for learning.

4.1 Academia Meets Rural Community for Architecture

During this first experience the site survey of meeting the rural community has totally changed the design result of students' work. Indeed, after the *contamination* of other disciplines' methods it was visible a shift of the architectural outputs. Students experienced for few days a research stay in Morino (Aq) in direct contact with the community. Each group could focus on different tool and method to conduct the inquiry: they had the opportunity to read extra disciplinary book of the library and were free to use the tool of investigation they preferred. Indeed, it was mainly an ethnographic survey, in which they could relate with people, place and objects within the rural community. The field research indeed has been the joint between ethnography (relating to people) and architectural design (relating to the place). Only after the inquiries they came back to the architectural core, proceeding to a self-revision of their own works (and we could spot the tremendous effort on self-judgment and re-elaboration).

Discipline, on the other hand, provides the theoretical framework and critical thinking tools necessary for architects to navigate complex design challenges. In rural are-as, the discipline of architectural pedagogy should embrace an interdisciplinary approach, integrating knowledge from fields such as sociology, anthropology, environmental science, and engineering. This interdisciplinary lens allows students to holistically understand the multifaceted dimensions of rural life and design solutions that address social, cultural, economic, and environmental aspects. It broadens their perspective and equips them with the tools to think beyond a conventional architectural definition through boundaries.

4.2   Rural Community Meets Academia Through Architecture

The central question that guides this architectural pedagogy experiments shows all is dealing with the problem of living in an inner area. The answer is multifaceted and for sure does not follow a singular trajectory and Vione envisioned itself as an experimental laboratory for exploring these possibilities.

Indeed, *Vione laboratorio permanente* [9]. is hosting students for its workshops to develop communication tools for a project that encourages reflection on the theme of living in a mountain village. By temporarily utilizing spaces in the borgo, the Municipality and the Parish of Vione are embarking on regenerative initiatives throughout the region, fostering new paths for development. An alliance of stakeholders has been formed to accomplish this ambitious objective against its inner area depopulation, including the Municipality of Vione, and other political administrations. Additionally, numerous other local and Alpine network institutions within the education field (universities and technical high schools) have been invited to actively support reshaping the community spaces through their disciplinary knowledge. The peculiarity of this experimentation lays in its community involvement in the architectural pedagogy process. The rural community first took part in informative meetings about architectural reuse, and then it was involved in hosting the participants of the workshops. Indeed, inhabitancy necessitates residing in a place, and VIONELAB's experimentation includes hospitality programs in private houses and municipal public spaces, targeting students, especially university students, and professionals. They are invited to engage in campus activities, sector workshops, training sessions, technical meetings, and team-building sessions. By gradually extending this hospitality initiative across the community, VIONELAB aims to reinvigorate underutilized or abandoned spaces, actively involving the local population. The presence of young people and their instructors is intended to open new horizons into the town, stimulate local people's curiosity, and spread awareness of design possibilities among the participants. This represents a continuation of Vione's journey that began in 2019, exploring the path towards its future. The Vione community seeks to establish dialogue with young individuals and educators from various universities and training institutions, engaging in discussions about their ideas and aspirations for change and transformation. The residency calendar is rapidly evolving and weaving a network of fresh relationships. With VIONELAB, Vione has already initiated the process of change for this small Upper Camonica Valley town. Progress is marked by incremental steps, focusing on the region's needs, intentions, projects, and resources. Collaborative efforts with universities, academies, and higher education institutions have revaluated the public and private spaces within Vione's historic centre. Though the aim was to fight against rural depopulation, learning and studying architecture had become the method through which students from different disciplines could propose new ideas for the town. At the same time, architectural pedagogy was a tool to approach the rural community's interest in achieving the program's primary goal.

5 Conclusion

5.1 Looking at the Margins to Redefine One Discipline's Core

The outcome of this reflection on architectural definition, whether as a discipline defined by a scientific status or as a modus operandi, stands on the idea that within the architectural pedagogy need not be a rigid constraint but rather a flexible framework that accommodates a diverse array of tools and methods. This approach accommodates a wave of positive transformation, particularly as we navigate the dynamic currents of contemporary global challenges [10]. Rural areas, for instance, often at the forefront of these challenges, an example of bridging methods between disciplines that have shown benefits from this dynamic approach. Understanding the interplay between method and discipline is crucial for effectively educating future architects who can address the unique challenges of rural context. Rural areas, marked by depopulation, limited connectivity, and a trove of historical and natural heritage, emerge as complex terrains where architectural pedagogy can make a substantial impact. To address the multifaceted challenges posed by these unique environments, architectural education must embrace a panoramic, interdisciplinary approach. By drawing insights from sociology, anthropology, environmental science, and engineering, students as well as rural communities, are equipped with a holistic perspective that extends beyond the traditional confines of architecture. In conclusion, the confluence of method and discipline within architectural pedagogy in rural areas emerges as a pivotal point of innovation. A participatory, community-oriented method beckons students to actively engage with rural communities, fostering co-creation of solutions deeply rooted in the local context. Simultaneously, a disciplinary methodology can be adopted in different disciplines to vehiculate community engagement in innovative processes. The interdisciplinarity interplay of methods and tools equips students (or better: anyone who wants to learn) with the theoretical knowledge and critical thinking skills necessary to navigate the intricate world of shifting conditions. By striking a harmonious balance between method and discipline, architectural education, especially in rural areas, emerges as a catalyst for sustainable development and positive change in one's community.

References

1. Agenzia per la coesione territoriale, Strategia nazionale Aree Interne. https://www.agenziaco esione.gov.it/strategia-nazionale-aree-interne/. Accessed 20 Feb 2024

2. Openpolis, Parole. https://www.openpolis.it/parole/che-cosa-sono-le-aree-interne/. Accessed 15 Oct 2023
3. OxfordLanguage English Dictionary. https://languages.oup.com/google-dictionary-en/. Accessed 15 Oct 2023
4. Alexander, C.: The Nature of Order: an Essay on the Art of Building and the Nature of the Universe. 1: The Phenomenon of Life. The Center for Environmental Structure, Berkeley (2002)
5. Pagano, G., Daniel, G.: Architettura Rurale Italiana, p. 6. Milano, Hoepli (1936)
6. Bilò, F.: Le Indagini Etnografiche di Pagano. LetteraVentidue, Siracusa (2019)
7. Lynch, K.: L'immagine Della Città (Ed.Trad. 2001). Marsilio Editori, Padova (1960)
8. Weyland, B., Attia, S.: Progettare Scuole Tra Pedagogia e Architettura. Guerini Scientifica, San Giuliano Milanese (2015)
9. Vione Laboratorio permanente. https://vionelab.it/#eventi. Accessed 15 Oct 2023
10. Settis, S.: Architettura e Democrazia: Paesaggio, Città, Diritti Civili, 1st edn. Casa editrice Einaudi, Cles (2017)

The Merits of Teaching Architecture as General Education

Arno Suzuki[✉]

Kyoto Tachibana University, Kyoto 607-8175, Japan
`suzuki-ar@tachibana-u.ac.jp`

Abstract. Teaching architecture as general education serves various purposes. Studying familiar buildings may become an intriguing introduction to liberal arts, including art, history, and natural and social sciences. Students can learn critical thinking, problem-solving, and interdisciplinary communication when studying architecture, which are vital higher education goals nowadays. Architecture students and other majors studying together may also improve the future mutual understanding between architects and clients. Housing and living environments are so essential that various efforts are made to teach them to children. Still, primary and secondary school teachers need more time, experience, or confidence to teach such a vast and technical subject. Architecture professors and practitioners can take it over in higher education. To examine the feasibility of this proposal, the author gave a 100-min online experimental lecture on introductory architecture to 1600 first-year students in different majors and conducted a quantitative text analysis on post-learning evaluation. The result proves that practical topics such as safety and environmental preservation interested students regardless of the major. However, a questionnaire survey revealed that students are much less interested in their living environment than in food and clothing. We ought to think about how to make them aware of the importance of this subject.

Keywords: Liberal Arts · Text Analysis · Questionnaire · Japan · Sustainability · Safety

1   Introduction

1.1   Purpose and Background

While Western universities often offer non-professional or minor architecture courses [1, 2], such courses are almost non-existent in Japan [3], probably because architecture is one of the engineering fields in the Japanese educational system. Japanese students seldom have opportunities to learn the subject, and the knowledge gap between professionals and the public is worrying.

Even though the living environment is one of the essentials for humans, "Home Economics" in primary and secondary schools only briefly touches upon the living environment, among other topics such as food and clothing. Teachers are not confident in teaching architecture-related topics mainly because few specialise in it at teachers'

© The Author(s) 2025
M. Barosio et al. (Eds.): EAAE AC 2023, SSDI 47, pp. 171–181, 2025.
https://doi.org/10.1007/978-3-031-71959-2_20

colleges [4]. "Technology" may cover design and drawing in woodworking or mechanics laboratories, but not in all schools. Moreover, "Home Economics" was only for girls, and "Technology" was only for boys until the 1990s.

"Our Home" was taught in mathematics in middle school from 1950 to 1952 but disappeared soon [5] because some educators complained that pupils should spend more time on abstract thinking and drills to improve their calculation skills [6]. Ironically, the mathematics and natural sciences scores of Japanese pupils have dropped significantly in 50 years after that. Less than 30% of high school students choose to study natural science courses now. However, "Home Economics" has recently started covering more practical topics such as safety and environmental issues, especially after the repeated natural disasters and COVID-19 pandemic [7].

We must teach architecture to Japanese people because they are more ignorant of their living environment than Europeans and Americans are. For example, many Japanese believe wooden houses last only for 20 years and reinforced concrete buildings for 50 years. They choose to scrap and build rather than maintain the old buildings. The construction industry has promoted this approach because it is easier and more profitable. The government supports this system through various tax laws. Homeowners do not bother about the maintenance of their houses and prefer easy-care and short-life synthetic materials over natural ones (see Fig. 1). Most believe in prefabricated semi-order homes sold by large nationwide "house maker" corporations. Others purchase ready-made homes or flats with finished interiors. They fear hiring architects and local builders, saying they do not know whom to trust. Still, those semi-order or ready-made homes hike the average mortgage to seven times their annual gross income. Growing up in such an environment, college students, including architecture majors, know almost nothing about their houses. Most students blindly believe in "Three Little Pigs" and do not even think if the story applies to Japan. They think the red-tile-clad reinforced concrete walls of their buildings are made of brick masonry.

Fig. 1. The upper row: traditional Japanese architecture with sustainable natural materials (Photographs by Arno Suzuki, taken in Japan, 2001). The lower row shows synthetic materials common in Japan (Photographs by Arno Suzuki, taken in Japan, 202).

Hypothesis

The rationale for teaching architecture as general education is as follows:

- Specialist teachers are available in universities—no teacher's license required
- More teaching hours can be given on architecture than in secondary schools
- Topics are related to students' everyday life regardless of the major
- Problem-solving and participatory learning opportunities
- Interdisciplinary teamwork with diverse students
- Citizenship cultivation—learning about social capital and responsibility

On the other hand, there are concerns as follows:

- Architects' communication skills to teach beginners
- Difficulties in teaching a group with different interests and backgrounds
- Lack of students' interest in housing or architecture

In this research, the author will investigate the last two concerns. First, she analysed students' responses to her experimental introductory lecture and proved it is possible to interest all majors with different backgrounds. Second, she proved that few students are interested in architecture before learning it.

1.2   Previous Studies and Practices

Architecture minors and introductory courses offered for non-majors focus usually on non-practical, non-technical subjects such as architectural history and philosophy. However, architecture may be ignored if it stays within the boundaries of humanities because the field tends to be unfairly disrespected by students and institutions. Instead, hands-on courses such as design-build studios [8] and construction laboratories [9] have gained popularity. They offer interdisciplinary education opportunities.

In Japan, teaching architecture in non-accredited schools or general education programs is a new idea; therefore, there was little previous research or mention. However, in recent years, general education has shifted from the traditional three areas of liberal arts, humanities, social sciences, and natural sciences to more cross-curricular and skill development courses such as communication and computer laboratories [10].

Various researchers on secondary education have proved that students will become interested in architectural topics after learning properly in school [11].

1.3   Findings from Teaching Experience

The author taught architecture to a student group that included non-majors at a land-grant research university in the US from 2000 to 2005. The author did the same in Japanese universities, from relatively unknown ones to one of the top national research universities with student bodies of various nationalities, from 2002 to the present. From this experience and the students' feedback, the author is assured that architecture and living environment studies can heighten interest in liberal arts and train communication and problem-solving skills. Comments such as "I did not like history at all in high school, but I found it interesting for the first time after learning about architecture that we can see" were noted numerous times at different schools. The students also enjoyed teamwork with strangers of various cultural and academic backgrounds in international education courses and small-class freshman seminars.

At Kyoto University, with academically top-of-nation students, the author conducted an open questionnaire every year from 2008 to 2019 before teaching to check their basic knowledge of architecture. The following is the summarised outcome from Japanese students, including architecture majors:

- Unaware whether their home is built of wood or reinforced concrete
- Unaware whether their home is priced at around 0.1 or 1 million Euro
- Their student housing, usually rental flats, is chosen by parents
- Ignorant of internationally famous Japanese architects such as Tadao Ando
- Ignorant of the names of construction and fittings
- No preference for the kind of house they want to live in, in the future

The most surprising was that almost none of the architecture students wanted to design their own house in the future. They said buying an industrialised house built by a nationally famous company would be safer.

2 Quantitative Text Analysis

2.1 Experimental Course Used for the Analysis

At Kyoto Tachibana University, the author delivered a 100-min online lecture to freshmen of 15 majors, including humanities, engineering, business and economics, and health sciences. The contents of the lecture were as follows:

Chapter 1: Introduction to Wooden Architecture
Which is the safest house: Rethinking "The Three Little Pigs" with earthquake area/The 1400-year-old wooden building in Japan/The 800-year-old wooden building standing in the sea/Difference between traditional and modern wood construction/Wooden buildings in Europe and the USA/Industrialised fake materials

Chapter 2: Structure *Traditional structure and construction methods/Columns not fixed to the ground/Fragile walls to absorb force/Flexible structure applied in contemporary buildings*

Chapter 3: Materials *Material and form/Why did ancient Greek temples collapse?/Invention of the arch/The circle of climate-materials-form-function-materials/Examples of passive design*

Chapter 4: Sustainability *Environmental performance of traditional houses/Biological structure of timber/Materials and function of tatami/Thatched roofs/Recyclable materials*

2.2   Research Method

Taking advantage of the large student body, the author conducted a quantitative text analysis of students' responses to the post-learning evaluation to the question "What have you learned from this lecture?" After omitting meaningless words such as particles and too-common words such as "architecture", "lecture" and so forth, the author analysed 5,709 words from 1,596 students' writings using KH Coder, a free text analysis application software by Koichi Higuchi [12]. All personal information was removed, except for their majors. The software shows only the frequency of appearance and relationship among words; reading the meaning depends on humans.

2.3   Co-occurrence Network of Words

The co-occurrence network visually represents the relationships among the occurring words, in which the size of the bubble indicates the number of times the word appears, and words tending to appear in the same sentence are connected by lines. Words were connected as shown in the diagram, summarising the students' writings. For example, the contexts were: "Society in the Edo Era was sustainable with no waste" and "Traditional construction without glue and possible to disassemble and recycle" (see Fig. 2). These are the same as the lecture content. It proved that non-major students could understand architectural topics well in one lecture, even technical aspects such as structure, materials, and environmental issues.

Fig. 2. Co-occurrence network to summarise students' feedback. (produced with KH Coder and traced for legibility by Arno Suzuki, 2023).

2.4   Co-occurrence Network of Words by Departments

The co-occurrence network cross-tabulated by departments as variables shows that only a few words showed a connection to specific majors: for example, "design" in the Department of Architectural Design (80 respondents), "history" in the Department of History (100 respondents) and "architectural style" in the Department of Historical Heritage (55 respondents). Otherwise, there were no significant differences between majors. Particularly, the departments of Economics (240 students) and Business (260 students), with many students and various future career paths, demonstrated more dispersion. On the other hand, the paramedical departments for professional training in Clinical Examination, Physiotherapy, Occupational Therapy, Paramedical, and Psychology (420 students in total) showed inclination toward environmental topics. Humanities such as Japanese Literature, History, and Historical Heritage (240 students), whose career paths seemed unpredictable, tended to show more interest in their specialization. The words "sustainable", "reason to study", "impression", "knowledge", "residence", "traditional", "environment", "climate", "common construction", "Kyoto", "form", and "passive design" are connected to more than one department. It means they appeared frequently in multiple departments, suggesting these are common interests across majors (see Fig. 3).

The author also cross-tabulated interested issues with departments by coding similar or related words into categories. The associations found were paramedical students discussing environmental issues, Japanese literature students discussing their own experiences, and information technology students showing no interest in any issues.

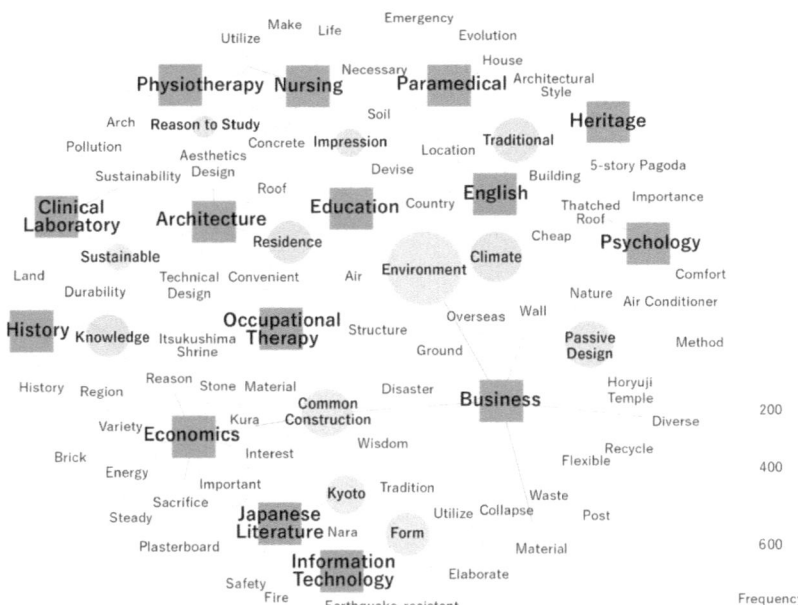

Fig. 3. Co-occurrence network of words cross-tabulated by departments (produced with KH Coder and traced for legibility by Arno Suzuki, 2023).

3 Questionnaire Survey

3.1 Purpose and Method

Some responses in the text analysis indicated that the student did not listen to the lectures. Additionally, the responses received to architecture were fewer than those to other immediately useful topics such as money, food, student life, and emergency medical care. Architecture may have seemed remote to young students compared to those other topics. To investigate this, the author and her students conducted questionnaire research focusing on housing and food. We chose these two topics because they are among the three essentials for humans (food, clothing, and shelter), and they also have some natural science aspects. Clothing was omitted because the result might be biased by young people's strong interests in fashion. Nakawaki, the author's undergraduate student, visited four universities in Kyoto City from June to October 2022 and conducted a face-to-face questionnaire survey. They gave the passing-by students two simple choices: architecture or nutrition, if the responding students had to take an elective course in their spare time. A response was received from 510 students; their profiles were 285 male and 225 female students; 153 were first year, 112 were second year, 123 were third year, and 122 were fourth year and older.

3.2 Results of the In-person Survey

Fewer students chose architecture (n = 235) than nutrition (n = 275). When cross-tabulated by the year, nutrition exceeded in the younger students, but the numbers almost equalized in the third year, and a slight reversal occurred in favour of architecture in the fourth. Nakawaki said, "Our interest in society increases as graduation and employment approaches, which may induce an interest in architecture, a field closely related to society." The cross tab by major field backs up his analysis by showing more interest in architecture by social science majors. The gender difference was very clear; male students chose architecture (n = 153) over nutrition (n = 132), whereas female students chose nutrition (n = 143) over architecture (n = 82) (See Fig. 4). Nakawaki thought that the Japanese custom, or the social pressure over the gender role, may have influenced students' choice even though "Technology" and "Home Economics" became co-educational in schools over 30 years ago.

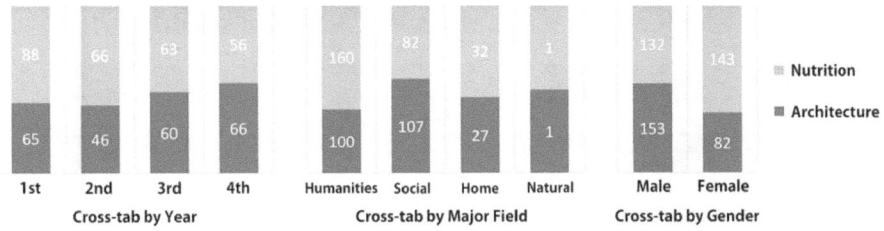

Fig. 4. Nakawaki's in-person survey results (graph by Arno Suzuki, 2023)

3.3   Results of the Online Survey

To further investigate Nakawaki's hypothesis, the author conducted a large-scale online survey in March 2023 to inquire about the reasons for the subject selection, which his team could not ask due to the time constraints of an in-person survey. From the 2000 responses from Japanese men and women of all age groups nationwide, the author extracted 447 valid responses from students aged 18–24 to compose a similar group to those of the previous in-person survey.

In the online survey, the author asked the respondents to choose a subject from six areas: nutrition, cooking, food safety, housing, construction, and real estate. Most female respondents (78.4%, n = 167) were in nutrition and cooking, and more than half of the male respondents (56.4%, n = 132) chose food-related subjects (see Fig. 5).

The gap between the food area and the housing area was wider than in the in-person survey, possibly because Cooking and Food Safety attracted more students. Nevertheless, an 18-year-old male respondent stated the reason for his choosing Cooking as "I want to be a good cook for my future children," suggesting that the gender gap is narrowing

Fig. 5. Choice of subjects by gender (graph by Arno Suzuki, 2023).

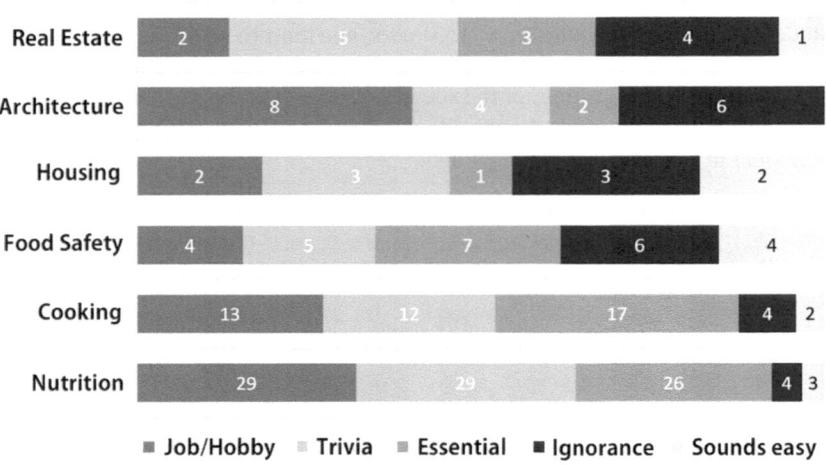

Fig. 6. Reason for the subject choice (graph by Arno Suzuki, 2023).

in younger minds. On the other hand, few respondents think learning about their living environment was necessary for their life (see Fig. 6).

4 Findings and Discussion

4.1 Students' Tendency Confirmed by the Text Analysis

The text analysis mostly satisfied the author's expectations and concerns. First, students have little knowledge of their living environment. They responded, "I had never heard of this before" (n = 197) and "I was surprised" (n = 560), even to some very basic information. In terms of traditional architecture, however, Kyoto Tachibana University students had a better understanding than Kyoto University students, probably because the former are mostly from areas with more historic houses remaining, whereas the latter are from big cities such as Tokyo. In addition, Tachibana students seem to have more familiarity with the on-site construction; at least 10 students stated that they had a carpenter in their family, and 40 said they knew someone in the industry. Connections with students' personal lives help them learn.

Second, the intention of the lecture was misconstrued sometimes. For example, there were many favourable comments on plasterboard (n = 263), although the author indicated that they were not biodegradable and caused environmental problems. Students' attention, however, was caught by the inexpensive, easy-to-assemble, fire-resistant, and sound insulation properties instead. It seemed "cheap and easy" was a value for the students. Another misunderstanding occurred when the author pointed out the loss of construction skills with the industrialisation of homebuilding. Many students thought the prefabricated housing was "safer" because it would not depend on the carpenter's skills. This exactly reflects the recent trends in the housing market in Japan.

Third, the result indicated that safety and environmental issues are the keys to sparking students' interests. "Disaster Prevention" and "Passive Design" interested students regardless of major. For example, "Earthquake" and "Traditional Architecture" appeared frequently and related to those two topics. Many students were impressed to learn about seismic isolation and vibration-damping principles in traditional Japanese construction with non-fastened posts and beams with deliberately fragile walls to absorb energy (See Fig. 2).

Finally, familiar topics raise students' interest in architecture. Names of nearby locations and local events such as *Kyoto* (n = 217), *the Kamo River* (n = 51), *the Gion Festival* (n = 47), and *Daigoji Temple* (n = 21) showed up a lot, and they all appeared in favourable ways. Other than local topics, everyday issues such as soundproofing, energy conservation, and disaster prevention were frequently discussed.

4.2 Conclusions and Further Discussion

The text analysis for the post-learning evaluation showed that all aspects of architecture and building sciences can be taught to all majors. It also suggested that we might raise interest in architecture regardless of the students' majors by discussing more critical topics, such as safety and environmental issues. Therefore, the author concluded that introductory architecture courses in general education should include more practical, natural, and social science aspects.

The questionnaire surveys revealed that students did not consider architecture a useful subject. Few people may enrol if architecture is an elective subject, therefore. Offering it as a part of a required lecture series was effective in making them listen. Another way to draw their attention and gain enrolment is to sprinkle information useful for their everyday lives, such as cost- and energy-saving. Once students are exposed to architecture, they will realise it is an interesting and important subject.

Acknowledgement. This research was funded by the Grant-in-Aid for Scientific Research by JSPS, Fundamental Research Category C, No.20K12529, 'How to Teach Housing in General Education'.

References

1. Barnett, P.M.: Elements of architecture. J. Archit. Educ. **30**(4), 11–14 (1977)
2. Jones, D.: Architecture as a discipline of the humanities. J. Archit. Educ. **34**(4), 18–23 (1981)
3. Suzuki, A.: Teaching Architecture as General Education for Non-Majors. In: 19th Architectural Education Symposium Proceedings, no. 19, pp. 41–48 (2019) (in Japanese)
4. Miura, M., Uno, H., Ohira, A.: Teachers' weaknesses in the area of housing and necessary support. Archit. Inst. Jpn Hokkaido Branch Res. Rep. Collect. **85**, 551–554 (2012). (in Japanese)
5. Takeuchi, K., Segawa, A.: Establishment and background of the unit 'Watashitachi-no-Jukyo' in the junior high school mathematics textbooks "Nichijo-no-Sugaku" in 1950. Bull. Inst. Archit. Des. **17**, 11–20 (2021). (in Japanese)
6. Morioka, J.: Hiraku Toyama's educational thought: focusing on his criticism of "experience unit learning." Hum. Sci. Bull. Osaka Prefect. Univ. Bull. **14**, 31–57 (2019). (in Japanese)
7. Hayami, T.: Housing education in school education: an investigation from a cross-curricular perspective. Res. Bull. Naruto Univ. Educ. **38**, 217–227 (2023). (in Japanese)
8. Middlebrook, J., Maines, K.: The introductory architecture studio revisited: exploring the educational potential of design-build within a liberal arts context. J. Archit. Educ. **70**, 154–164 (2016)
9. Garlock, S.: Architecture as Liberal Art. John Harvard's Journal January–February 2015 (2015). https://www.harvardmagazine.com/2014/12/architecture-as-liberal-art. Accessed 8 Feb 2024
10. Yoshida, A.: Liberal education in the university in post-war Japan: In: Search of the Place to Fit. Iwanami Shoten, Tokyo (2013) (in Japanese)
11. Hayami, T., Seto, A.: Senior high school students' awareness of home economics housing education and how that awareness changes as a result of taking classes. J. Home Econ. Jpn. **71**(3), 182–192 (2020). (in Japanese)
12. Higuchi, K.: New quantitative text analytical method and KH coder software. Jpn. Sociol. Rev. **68**(3), 334–350 (2017). (in Japanese)

Roots of Architecture: Ways of Research

Roots of Architecture: Ways of Research

Elena Vigliocco[✉]

DAD Department of Architecture and Design, Politecnico di Torino, Turin, Italy
elena.vigliocco@polito.it

Can Architecture be produced without an architect? Yes. It has always been like this. The historic centers of our cities are the result of designs whose authors are often unknown, not because they are not essential but because they are unnecessary [1]. The interest of architects, but not only in authorship, dates back to the nineteenth century, a period of excellent building ferment in which incredible urban transformations sanctioned the idea of the house as a consumer good [2].

If today, we ask ourselves whether Architecture belongs to the architect who designed it or to the community of its users, it makes us understand how much the importance of authorship was a parenthesis that lasted just over a century. The weakening of the theory of knowledge (weak thinking) [3] is closely linked to the weakening of the subject (the architect, who is de-responsibilized) and to the weakening of being (the discipline, considered porous, contradictory, polycentric, devoid of univocity) [4]. Therefore, it is logical to ask whether architecture schools should be configured as places of experimentation aimed at building new hypotheses for the future rather than service-oriented places according to the demands of communities and the market.

To try to contribute to the debate actively, it may be helpful to highlight the trends and changes taking place in the architectural research sector, broken down according to three key issues: themes, methods, and tools.

Concerning the themes, research increasingly tends to investigate the topics that concern civil society [5–7]. Climate change, concerning energy issues rather than the impact of human settlements on ecological systems, and the fragility of territories, conceived as a response to natural rather than social and economic disasters, are some of the investigated topics that lead Architecture to question itself on the civil role it is capable of assuming.

Regarding methods, the research approach is increasingly multidisciplinary and trans-scalar. The complexity of the systems investigated requires both the involvement of other and new disciplines, capable of analyzing the areas that Architecture does not reach and reasoning at multiple levels of in-depth analysis to increase the effect of the research results [8, 9].

Lastly, the tools are increasingly sophisticated and require researchers/teachers and students to update their skills continually. In a world that has become increasingly smaller, traditional research tools are being superseded by digital applications (an example for all is represented by AI), which open new research opportunities by simplifying routine operations.

© The Author(s) 2025
M. Barosio et al. (Eds.): EAAE AC 2023, SSDI 47, pp. 185–186, 2025.
https://doi.org/10.1007/978-3-031-71959-2_21

Concerning these considerations, a complex and highly challenging picture emerges in which Architecture Schools must question themselves and, above all, renew their practices.

References

1. Rykwert, J.: The Idea of a Town. The Anthropology of Urban Form in Rome, Italy and the Ancient World. Princeton University Press, Princeton (1976)
2. Londei, E.F.: La Parigi di Haussmann. Kappa, Bologna (1982)
3. Vattimo, G.: La fine della modernità. Garzanti, Milan. Translation by John R. Snyder (1991). The End of Modernity: Nihilism and Hermeneutics in Post-modern Culture. Polity Press (1985)
4. Borradori, G.: "Weak Thought" and postmodernism: the Italian departure from deconstruction. Soc. Text **18**, 39–49 (1987). https://doi.org/10.2307/488689
5. Space10: The Ideal City. Exploring Urban Futures. Gestalten, Berlin (2021)
6. Lokko, L. (ed.): The Laboratory of the Future. Biennale di Architettura di Venezia, Venezia (2023)
7. Farrell, Y., McNamara, S. (eds.): Freespace. Biennale di Architettura di Venezia, Venezia (2018)
8. Segapeli S (ed.) Vous Avez Dit Espace Commun? (2022)
9. Bruxelles, P.L., Marzo, M., Ferrario, V., Bertini, V.: Between Sense of Time and Sense of Place. Siracusa, LetteraVentidue (2022)

Practicing Care Through Architecture: Participatory Research as a Tool to Subvert Power Structures

Nadia Bertolino[✉]

Architecture and Urban Design Research Lab, University of Pavia, Pavia, Italy
nadia.bertolino@unipv.it

Abstract. The paper raises questions about the extent to which architectural research can function as a tool for challenging power dynamics and facilitating a critique of capitalism. This aligns with the assertion by Meiksins Wood that such a critique of capitalism "requires a constantly renewed critique of the analytic instruments designed to understand it" [1]. In the final chapter of "Critique of Architecture," Spencer, in an interview with Kosec, asserts that architecture serves as the central and foundational arena for education, surveillance, and discipline [2]. Within this context, the paper seeks to explore this question by offering a critical examination of the recent participatory research project "Architectures of Care". This project involved the mapping of social and environmental care practices within three self-organized communities in Italy, Turkey, and the UK. The paper will particularly focus on the challenges associated with conducting research through participatory methods, wherein participants become co-researchers. Additionally, it will address the controversial role of the "neutral researcher" within this framework. The "Architectures of Care" project provides valuable insights that contribute to the discussion of whether architecture can foster caring environments. Furthermore, it prompts reflection on the significance of involving self-organized communities in our research agenda. These communities are perceived as non-commodified agents that have the potential to challenge prevailing power structures, establish networks of care, and facilitate radical forms of social emancipation [3].

Keywords: Participatory research · Care · Spatial Production · Self-Organized Communities · Power Structures

1 Introduction: Research Question and Methodology

This paper introduces and discusses themes and findings of the research project "Architectures of Care", which is a pilot research project, funded by Northumbria University Research and Innovation Services, under the "Participatory Research" funding scheme. The project aimed to provide a reflection on what architects can learn from self-organized communities in terms of social inclusion and environmental care.

M. Barosio et al. (Eds.): EAAE AC 2023, SSDI 47, pp. 187–196, 2025.
https://doi.org/10.1007/978-3-031-71959-2_22

To address this, the primary objective of this research project was to conduct a pilot study that delved into the effectiveness of informal practices within three distinct eco-communities: Le Piagge in Florence, Guneskoy in Ankara, and Old Hall in Colchester. These communities served as invaluable sources of inspiration, as they presented potential responses to the pressing issues of our times, namely social inequalities and the climate emergency. In contrast to mainstream, commodified approaches, these eco-communities offered models that prioritize values such as integration, circularity, durability, and resilience. Building on the foundations of previously collected data, this project enhanced its dataset through direct engagement with community members. By facilitating group conversations within these communities, the research project achieved two critical objectives. First, it raised awareness regarding the informal practices already in place within each community. These practices can range from innovative methods of social integration to creative environmental conservation efforts. Second, the project endeavored to establish an international network that connects these three eco-communities. This network served as a platform for the exchange of knowledge and, over time, the development of essential skills for addressing shared challenges.

The project engaged eight representatives from each community, designating them as "co-researchers," through the organization of three facilitated group conversations held on February 2023. The principal goal was to collaboratively curate a repository of textual and visual materials that capture local practices of social inclusion and environmental care. These facilitated group conversations were designed to encourage participants to both verbally and visually respond to specific prompts, aimed at raising awareness and delineating potential areas of improvement within their respective communities. The textual content and drawings generated during these group conversations played a crucial role in shaping the pilot version of the online catalogue known as the "Archive of Informality". Subsequently, a comprehensive report was crafted for each community, involving an in-depth analysis of the data collected and a delineation of the key points and discussions that transpired during the group conversations. These reports have been shared with stakeholders in preparation for a final online event that took place on March 2023. This event focused on discussing networking opportunities and provided a platform for the collective formulation of a research question as a prelude to the submission of a standard grant proposal to the Arts and Humanities Research Council.

2   Theoretical Framework: Design, Participation, Care

2.1   Architecture as a Relational Process

"Architectures of Care" research project draws on the definition of architecture as a relational process and, consequently, delves into the role of the architect-researcher, who becomes an active agent and, in certain cases, a catalyst for change. In his essay "The Negotiation of Hope" within "Architecture and Participation," Jeremy Till describes spatial production processes as complex and not rigidly confined to the simplistic dichotomy of "bottom-up" and "top-down" design [4].

This approach highlights the close connection between the design process and what Rousseau terms "transformative action", which denotes participatory actions enabling the exploration and construction of alternative scenarios to conventional space production through active citizen involvement in public life. In response to Rousseau's model of transformative action, Pateman underscores the potential threat that a genuinely collective process of space production poses to the stability of established political systems [5]. Thus, it is considered acceptable, only if it can be manipulated to endorse pre-determined decisions. It is evident that, for Pateman, the theme of participation is an integral part of defining not only the space production process but, more broadly, the creation of political life. Referring this back to the realm of architectural practice, Till emphasizes that complete participation is an ideal that is challenging to realize in architectural design, as it presupposes a shared and communicable culture between the architect and the end user. This condition is typically not met in design, where the architect's technical knowledge and the empirical (not necessarily explicit) knowledge of the user operate at different levels. Additionally, communication channels are inherently characterized by codes, conventions, and authority dynamics that make the design process less accessible [4]. In fact, in 1970, scholars in the fields of architecture, design, and computer science highlighted one of the primary barriers to participation in architecture: the opacity of the design process. Moreover, in the same year, during his concluding remarks at the Design Participation conference, Cross emphasized how the introduction of new technologies and techniques could potentially circumvent the conventional political control over the design process [6]. He regarded it as a systemic issue that the conference had to confront political problems. However, it is clear that participation is intrinsically a political matter, not in terms of allegiance to a particular party but rather due to its integral role in a process that inevitably impacts public life. In those same years, the debate revolved around the contentious role of the architect in participatory spatial production. On one hand, the architect, possessing specialized knowledge, was viewed as the sole figure capable of dictating the rules of the participatory process, a position that inherently reinforced a hierarchical structure. On the other hand, following a more radical perspective, the architect was divested of authority, and their professionalism was exercised through the role of a technical facilitator, able to translate the community's desires into spatial realities without imposing their own position. Nevertheless, it must be observed that within this latter realm of action, the architect cannot harness their knowledge as a transformative tool. On the other hand, their technical expertise is insufficient to support the user in defining new visions of the built environment. As Rose succinctly summarizes: "The architect steps back, but the citizen does not access power" [7].

It is, therefore, pertinent to consider whether there exists an alternative path for defining the role of the architect, starting from a profound and comprehensive understanding of all the variables involved in the definition of a new spatiality. This essay takes the position that the process is based on the architect's act of critical positioning as an active citizen within the context they are called to intervene in [8, 9], triggering an alteration of the status quo through creative action [10, 11]. This process is then completed through the collective recognition of a new meaning of creating space and inhabiting it. In particular, Agamben's reflections on creative action introduce the concept of human potential as the individual's capacity to act creatively [10]. The theme of "potency" and the distinction between "potency" and "act" are particularly relevant here. According to Agamben, human essence lies in the capacity-in-potency to act, rather than in the concrete and realized action, which could be viewed as a structural element of the previously mentioned technical design knowledge. Equally relevant is the need to redefine the scope of project action and the professional role of the architect, as discussed by the Spatial Agency group, founded at the Sheffield School of Architecture in 2010. According to Till, Awan, and Schneider, we should refer to a much broader field of opportunities where architects and non-architects can collaborate, suggesting "other ways of doing architecture" [8]. This provocative manifesto also suggests that a building is not always necessarily the best solution for resolving problematic spatiality.

2.2 Participation in Architectural Research

Participatory architectural research is a foundational paradigm within the domain of architectural design, illuminating the central role of community engagement and collaboration in shaping the built environment. This approach, recognized by scholars such as Petrescu and Blundell Jones [12], Patsy Eubanks Owens [13] and Nabeel Hamdi [14], holds significant potential and grapples with various challenges, making it a complex yet rewarding avenue for architects to explore.

Owens, in her work "Community-Centered Design," underscores the power of participation in the design process. She argues that engaging communities in decision-making not only results in architecture that is more aligned with their unique needs and aspirations but also has the capacity to empower marginalized populations [7]. This approach fosters a profound sense of ownership and agency within their surroundings, ultimately democratizing the design process. However, realizing this potential is no small task. The challenges are multifaceted, encompassing issues of representation, access, and cultural sensitivity. Achieving meaningful participation requires architects to navigate these complexities while respecting the diverse perspectives and lived experiences of the communities they serve.

Drawing from the field of urban sociology, the works of scholars such as David Harvey [9] and Richard Sennett [10] further inform the challenges and potentials of participatory architectural research. Harvey's insights on the right to the city emphasize the importance of participatory approaches in urban development, as they contribute to more equitable and inclusive cities. The notion of the right to the city implies that citizens should have an active role in shaping the urban environment. This aligns with the principles of participatory architectural research, where community engagement is central to design decision-making. Sennett's exploration of the open city and the role of the public realm in urban life sheds light on the potential of participatory design in enhancing urban spaces [10]. He argues that a well-designed public realm, one that emerges from collaboration and community input, fosters social interactions and a sense of belonging among residents.

In addressing the challenges of participatory architectural research, architects must prioritize inclusivity and equity. Owens' work reinforces the need to recognize the voices of marginalized groups who have historically been excluded from the decision-making process [13]. Communities vary in terms of culture, socioeconomic status, and background, and architects must adopt an approach that values these differences. Bridging the gap between the lived experiences of community members and the architectural design process is not a straightforward task, but it is essential for creating spaces that truly serve and inspire their inhabitants. Nabeel Hamdi's emphasis on small-scale, community-driven interventions calls attention to the potential of grassroots initiatives [14]. These local, community-focused projects can serve as catalysts for broader urban changes. However, architects and researchers face the challenge of scaling up these interventions to create city-wide impact. This necessitates navigating bureaucratic structures, securing funding, and aligning projects with larger urban development strategies. Participatory architectural research is, at its core, a cooperative endeavor. Architects, communities, and stakeholders must come together to explore new dimensions of architectural design. This approach not only engages the public but also promotes a deeper understanding of the social, cultural, and contextual intricacies of architectural projects. However, realizing its full potential depends on overcoming the challenges it presents (Fig. 1).

Fig. 1. Re-cycling site at Comunità delle Piagge, Florence (Italy), 2023. Photograph by Nadia Bertolino.

One of the challenges faced in participatory architectural research is the need for a shared language and a common understanding of design principles between architects and community members. It is crucial to facilitate effective communication, as architectural concepts and terminology can be daunting for those unfamiliar with the field. Architects should employ accessible and visual methods, such as diagrams, sketches, and models, to bridge this knowledge gap and create a more inclusive design dialogue. Another challenge is the potential for conflicts and disagreements within the community or between architects and residents. Differing opinions and priorities can arise during the decision-making process, potentially leading to tensions. Architects need to act as mediators, facilitating constructive discussions and finding common ground. This involves patience, active listening, and the ability to adapt the design to accommodate various perspectives.

2.3 Making *Other Worlds*: Care and Architectural Design

Care assumes a pivotal role in participatory research. Examining the three communities through the framework of care facilitated the reflection on various forms of nurturing relationships that exist and bolster the operation of these societal institutions and allowed to determine the spatial unfolding of such practices. Care is mutually experienced, by the ones providing the caring deed and the others that are the recipient of this deed. It is an empathetic approach that helps to meet needs and think beyond that. As confirmed by most participants in the research project, care can be described as a way of being that makes life habitable for other people, animals and the planet [15]. A caring attitude ensures that a particular lifestyle does not exploit another person or thing. It has been defined as a "social practice that is essential to the maintenance and reproduction of society" [16]. In fact, Gilligan's approach to understanding caring practices goes beyond conclusions based on justice, rules and regulations. It is rather perceived with emotions and empathy. This creates the basis for building relationships between people. The "conception of morality reflects the understanding of social relationships" [17], defining the role played by humans in society.

Aligned to Gilligan, Noddings' approach establishes a connection between justice and care, delineating two distinct stages within the caring process: "caring about" and "caring for" [18]. These stages constitute essential components of any caring relationship. During the act of caring, one relinquishes self-interest and engages in selfless concern for others, a state Noddings terms as being engrossed in care. Empathetic caring occurs naturally and represents a more active form of caring, albeit less receptive compared to engrossment. Noddings defines caring-about as the emotions we experience within a broader and more generalized context, primarily focused on recognizing the need for care. This stage contemplates care from a theoretical standpoint rather than addressing it practically. The caring-for phase refines our focus, translating intentions

Fig. 2. Guneskoy eco-community site, Ankara (Turkey), 2023. Photograph by Nadia Bertolino.

into tangible actions and makes evident how caring practices contribute to the creation of alternative, other worlds, where power structures are challenged and subverted.

From this perspective, it becomes apparent that the project could indeed absorb the intangible dynamics of the human relations, devising strategies capable of translating these into spatial constructs and providing environments that can nurture caring practices. It is an agency that requires annexing, selecting, or abandoning, and ultimately engaging in dialogue with what is discarded rather than categorically excluding it as inappropriate material. Thus, new operational scenarios emerge, new territories that already existed but were previously overlooked by our gaze, embracing the "mutable identity of places" [19]. The future of these blank areas, perennially awaiting a defined role, may therefore involve forming zones of change: ready to re-enter the game in the event of necessary revisions to the existing structure, thereby accommodating the uncertain and the undefined. Making *other worlds*, eventually (Fig. 2).

3 Discussion: Spatial Unfolding of Caring Practices

The three communities involved in the "Architectures of Care" project were selected due to their radical agenda and grassroots approach to sustainability. While this article doesn't delve into the specifics of these communities, it's more relevant to recognize that, despite their differences in location, context, and governance models, there exists a common ground that elucidates their operational principles and the spatial provisioning processes they entail: that is putting "care" at the very core of their agenda and everyday community life. The research findings from the eco-communities in Florence, Ankara and Colchester unveil a compelling set of principles and practices that have played a pivotal role in nurturing inclusive and sustainable environments. All the points raised by the research participants are extremely relevant to an architectural discourse and foster a reflection on new design paradigms.

Firstly, the principle of low thresholds and blurred boundaries explains the critical role of open and inclusive spaces. An issue highlighted by most participants across the three communities is that, by minimizing barriers to access and participation, these communities cultivate an environment that welcomes diverse contributions, fostering a sense of belonging and shared responsibility among inhabitants. This philosophy of inclusivity not only values the input of every individual but also serves as a cornerstone for nurturing a cohesive community fabric.

Moreover, the emphasis on designing spaces with undefined character and flexibility amplifies the notion of empowerment within these communities. By providing inhabitants with the agency to shape their environment according to their needs and preferences, these spaces become catalysts for fostering deeper connections and a heightened sense of ownership and place attachment among residents. Such flexibility allows for the organic evolution of spaces, ensuring they remain responsive to the evolving dynamics and aspirations of the community.

Additionally, the activation of networks and territorial connections extends the influence of these communities beyond their immediate boundaries. Through collaborative efforts and alliances with neighboring communities, they not only amplify their impact but also underscore the interconnectedness of social and environmental ecosystems.

This broader engagement enables communities to pursue collective goals and address shared challenges, thereby reinforcing the notion of collective responsibility and solidarity. Furthermore, the adoption of sustainable practices, such as recycling spaces and embracing low-tech building technologies, embodies a commitment to environmental stewardship and resource efficiency. By repurposing materials and engaging in hands-on construction processes, these communities not only minimize their ecological footprint but also foster a culture of innovation and resilience. Moreover, these practices serve as educational tools, empowering residents with valuable skills and knowledge related to sustainability and construction.

In essence, these eco-communities epitomize the harmonious integration of principles such as inclusivity, flexibility, networking, recycling, and self-construction. These principles not only inform the physical design of spaces but also permeate the social and cultural dynamics within these communities. As architects, urban planners, and policymakers endeavor to design environments aligned with these principles, they must navigate various challenges, including balancing diverse interests and addressing power dynamics. Yet, by embracing transparency, collaboration, and a profound respect for local contexts, they can unlock the full potential of participatory design in shaping vibrant, resilient, and inclusive urban environments.

4   Conclusion: Subverting Power Structures Through Participatory Research

Drawing on the research project mentioned above, this paper advocates for care as a fundamental and critical action, serving as a potent tool for critiquing the prevailing capitalist modes of spatial production. The three communities involved allowed to introduce and explore various dimensions of *urban care*, *care institutions*, and *care as an agency* to underscore the significance of care as a form of design paradigm and critical action. In this paper, it is believed that prioritizing care for "the other," for individual and collective well-being, for the planet, and for the city is imperative. To navigate the challenges of our times, fostering a sense of care and interconnectedness should be at the forefront of our intellectual and societal discourse, as this aligns with the human geography paradigm of examining the dynamics of space, place, and society [20–22]. By emphasizing the value of care, it is provocatively proposed a reevaluation of thought processes and the restructuring of societal relations that can ultimately steer towards a more compassionate, inclusive, and sustainable future. The objective is to reinstate the concepts, principles, and a commitment to novel civic organizations that resonate with the continuous battle concerning societal, political, and democratic existence. The endeavor lies in discovering techniques and actions fostering solidarity, collaborative efforts, and a moral framework grounded in compassion, as essential components for an efficient response to the deterioration of our urban centers and the environment. The mission is to redefine care as a pivotal form of engagement that operates on multiple levels, spanning from individual responsibilities to societal interactions and even planetary interconnectedness. Care, often confined to the private realm and connected with marginalized groups, assumes significance when harnessed as a critical tool for evaluating the impacts of capitalist spatial production. Chiara Bottici, in "Anarchafeminism"

[23], advocates an intersectional approach aimed at emancipating all from capitalist exploitation and male-centric dominance. This perspective highlights the essential role of care in understanding the intersections of various oppressions and nurturing anti-state solidarity and resistance.

The division between "work" and "labor," proposed by Hannah Arendt in "The Human Condition" [24], distinguishes productive work from the seemingly unproductive labor of care, crucial for sustaining the world. However, care work is often undervalued, even though it encompasses caring for others, spaces, everyday life, and the planet. In an era marked by democratic fragility, civic culture erosion, and the neglect of collective life, care for others has waned, replaced by regressive individualism that erodes the social imagination and civic institutions. Hélène Frichot's concept of "Infrastructural Love" positions care as a relational approach that connects pedagogy, practice, and theory [25]. Ellen Meiksins Wood, in "Democracy Against Capitalism", argues for perpetual attention and reconfiguration of democratic life [1]. Care, as a critical action, provides a lens for critiquing capitalist spatial production and exploring meaningful ways of living and acting. Amidst the backdrop of capitalist urban space, new institutions and communities are needed to align with the broader quest for an alternative collective life [26]. Critical thought and civic action spaces must be established, fostering a social imagination that encompasses care for all beings, human, non-human, and more than human. Articulating care's centrality in shaping agency, values, social relations, and future visions of collective life is essential. Ultimately, care as critical action emerges as both a method and guiding principle within the care framework underpinning the research project discussed in the paper.

The research project discussed here embodied a collaborative and inclusive approach, positioning community representatives as co-researchers, thereby underscoring the significance of local knowledge and experiences. The aim was not merely to document and disseminate these insights but also to foster meaningful connections and sustained dialogues among eco-communities. The research project recognizes the value of informal practices to challenge and, to some extent, subvert power relations in developing the urban realm. By shedding light on the innovative, localized, and community-led responses to our contemporary challenges, the project contributed to a broader discussion on sustainable and socially inclusive practices. Furthermore, it fostered a sense of shared purpose and knowledge exchange among eco-communities across borders. The collaboration among these communities enabled the development of skills, strategies, and best practices that can be applied not only within these communities but also on a global scale, providing practitioners and policy makers with a radically new set of paradigms to design spaces that enable and nurture practices of mutual care.

References

1. Meiksins Wood, E.: Democracy Against Capitalism: Renewing Historical Materialism. Cambridge University Press, Cambridge (1995)
2. Spencer, D.: Critique of Architecture Essays on Theory, Autonomy, and Political Economy. Birkhäuser, Basel (2021)
3. Bollier, D.: Commoning as a force in urban design. In: Mahdavi, M., Wang, L. (eds.) New Geographies 12 Commons. Harvard University Press, Cambridge (2022)

4. Till, J.: The Negotiation of Hope. In: Blundell Jones, P., Petrescu, D., Till, J. (eds.) Architecture and Participation. Routledge, London and New York (2005)
5. Pateman, C.: Participation and Democratic Theory. Cambridge University Press, Cambridge (1970)
6. Cross, N.: The Automated Architect. Pion Press, London (1977)
7. Rose, G.: Athens and Jerusalem: a tale of three cities. Soc. Legal Stud. Int. J. **3**(3), 333–348 (1994)
8. Awan, N., Schneider, T., Till, J.: Spatial Agency: Other Ways of Doing Architecture. Routledge, London and New York (2011)
9. Rendell, J.: Critical Architecture: Introduction. J. Archit. **10**(3), 227–228 (2005)
10. Agamben, G.: Creazione e Anarchia. L'opera Nell'età Della Religione Capitalista. Neri Pozza, Vicenza (2017)
11. Boden, M.A.: The Creative Mind, Myths and Mechanism. Routledge, London and New York (2004)
12. Blundell Jones, P., Petrescu, D., Till, J. (eds.): Architecture and Participation, 1st edn. Routledge, London and New York (2005)
13. Owens, P.E.: Community-Centered Design. Taylor & Francis, London (2007)
14. Hamdi, N.: Small Change: About the Art of Practice and the Limits of Planning in Cities. Earthscan, London and Edinburgh (2004)
15. Voelcker, B.: Care. In: Mould Collective (eds) Architecture after Architecture - Spatial Practices in the Face of the Climate Emergency (2022)
16. Struening, K.: New Family Values: Liberty, Equality, Diversity. Rowman & Littlefield, Lanham (2002)
17. Gilligan, C.: In a Different Voice: Psychological Theory and Women's Development. Harvard University Press, Cambridge (1993)
18. Noddings, N.: The language of care ethics. Knowl. Quest **40**(5), 52 (2012)
19. Marini, S.: Nuove Terre. Architetture e Paesaggi Dello Scarto. Quodlibet, Macerata (2018)
20. Smith, N.: Contours of a spatialized politics: homeless vehicles and the production of geographical Scale. Soc. Text **33**, 54–81 (1992)
21. Lefebvre, H.: The Production of Space. Wiley-Blackwell, Hoboken (1991)
22. Harvey, D.: Justice, Nature & the Geography of Difference. Blackwell, Hoboken (1996)
23. Bottici, C.: Anarchafeminism. Bloomsbury Publishing, London (2021)
24. Arendt, H.: The Human Condition, 2nd edn. University of Chicago Press, Chicago (2018)
25. Frichot, H., Carbonell, A., Frykholm, H., Karami, S. (eds.): Infrastructural Love: Caring for Our Architectural Support Systems. De Gruyter, Berlin (2023)
26. McEwan, C., Bertolino, N., Matteucci, C.: (eds) Editorial, Care and Critical Action. In: "Lo Squaderno n.65, Explorations in Space and Society" (2023)

Who is in?: Non-Living and Hybrid Constituents in More-Than-Living Ecosystem of the Studio

Erenalp Büyüktopcu[(⊠)] and Ayşe Şentürer

Istanbul Technical University, Istanbul, Turkey
{buyuktopcu,senturer}@itu.edu.tr

Abstract. This research text explores the miscellaneous roles of more-than-living (non-living and hybrid) constituents within the pedagogical universe of architectural studios, highlighting their progressive potential to transform educational practices towards more inclusive, emancipatory, and experimental directions. Through an in-depth analysis of diversified pedagogical experiments, from the Radical Pedagogies selection of the late 20th century to the contemporary approaches presented in the 2022 ABC-Architecture Beyond Capitalism School, the research illuminates how more-than-living constituents act as catalysts for new subjectivities and pedagogical methods. The research endeavors to reveal the transformative and progressive potential of more-than-living constituents through their contextual and performative actions in their political, spatial, and networked positions. Unlike the majority of studies on studio pedagogies that primarily focus on the power dynamics and hierarchical tendencies between participants/students and facilitators/instructors, this research underscores the overlooked significance of spatial arrangements, collective experiences, open-source strategies, and other factors. By integrating these constituents, the text aims for a pedagogical approach that embraces dissolved physical/mental boundaries and all modes of action, facilitates equitable power dynamics, fosters a critical engagement with pressing contemporary issues, opens up new avenues for critical-creative exploration, and promotes acts of resistance and resilience within the studio atmosphere.

Keywords: living and non-living constituents · studio space · pedagogical universe of the studio · contextual and performative actions · studio pedagogy

1 Intro and Warm-Ups

As a place centered at the architectural pedagogy, the studio space generously hosts living and miscellaneous non-living constituents: students/participants, instructors/facilitators and invited individuals; spatial/temporal elements and components, physical/digital agents or representatives, owned/hacked/open-sourced tools, mediums and strategies, and others. Inside the studio, pluralistic and inclusive atmosphere and egalitarian practices hold strong potential to deconstruct the conventional education models and their hierarchy-prone power dynamics established through the giver and receiver of knowledge. The giver, in other words professor are positioned as "the privileged transmitters of

© The Author(s) 2025
M. Barosio et al. (Eds.): EAAE AC 2023, SSDI 47, pp. 197–207, 2025.
https://doi.org/10.1007/978-3-031-71959-2_23

knowledge" [1] in the conventional models. Contrary to the conventional models, pedagogies on the studio fosters an environment of mutual communication and exchange that stimulates critical discussions and provokes creative experiments. Contributions of the non-living and hybrid constituents, strengthen the opportunity of experiencing a collective existence and solidarity between the living members of the studio. Within the context of this research text, we will illuminate the non-living and hybrid constituents, their characteristics and contributions to the design studio and its more-than-living ecosystem. For warming up to the context and employing experimental, progressive practices to make critical interrogations and interpretations, we select "Radical Pedagogies" and "Architecture Beyond Capitalism (ABC) School 2022" as the sources of research. While practices in "Radical Pedagogies" give also substantial clues about the political climate arisen on the post-war period of the 20th century, current studio practices from the "ABC School 2022" reveal the alternative, emancipatory approaches on the contemporary politics.

1.1 Warm-Up1: The Selection of Radical Pedagogies, Exhibition and Book

The "Radical Pedagogies" research project, led by Beatriz Colomina, Britt Eversole, Ignacio González Galán, Evangelos Kotsioris, Anna-Maria Meister, and Federica Vannucchi, was exhibited at the 2014 Venice Biennale under the same names as curators and with the subtitle "Action-Reaction-Interaction". The catalog of pedagogical practices displayed was also shared on the project's website (www.radical-pedagogies.com) and published as a book in 2022 [2]. The selection compiles pedagogical approaches and experiments considered radical, mainly centered in Europe-America and from various universities around the world, including workshops, conferences, protests, tours, research, exhibitions, or biennials focused on architecture, planning, design, and art education, totaling 89 practices. The reasons for cataloging and examining these practices as radical pedagogical experiments and trials vary from their political underpinnings to the pedagogical atmosphere they create and the methods and tools they utilize. However, they share a common feature: all bear strong traces of the inclusive, emancipatory, interactive, critical-creative, and investigative pedagogical universe of the design studio in various aspects.

The structure of the warm-up exercise can be summarized as exploring the practices that constitute the selection in the "Radical Pedagogies" exhibition and book in terms of their political, spatial, atmospheric contexts, and their atypical, non-orthodox, experimental, emancipatory methods, and critical-creative representational productions; then creating a chart of the selection through specific criteria and generating mappings and graphs; and finally, conducting interrogating, critical comparisons, and speculative, progressive evaluations on them. We are moving to the Warm-up2 exercise to explore studio practices from the current time range and to make comparisons and evaluations on them.

1.2 Warm-Up2: 2022 ABC-Architecture Beyond Capitalism School, Sessions

The Architecture Lobby, a grassroots organization, was established by architecture laborers from various fields such as architects, landscape architects, planners, designers, students, and academics to build a critical-based solidarity and advocate for fair labor practices and inclusive work environments in the USA. Within the organization, the Academy Working Group has been conducting a series of workshops titled "Architecture Beyond Capitalism (ABC) School" since 2021. The school in 2021 advanced through a detailed exposition of concepts such as capitalism, labor, and commons. The series of workshops in 2022 (ten days, thirteen workshops, and opening-closing sessions) was organized with the theme of examining these three concepts directly through educational practices in the studio, discussing and developing alternative, experimental, progressive methods for pedagogical practices in the studio. The 2022 school focused on the studio environment and education, considered the dynamo of architectural culture. Issues such as more-than-living studio constituents and their (equatable) power dynamics, the construction of a more transparent, egalitarian, and participatory studio culture through program and studio curricula, political, economic, and physical/digital factors affecting the operations and outputs of the studio, as well as inter/transdisciplinary practices and impressions from studio practices in other disciplines were opened for discussion.

The 2022 school featured the ability for anyone registered on the web/cloud portal to access workshop recordings at any time; the drafting and openness for editing of a common text on studio culture and alternative, transformative action proposals during the closing plenary of the school, which was held online for all sessions; and the establishment of an active learning, sharing community where individuals we see as facilitators in one workshop could easily be seen as participants in another. These aspects, along with the presentations and discussions in the workshops, are considered experiments and trials for building studio practices.

2 The Pedagogical Universe of the Studio: Contextual and Performative Actions

We will attempt to construct the "Pedagogical Universe of the Studio" as a networked, and nebulous 'carrier bag' (referencing Ursula K. Le Guin and Donna Haraway), drawing from the practices we selected and quoted from the "Radical Pedagogies" selections [Warm-up1] and the workshops we decrypted from "Architecture Beyond Capitalism School 2022" [Warm-up2] as well as readings on pedagogy. After explaining the conceptual foundation that will form the carrier bag, we will start to fill its pouch with practices and trials from Warm-up1–2 and insights from readings.

We can evaluate the actions occurring within the "Pedagogical Universe of the Studio" under two main headings: "contextual" and "performative". Positions in terms of politics (serving as an incubator where collectivity, solidarity, and equality are observed; being a source of activism and resilience/resistance; creating awareness; and more), spatial (disconnected, dissolved, or blurred boundaries of the studio; engagement with or outflow into the urban environment; and more), and networked (inter/transdisciplinary research subjects and exploration methods; non-hierarchical, horizontal power dynamics among all living constituents of the studio including students/participants, facilitators/instructors, invited individuals, etc.; and more) refer to "contextual actions". Decision mechanisms and practices directed at the functioning of the studio through these contexts are indicators of "performative actions". Performative actions can be characterized through a broad adjective pool such as provocative, speculative, triggering, radical, alternative, off-line, atypical, non-orthodox, experimental, proactive, progressive, participatory, egalitarian, emancipatory, inclusive, pluralistic, heterogeneous, investigative, and creative.

[Warm-up1]-1. Cedric Price [London, UK; 1959–65] focused on developing flexible learning modes and methods that would require mobile/transportable infrastructures to process his productions as physical agents in continuous flow. In the Potteries Thinkbelt, Price designed not only as a physical but also as a 'contextual' stimulant towards the dream of an educational environment/atmosphere that dissolves into daily life, utilizing the rail system infrastructure connecting abandoned factories/production facilities that once produced pottery. His envisioned educational system's physical transfer centers and transportation modes (consisting of wagons/cabins on rails) also 'perform' as permanent constituents and temporary agents of knowledge transfer. In the plan&section sketches for the Pitts Hill Transfer Area, coded as a permanent constituent of knowledge transfer, there are 'large' and variable structural volumes supported temporarily or self-sustaining, formed by the congregation of 'small' service provider units within a fixed skeleton with movable platforms, in a multi-stacked manner.

During the same periods of production and through partnerships such as running studio/workshops at the Architectural Association School of Architecture (AA) in London, the 'kit-of-parts' logic frequently encountered in the designs of Archigram and Price can also be seen in Price's Potteries Thinkbelt proposal. Through a dynamic kit-of-parts that would trigger volumetric configurations of unit mobility, Price proposes a new architectural 'context' response to what he considered archaic and static educational environments. The rail system network, reinforced by the support of the existing/added service roads and the nearby airport, connecting the pottery factories, becomes transformative in character with its 'performance', transforming pedagogical practices. The time spent on the rails between centers turns into parameters that could redefine the duration of the class/studio, while the wagons, possessing a moving, shaky, narrow but elongated interior space with a continuous flow of exterior space, become parameters that could redefine the class/studio atmosphere. Units and volumes at the transfer center and rails and wagons in the transport network are important non-living constituents of Price's proposal, included in the more-than-living ecosystem of the pedagogical universe.

[Warm-up1]-2. The Institute for Lightweight Structures (IL-Institut für Leichtbau Entwerfen und Konstruieren-ILEK) [Stuttgart, Germany; 1964–90], was established under the leadership of Frei Otto and utilized one of Otto's tent structures for shelter. Positioned in a wooded area within the new campus of the University of Stuttgart, the tent structure that formed the architectural and provocative 'context' of IL was described as "an anomaly (deviation, disorder, oddity) next to science-focused neighboring institutions/institutes, a thorn that arouses curiosity among the buildings belonging to the sciences." Starting with only "six students, IL reached a population of seventy students from various parts of the world, attracted by the open teaching approach and the reputation for experimentalism at the institute, and interested in alternative pedagogical practices and young architects" by 1971 [3].

IL was designed with a focus on total space usage that would not contain "a sharpened division/separation of functions/interior spaces" but would offer spatial flexibilities accommodating different scales of work, activity modes. This arrangement allowed for 'performative' variability in the programmatic setup, such as "expansion and contraction of experiment/discovery groups within their research/production processes" using dividing equipment, level differences, mezzanines, and "the possibility of collective events like seminars/meetings to spill outside (frequently) whenever desired." "The laboratory environment at IL, focusing on physical model/prototype production and their documentation through various methods, was predicated on a direct experience of playful experimentation using unusual materials such as eggs, balloons, shaving foam, and technical equipment like cameras, small gauges, fostering a fun/playful atmosphere observed during tight collaboration" [3].

[Warm-up1]-3. Architects Alberto Cruz, poet Godofredo Iommi, and sculptor Claudio Girola from the School of Architecture at the Pontifical Catholic University of Valparaíso (Pontificia Universidad Católica de Valparaíso-PUCV) [Valparaíso, Chile; 1952–72], who wanted to explore the urban environment with creative methods and various dérives through the city, designed a truly interdisciplinary studio/pedagogical practice that would "use language performed in a poetic and collective manner through poetry as a tool," [4]. In the practices at PUCV, language was used in a way that aligns with Kasia Nawratek's definition that it "acts as a catalyst for the emergence of images and consequently meanings, inherently containing the power to facilitate the phenomenon of designing" [5]. They explored urban landscapes as a playground without defining any specific strategy or purpose, other than through poetry and the language reconfigured by poetry's unique methods (re-structuring, distorting, playing with layout). This method of make instrumental poetic language also entailed a complete liberation from historical layers within the 'context' of the urban environment, associations related to memory/remembrance, and physical/spatial stimuli and directions.

The dérive practices at PUCV differ in some aspects from those initiated by Guy Debord and Asger Jorn within the Lettrist International during a similar period (between 1952–1956) and later transferred to Situationist International practices in 1957. According to a passage Greil Marcus cites from Lettrist's publication Potlatch, in their dérives, "they potentially used the existing urban infrastructure, from metro corridors and tunnels to dark parks, gardens, fire escapes, and roofs" [6]. While Debord and Jorn's psychogeographic explorations in parts of Paris, Copenhagen were conducted with the active, intertwined guidance of memory and senses; PUCV set aside memory and the past, framing city dérives under the guidance of emotions evoked by poetic language. Debord and Jorn transformed their dérives and vague encounters in familiar parts of the city they were deeply connected to into psychogeographic representations (collages, mappings, fanzines, pamphlets) [7]. PUCV that foregrounding the auditory impact of poetic language, experienced proactive and atypical 'performances' of close interactions and playful encounters with urban landscapes through 1:1 apparatuses like body extensions, hybrid skeletons, and prostheses which enable to engage without leaving footprints in the sand, walk on the beach without sinking and getting wet, dérive around the city without staying in the sun, being affected by the wind, and play games by gliding without touching the ground.

The pedagogical practices that focus on exploring the city under the control of emotions provoked and triggered by poetic language, including close encounters with the city and the body, and playful encounters (tournaments), allowed the participant/student-facilitator/instructor living constituents at PUCV to be "open to producing new subjectivities." In their dérives not only within the city and its immediate surroundings but also to various points in South America with an interdisciplinary team, we can observe "a behavior plane where all participants are equalized and hierarchical, atypical power dynamics" are non-existent. Collectively performed, pleasure and enjoyment-focused 'performative' actions (papillons-phalènes) deeply shook pedagogical conventions and desired to "blur, dissolve, and even break the boundaries between learning-working-living," bestowing a "new erotic character" to the pedagogical universe of the studios at PUCV [8].

[Warm-up2]-1. Will Martin explored the trail of emancipatory agents that could be integrated into the studio as non-living constituents, in his workshop "Ours to Hack and to Own: Open-source Strategies and the Pedagogy of Potential" [9], aiming to trigger pluralistic, inclusive, co-evolving, and autonomous architectural pedagogies. The preconditions for the agent's 'context' within the studio were first defined as "not serving capitalist accumulation and naturally not training worker for it, and then not forming its strategies and production tools over orthodox, patriarchal, and reductionist frameworks" [9]. The proposed non-living constituent to permeate/settle in the studio and 'perform' was the GitHub service, a cloud-based and open-source version control system frequently used by software developers.

GitHub's repository can host content in various formats, including visuals, text, coding, etc. Software developers, defined as 'contributors' to the repository, have access rights to view, edit, and develop content. The version control system allows developers to simultaneously follow the entire process of the main project or its branches (semi-autonomous parts that don't affect the main project but can transfer changes, edits, and improvements back to the main project if necessary) and revert any errors that occur during the process. In summary, GitHub is scrutinized as an open-source, user-friendly, and participatory system/interface that facilitates the project process to progress in a negotiator manner and in collaboration by reducing the fear of making mistakes. The participatory, alternative studio pedagogy development tool or the performance of GitHub as a pluralistic and emancipatory agent within the pedagogical operation is examined, thanks to its open-source nature, which allows it to be owned or hacked by everyone.

[Warm-up2]-2. In the "Collectivized Pedagogies" workshop [10], constituents of the Dark Matter University (DMU) jointly presented their pedagogical universes with inclusive and progressive studio themes and course contents. Studio facilitators from various universities across the United States, who came together during the COVID-19 process, designed a curriculum for political design studios focused on "Design Justice," developing a participant and disputative approach through online programs (zoom, mural, miro, etc. for video conferencing over the internet, information sharing, enabling joint discussion and development like) and interfaces. As a positive outcome of the distance education model, which became part of contemporary pedagogical practices due to the compulsory long-term, intense lockdown, and quarantine measures during the pandemic, DMU constituent studio facilitators had the opportunity to apply and observe their curriculum and 'context' synchronously in various universities and through differing internal operations and dynamics. This situation provided a foundation for 'performative' activities such as an unprecedented permeability between studios (studio collaborations established with facilitators from different states and virtual visits between them), mutual interaction and sharing (discussing 'contextual' differences that emerge when the same theme is applied in studios at other universities). The main themes of the studios focused on principles like "working together and exploring the city/state's infrastructure and current neighborhoods"; "sequential modules aimed at comprehensively researching, exploring, and understanding socio-economic/political topics such as housing, economic developments, and labor" [10] were applied in a collaborative manner across different universities.

Implementing their pedagogical experiments in this 'context' and striving to conceptualize a political design studio, Dark Matter University (DMU) aims to define an approach that can critically and inquisitively focus on commonly overlooked, normalized issues avoided in confrontation in the current context, such as xenophobia and "anti-racist pedagogies; commons, collectives, and marginalized communities; fragile identities, decolonization, and care practices; isolation, surveillance spaces, and the dynamics of power and labor" [10]. This approach aims to combat existing inequalities and adopt a progressive stance towards transformation because, as Bhabha puts it, the "condition of being a minority and the culture of the migrant being in-between/in-limbo" [11] creates transitory sensations capable of gathering and integrating all constituents of the studio into acts of resistance and resilience. In later stages, DMU plans to enrich its inclusive and emancipatory pedagogical universe by adding courses/studios related to the personal research and experience areas of its constituents that align with the universe.

3   The Pedagogical Universe of the Studio and More-Than-Living Constituents: Conclusions

From [Warm-up1]-1. Cedric Price harbored the dream of reaching a pragmatic utopia through a revolutionary set of proposals he designed by himself, without the request of any person/institution. He believed that everyone involved in the education/learning/thinking network, which is "embedded/integrated into human actions, including everyday life," would also be freed from the shackles of dogmas and "behavioral patterns based on patriarchy" [12]. Price stated that the Potteries Thinkbelt was a proposal that would "facilitate a state of social justification where students and the public, instead of being separated, come together, which architecture education (and universities in general) needed" [13]. Although not a direct counterpart of emancipation and unification in his proposals, he utilized tools in the studios/workshops he conducted within the AA to empower students by authorizing them. "According to Price, being a student also encompassed the responsibility of acting as an independent individual. In this direction, the 'Taskforce' program he initiated at the AA aimed for a contract to be signed between the student/participant and the facilitator/instructor, leaving the task of defining what the student wanted to achieve throughout the year to the student" [12]. The contract, as a non-living constituent added to the studio, becomes a catalyst for both making the students a more responsible and at the same time more independent constituent of the studio and transforming the archaic and hierarchy-prone power dynamics in the studio. Our critique that signing a contract does not break the dual structure between student-facilitator sides and that aiming for the student's independence paradoxically results in a document that might pressure the student is a note on the method applied.

From [Warm-up1]-2. Like Buckminster Fuller's domes [14], we can see in IL under Frei Otto's coordination the use of the total space defined by a shell including alternative structural systems and materials as a non-living constituent to transform the pedagogical atmosphere in a non-orthodox manner. The difference in IL lies in the expectation not for the transformation to be as spatially focused as on the domes but for the space to be positioned as a facilitator and intermediary constituent of this transformation. The space is successful as a facilitator insofar as it can open up room for experimental and participatory pedagogical practices, allow flexibility in operations, and beyond that, transform the hierarchy-prone power dynamics among living constituents towards a progressive and egalitarian direction. Finally, the almost sacral position of space in the domes has given way to a more sarcastic approach in IL, where space is described as strange and even alien-like.

From [Warm-up1]-3. In previous practices, we observed structural (and non-living) constituents with their own tectonics, such as domes, transfer centers, transportation modes, and tent structures, directly or indirectly transforming the studio atmosphere and operation in a progressive direction. In the practice at PUCV, where the illusion of a binary ecosystem defined by living and non-living constituents is dismantled, we start encountering constituents with intermediate forms and hybrid structures. Language conveyed as poetry gains a hybrid constituent nature because it can be 'performed' through living constituents while also preserving its ontological 'context'. The emphasis and intonations, the words, and lines awaken different meanings, images in the mind of each listener, producing a collective reaction while being listened to and perceived collectively in a form that has turned individual, almost ritualistic. This reaction then aligns with Donna Haraway's description of an atmosphere "where a critical and joyful uproar is occurred; a trouble is produced through productive joy, terror, and collective thinking" [15], also describing a practice that can become autonomous from the dominant and surrounding system. Besides directing speculative, emancipatory actions of the studio/pedagogical practices, it also encourages the design of complex skeletons, prosthetic-like body extensions as representational productions, and their development using progressive and experimental structures and materials.

From [Warm-up2]-1. The practices discussed from the Radical Pedagogies selections compiled from the second half of the twentieth century (1940–1990) [Warm-up1], defined non-living and pedagogically transformative constituents of the studio through structural formations and spatial qualities, encounters/explorations/close encounters in urban landscapes, and mediated actions. When the studio gained a more holistic, flexible, and variable 'context' through spatial arrangements, and when 'performative' actions dissolved its boundaries into daily life and the city, the perceived physical atmosphere within the studio was mentioned. In the workshop by Will Martin at the 2022 ABC-Architecture Beyond Capitalism School, we see that technological equipment, systems, and interfaces, which were previously applied mainly for recording studio productions, documenting parts of the city forming the studio context, or archiving, have now evolved from being supportive constituents to being game-changers, fundamental building blocks. The impact of technological leaps over the long interval, digital interfaces and applications that have become pervasive in our lives, is evident in the change of role distribution within the studio. In here, instead of choosing a digital system for pragmatic reasons to facilitate studio operations, we see it uniquely involved in the preparation phase of the studio's pedagogical universe through the program, positioned against the capitalist system, breaking down hierarchies prone to formation within the studio, and mediating an active interaction and feedback mechanism as a non-living, moreover virtual constituent.

From [Warm-up2]-2. in [Warm-up1], we had defined the transformative powers of non-living or hybrid constituents on the physical atmosphere of the studio's pedagogical universe. In the previous practice of the 2022-ABC Architecture Beyond Capitalism School, we evaluated the non-living (and virtual) constituent for directly focusing on emancipating the studio's power-relationship dynamics from hierarchization and even preventing the reflections of the neoliberal political and economic system from leak the studio. The difference in the DMU practice similarly aimed at studio operations, beyond focusing on the preparation process through the curriculum for conceptualizing the studio 'context', also generates counterparts for 'performative' actions actively taking place during the studio's operation. Topics and issues that studio focuses on, and the key cornerstone of this holistic approach emerge in a hybrid structure with the non-living qualities of the spaces it concentrates on, and the living (and both physically perceptible or mentally apprehensible) qualities (From [Warm-up1], similar to the practice at PUCV) of the communities and commons in DMU. Communities and concepts marginalized in our current context are deeply scrutinized, confronted through interrogating, and addressed for designing proactive and progressive proposals in DMU's political design studio. As a natural consequence of this approach, the internal power dynamics of the studio and modes of research/discussion/production are redefined specifically for DMU. Consequently, the pedagogical universe of the DMU is planned through a system of in-depth disputative and critical-creative thinking to leave prejudices and dogmatic beliefs outside the studio, embrace and welcome all modes of action, including progressive and speculative, and to not provide an opportunity for "knowledge to be taught implicitly as part of the hidden curriculum," as Illich puts it [16].

Another significant (and non-living) constituent observed in the DMU practice, online programs, unlike the previous practice, do not undertake any triggering or transformative role in the studio's pedagogical universe. Rather than, it only function as a technical interface that in facilitating transitions and mobility or mediating gatherings virtually. We would like to conclude the examination of DMU with a somewhat speculative evaluation. In the practices selected from the second half of the twentieth century in [Warm-up1], we generally saw the reflection of universal political fluctuations, such as the events of '68, in pedagogy, usually in the form of pressures and rebellions from student groups towards the (undoubtedly positioned more privileged) facilitators. In [Warm-up2], especially in the DMU practice, both studio facilitators and the participant/student composition in the universities where the practices are applied are mostly positioned as constituents that are directly the subjects of the issues and problems addressed by the studio or fit various in-between/marginalized definitions. We see this transformation as important also in terms of evolving relationships that are eager to hierarchize among living constituents (with a bit provocative interpretation, from privileged to needy) towards a more inclusive, egalitarian, and emancipatory ground.

In light of the above conclusions from the section of contextual and performative actions, in this research text, we have focused on the miscellaneous roles and power of the studio's non-living and hybrid constituents in transforming the pedagogical atmosphere and operations in a participatory, inclusive, and emancipatory manner. We chose to underscore the unique/quirky/eccentric, troublemaking/pesky, fickle/inconsistent, triggering/provocative, playful/pleasure-seeking/erotic qualities of these constituents, which studies on studio pedagogy focusing on the student-facilitator duo (sometimes turning into a trio with the addition of invited individuals) have not examined much, and preferred to explore and interrogate the studio's pedagogical universe with an unconventional approach guided by these adjectives. We also discovered that the experimental and proactive structures of non-living and hybrid constituents contribute to the transformation from hierarchical power dynamics towards more engaged pedagogical practices among living constituents.

References

1. hooks, b.: Teaching to Transgress: Education as the Practice of Freedom. Routledge, New York (1994)
2. Colomina, B., Eversole, B., Galán, I.G., Kotsioris, E., Anna-Maria Meister, A.M.: Radical Pedagogies. MIT Press, Massachusetts (2022)
3. Fabricius, D.: A Spinner in His Web. In: Pedagogies, R. (ed.) Colomina, B, pp. 318–323. MIT Press, Massachusetts (2022)
4. Galán, I.G.: Autonomy...to Join Life, Work, and Study. In: Pedagogies, R. (ed.) Colomina, B, pp. 154–160. MIT Press, Massachusetts (2022)
5. Nawratek, K.: Introduction. In: Nawratek, K. (ed.) Space and Language in Architectural Education: Catalysts and Tensions, pp. 1–17. Routledge, London (2022)
6. Marcus, G.: Lipstick Traces: A Secret History of the Twentieth Century. Harvard University Press, Massachusetts (1989)
7. Büyüktopcu, E.: Mimari Tasarımda Hayali Gelecek Kurguları: Bir Zamanlar Arası Söküm Denemesi-Imaginative Future Speculations in Architectural Design: An Experiment of the Intertemporal Disassembly [Master Thesis, Istanbul Technical University] (2017). https://tez.yok.gov.tr/UlusalTezMerkezi/tezDetay.jsp?id=cNKnw1PEgNbq4aA qzunQiw&no=A2AyuClciw3E2PsgolgE6w, last accessed 2024/02/24
8. Galán, I.G.: Revisit: Ciudad Abierta in Ritoque, Chile. Architectural Review. https://www.arc hitectural-review.com/essays/revisit/revisit-ciudad-abierta, last accessed 2024/02/24 (2022)
9. Martin, W.: Ours to hack and to own: open-source strategies and the pedagogy of potential. 2022 ABC-Architecture Beyond Capitalism School. https://youtu.be/_AWaaEjNGFA, last accessed 2024/02/24 (2022)
10. Riano, Q., et al.: Collectivized pedagogies. 2022 ABC-Architecture Beyond Capitalism School. https://youtu.be/CfrR_kLhs3A, last accessed 2024/02/24 (2022)
11. Bhabha, H.K.: The Location of Culture. Routledge, London (1994)

12. García-Germán, J.: Cedric price the architectural association and others. Radical Pedagogies. http://radical-pedagogies.com/search-cases/e18-architectural-association/, last accessed 2022/03/01 (2014)
13. Price, C.: Potteries Thinkbelt. New Society:192 (1966)
14. Wigley, M.: Parasitic Pedagogy: The Buckminster Fuller Teaching Machine. In: Pedagogies, R. (ed.) Colomina B, pp. 219–225. MIT Press, Massachusetts (2022)
15. Haraway, D.: Tentacular Thinking: Anthropocene, Capitalocene, Chthulucene, in Staying with the Trouble: Making Kin in the Chthulucene. Duke University Press, London (2016)
16. Illich, I.: Deschooling Society. Harrow Books, New York (1973)

Action Based Research for Capacity Building of Neighborhood Communities

Daniela Calciu, Vera Marin, and Oana Pavăl[(✉)]

Association for Urban Transition, "Ion Mincu" University of Architecture and Urban Planning, Bucharest, Romania
daniela.calciu@uauim.ro, vera.marin.100@gmail.com, sooana.paval@gmail.com

Abstract. This paper aims to present action research experiences in Romanian context and reflection on possible contributions from architectural education to helping neighborhood communities. For the Bucharest team of ATU- Association for Urban Transition, the project entitled Urban Education Live - Innovative Urban Education in Live Settings (2018–2020) was about applying social mapping tools for helping a civic initiative group to define a common agenda that was presented to the local authorities as the first neighborhood development plan in Romania. A new project entitled CoNECT - Collective Networks for Everyday Community Resilience and Ecological Transition was initiated in 2022 and it is aiming to identify collaboration mechanisms among various stakeholders for helping grass-roots initiatives addressing transition challenges to have more visibility and impact. Both projects are implemented in the framework of the Urban Europe Joint Partnership Initiative as a research program demonstrating a solid concern for the social impact of research on communities. Research on connections between space and communities means testing ideas and tools coming from the training of built environment professionals and helping neighborhood communities and master students to work together on concrete objectives and formats. By doing so, architects and other professionals learn how to offer their support in building bridges between grass roots initiatives and decision-making processes since these formats are the ones used to plan and design interventions for transforming spaces.

Keywords: Collaboration Mechanisms · Built Environment Professionals · Education · Action Research · Live Pedagogy · Ecological Transition

1 Introduction

Over the past two decades, an increasing number of architecture schools across Europe have allowed the space for experimenting with new roles of the engaged architect, roles that have strayed away from the consecrated ways of practicing architecture and have opened the panoply of clients of the professions of the built environment. Such education practices are in line with new professional strands built on the integration of the premises that the end users of architecture and the city are not only passive and compliant beneficiaries, but rather active participants in processes in which they bring

M. Barosio et al. (Eds.): EAAE AC 2023, SSDI 47, pp. 208–218, 2025.
https://doi.org/10.1007/978-3-031-71959-2_24

unique and valid perspectives. This acknowledgement alone is not enough to foster real change in how architects can serve society. However, it has prompted new ways of thinking and doing architecture, expanding the collaborative approach and recognising and integrating the agencies of others over the built environment. It has also prompted new research objects and methods, carefully connecting academia with the grassroots initiatives looking to improve the spaces of the everyday life of urban communities.

"Spatial Agency: Other Ways of Doing Architecture" [1] was a first extensive collection put together by Nishat Awan, Tatjana Schneider, Jeremy Till, to showcase "new ways of doing architecture", which has served as an inspiration for educators and students concerned with the uncertainties of our future. The online database [2] has thus become a destination for those interested in the process rather than the results, and in the opportunities that arise from venturing outside the established path of shaping the environment towards reflecting on questions of appropriation, dissemination, empowerment, networking, subversion, knowledge, organizational structures, physical relations, or social structures.

Among the more recent collections of alternative practices, the editorial project coordinated by Mathias Rollot verifies the "collaborative hypothesis" [3] while suggesting even the emergence of a new profession, which is coming to life in-between the constellation of actors that shape the built environment, and which is developing under the conceptual auspices of "making the city differently". This "profession" would actually be the recognition of a new competence - that of working collaboratively within a knowledge producing system including and intertwining all sectors of our society. In fact, the various taxonomies and analyses of "other practices" overlay in several points: (1) that interdisciplinarity is a hub for different fields learning from each other; (2) that action research is a continuous learning and acquirement of skills, values and attitudes; and (3) that collaborative action research is a territory of mutually beneficial encounters of the various actors and stakeholders.

As urban stakeholders are redefining their roles within collaborative networks, architectural education is prompted to emphasize community involvement and partnership. Consequently, there is a burgeoning interest in cultivating learning experiences that intertwine with research on community-based initiatives and urban innovation ecosystems involving all stakeholders to co-create and co-design sustainable developments in support of more liveable, just, inclusive and attractive neighbourhoods. Simultaneously, these learning experiences have demonstrated their enhanced value when extricated from the confines of the "ivory tower" of academia and taken into the "real world" - by changing the place of learning, but also by changing the agents and the relationships of the pedagogical act.

The text will discuss a heuristic process of non-formal education developed by a Romanian NGO that gathers students and teachers from various universities to collaborate on live projects aimed at enhancing the neighbourhood practices for a better and more sustainable life within the city. First, we will place the program within the theoretical framework of the Multiple Helix Collaboration (MHC), [4] which fosters a holistic approach to addressing complex urban challenges and advancing architectural education. Second, we will present our way of blending interdisciplinary engagement and collaborative action research. Although we are also teachers within the "Ion Mincu" University of Architecture and Urban Planning (UAUIM) Bucharest, we are developing this model through the Association for Urban Transition (ATU) as a platform where more university agents can meet in a transparent governance structure, inclusive decision-making processes, and effective communication that allows an equitable exchange among all the participants. Third, we will discuss the implications of such programs for curriculum development, research, and community engagement, highlighting the transformative potential of collaborative approaches in reshaping architectural practice and education.

2   The Multiple Helix Collaboration (MHC) for Urban Innovation at the Neighbourhood Level

Over the past few decades, innovation has proven extremely effective when understood as a multifaceted and collaborative process that involves interactions among a variety of actors across different sectors [5]. The first conceptualisation of this idea, as well as the first comprehensive framework for analyzing the dynamics of innovation in this context, was the Triple Helix Model (THM), proposed by Etzkowitz and Leydesdorff in the 1990s. They argue that the three primary institutional spheres - academia, industry, and the government - represent distinct yet interdependent actors, each contributing unique resources, expertise, and incentives to the innovation process. As the benefits of this three-way collaboration have grown, innovation ecosystems have developed to include civil society and the natural environment. These two additional helices were incorporated by Carayannis and Campbell in 2009, who coined the Quintuple Helix Model (QHM) [6] as an extension of the THM, broadening it on different collaboration plans, to create a more inclusive and holistic approach to innovation.

MHC can take various forms, including collaborative research projects, innovation clusters, public-private partnerships, and multi-stakeholder platforms. They facilitate knowledge exchange, technology transfer, and capacity building across sectors, fostering innovation-driven growth and socio-economic development. By engaging with civil society organizations, grassroots initiatives, and environmental stakeholders, MHC seeks to promote inclusive innovation that addresses societal needs and environmental challenges, and to enhance the resilience and adaptive capacity of innovation ecosystems, enabling them to respond effectively to changing socio-economic and environmental conditions.

Such initiatives can also offer numerous benefits for architecture education, including enhanced interdisciplinary learning, real-world experience, and networking opportunities for students. While the THM prompted the involvement of students in projects developed together with the industry, the QHM extended the realm of live projects to diversify the forms of collaboration between all categories of stakeholders, and also to explore more widely the role of instruments of design as a public service. The implications for curriculum development and research in architecture schools can be profound, given that by integrating interdisciplinary approaches, community engagement, and sustainability principles into the curriculum students can become more socially responsible and environmentally conscious architects. Moreover, by being part of collaborative research projects with external partners, architecture schools can generate new knowledge, promote innovation, and contribute to the advancement of architectural practice and theory.

Despite its potential benefits, MHC faces several challenges, such as institutional barriers, divergent interests, or power dynamics among stakeholders. Herein lies the need for innovation of the governance structures, as well, by crafting collaborative mechanisms that promote transparency, accountability, engagement and cohesion. An opportunity in this respect comes from the shift in the modes of organization of professional practice corresponding to the definition of new roles of architects and approaches to architecture. Within the Romanian context, the nonprofit sector has emerged as a prevalent form for alternative ways of practicing architecture, but also for innovative models of collaboration and community engagement that can complement and enhance MHC efforts. One example is ATU, [7] which stands as the oldest non-governmental organization dedicated to the built environment in Romania.

3   A Think Tank NGO as a Platform for Multiple Helix Collaboration

Established in 2001 as a spinoff of UAUIM (the Integrated Urban Development Master Program), ATU has sought to promote a more democratic approach to urban studies and practices, initially focusing its efforts in Bucharest, then expanding to Sibiu. Its foundational principles rest on two pillars: facilitating dialogue among various actors and stakeholders through negotiation-based decision-making, and advocating for an interdisciplinary approach to urban planning. These guiding principles were born out of the disparities observed during Romania's transition following Communism, but are still valid in the present state of transition towards ecological sustainability. Although the projects have shifted their focus and discourse in accordance with the global turning points at all levels, ATU has continued to develop their methods of collaborative action research, still exploring ways for bridging academia and local communities, engaging with diverse stakeholders to navigate this ongoing transition.

Over the span of two decades, ATU has cultivated a membership base comprising approximately seventy individuals, forty of whom are professionals of the built environment, a significant portion being UAUIM graduates, including seventeen who now serve as faculty members within the same institution. This representation has fostered a robust collaboration between ATU and UAUIM, but ATU has extended its reach to establish solid partnerships with various other universities, facilitating collaboration across disciplines such as construction engineering, art, anthropology, sociology, human and urban geography, law, political sciences, and public administration studies. This deliberate focus on interdisciplinary engagement has been ingrained in ATU's ethos from its inception, defining an independent platform for collaboration that transcends the constraints often associated with traditional disciplinary boundaries within the academic curricula.

In addition to serving as a versatile research unit registered as an NGO, facilitating collaboration among researchers from diverse academic backgrounds, ATU plays an interesting role in providing non-formal educational opportunities for students. These opportunities encompass a spectrum of engagement, from active involvement in events through volunteering to participation in structured internships or fellowship programs. Notably, within the framework of fellowship initiatives, the organizational structure fosters a horizontal dynamic between mentors and fellows, contrasting with the hierarchical mentorship models typically prevalent in university-based professional training programs. This emphasis on a more egalitarian relationship dynamic fosters collaborative and inclusive learning environments, which stand as a departure from conventional educational paradigms. Through fellowship programs embedded within ATU's projects, a methodological framework is being developed, which holds the potential to serve as a replicable model for fostering collaborative endeavors in other contexts of partnership.

This initiative aligns seamlessly with academia's third mission, [8] which emphasizes the generation of knowledge to address societal issues and contribute to societal development. Consequently, action research becomes intricately intertwined with societal engagement. ATU has been adjusting its discourse and methods by exploring collaborative mechanisms capable of addressing critical urban challenges, such as the preservation of public space, heritage protection, informal settlements, and the development of green and blue infrastructures. These efforts are underscored by the imperative of negotiating the public good and establishing a common ground among stakeholders within neighbourhoods or cities.

Both within ATU and UAUIM, an important and persistent inquiry revolves around the responsibilities of architects and urban designers in shaping the goals of spatial transformations, particularly concerning public spaces and amenities. Fundamental considerations include how project requirements are delineated and negotiated, how the professional responsibilities are shifting under the influence of the changing societal and environmental conditions, and how professionals can be more relevant within the decision-making processes and in relation to public institutions. This is especially pertinent in contexts where existing spaces are in use, and where the determination of public interest requires active engagement with the community. Public spaces and facilities mirror most accurately the challenge of defining the roles of architects, urban designers, or landscape architects in accurately identifying public interest. Thus, how can we structure commissions for interventions funded by public resources in a manner that enables meaningful public input?

In response to these inquiries, our projects have been experimenting with novel forms of collaboration between academia and urban communities, in a form of live pedagogy (the fellowship) that prompts complex and interdisciplinary learning experiences based on direct contact with socio-spatial situations and collaboration among students, mentors, and civic partners from various fields. The fellowship stands on the principle of mutual benefit: the collaborative milieus contribute to empowering residents to voice their needs and take action to improve their living environments; concurrently, they inspire students to evolve into empathetic professionals who harness their skills creatively for the betterment of urban life.

4   The Fellowship as a Pedagogical Framework for Collaborative Action Research

4.1   An Interactive Mobile Workshop for Public Participation

The most consistent program developed by ATU is URBOTECA, a set of instruments and methods for the research, mediation and design of frames of exchange and collaboration between the top-down and the bottom-up. Started in 2014, it stemmed from the ambition to disseminate academic knowledge and perspectives concerning effective urban governance and participatory democracy in urban planning and design processes in Romania. The need rose from the Romanian context, where there is no history of practices based on community engagement and urban development projects have a top-down approach, based on quantitative studies and generic surveys which lead to a birds-eye level context understanding. At the same time, both formally established or informally self-organized civic groups usually come together to react to stringent problems that affect them directly. Residents and communities are not informed about, or included in, processes and projects from the beginning, so their primary action is often to contest un-negotiated market driven urban developments that would negatively impact their living environments, or to demand fixing a problem or undoing what they consider to be wrongdoings.

Our initial scope of inspiring residents to also gather pro-actively and ask for more transparency of decision-making processes was followed by the intent to open a career pathway for graduates in architecture, urban planning, landscape architecture, along with professionals in sociology, anthropology, and communication, who would collaborate to facilitate negotiations among the various urban stakeholders, to better balance the forces shaping our cities. Thus, the students were our partners in testing and drafting guidelines for establishing impartial environments for constructive and efficient dialogue, wherein public officials, local representatives, and members of the community could express and understand conflicting views and common interests.

After this first experience blending action research and non-formal education, we felt that our efforts had been too dissipated in one-time actions across the city. Although we had much to learn from the occasional exchanges with residents, we had no way of verifying if our encounters had left any impact on our conversation partners. Consequently, we considered that a partnership with a civic initiative group would allow us to ask more complex questions and explore more ways in which academia might serve urban communities.

4.2   A Mobile Lab for Urban Education in Live Settings

In 2018 ATU designed the URBOTECA Fellowship programme [9], within a research project conducted together with the Sheffield School of Architecture, The Tampere School of Architecture, The Institute for Spatial Policies (IPOP), and The Centre for Spatial Sociology, University of Ljubljana. Between 2017 and 2021, Urban Education Live (UEL) [10] created and tested models of collaboration between universities and urban communities, where universities acted as catalysts of urban change through trans-educational urban capacity building. The objectives of the Fellowship were multi-faceted. Firstly, it aimed to enhance community engagement by broadening the participation of neighborhood residents, thereby fostering a more diverse and active neighborhood life and increasing residents' attention level towards the possibilities for further development of their environment. Secondly, it sought to establish a meaningful connection between academia and the local community, facilitating mutual trust and creating the context for the residents to benefit from the professional expertise of the fellows. Thirdly, the initiative aimed to gather pertinent data and information through field research, and to present it in a user-friendly format accessible to both professionals and community members. This approach aimed to promote understanding and long-term engagement among residents, fostering a sense of ownership and contribution to the neighborhood's development.

From the pedagogical standpoint, the program's focal point revolved around the comprehensive consideration of the three domains of learning as delineated by Bloom's Taxonomy [11]: cognitive, affective, and psychomotor. The targeted knowledge acquisition encompassed a profound understanding of various concepts, such as neighborhood communities, interdisciplinarity, live project cooperation (living lab), design thinking, participatory planning, as well as broader topics such as common agendas, public interest, decision-making processes, and indicators of change. The acquisition of skills, or the methodological approach to defining common interests at the neighborhood level, constituted a fusion of anthropological tools and those utilized by spatial planners, urban designers, or architects.

This interdisciplinary fusion was implemented over a duration of six months, during which fellows and the project team, as well as the community partners, engaged extensively in informal exchanges which proved more enriching than structured research methodologies like guided interviews or questionnaires. This "social mapping" was an ongoing process integrating an online mapping tool with a series of multimedia-recorded street interviews and interventions. These activities were documented in visual and written formats, and uploaded onto an online map of the study areas. Beyond the mere data collection, this process aimed to engage the residents in discussions about shared values, such as the public's right to be informed about impending changes to public spaces or common amenities, thus fostering a sense of collective ownership and engagement in the spaces of the everyday.

The 2018 Fellowship was intended for master's students and recent graduates (within five years of completing their master's degree) in various fields including architecture, urban planning, landscape architecture, geography, engineering, information technology, arts, anthropology, sociology, economics, psychology, law, and journalism in Bucharest. We received fifty-two applications, selected an initial cohort of seven fellows, and finally worked with five (an architect, a landscape architect, an urban designer, an artist, and a communicator).

The program drew inspiration from the "Collective Impact" [12] concept, aiming to cultivate a unified urban agenda with shared objectives. This approach fostered novel collaborations across the third sector, engaging a spectrum of stakeholders from both public and private domains. By integrating both hyper-local and extensively networked strategies, the initiative sought to address urban fragmentation. Embracing a bottom-up approach to comprehending urban challenges, the program charted a course toward a novel modus operandi, that involved harnessing local resources, including human capital and expertise, to bolster capacity, cultivate new networks, and spur innovation within the urban landscape.

The program was crafted to meet the time constraints and format criteria necessary to be recognised by (and possibly integrated) into the official curriculum of several universities that might become keen to leverage their resources for the betterment of local communities while providing students with opportunities for applied learning. At the same time, the program was designed in alignment with the primary objectives of the Urban Education Live research project, which entailed immersive engagement with neighborhood communities, employing novel social mapping methodologies to discern what people consider to be an asset or a liability in their living environments.

The "Super Site Specific" defined within the international research project echoed locally in the determination of the local agenda, in choosing and adapting the methods and choosing the essential interventions that were considered to be transformative as addressing the socio-spatial complexities and contemporary challenges prevalent at that moment. The trans-educational approach was supported by mapping techniques and technologies which focused on emerging patterns of use, desires and needs, employing a keen sensitivity to the social dynamics at play.

The outcome was not a project, but a community-led development plan for the neighbourhood; more specifically, a handbook that gathered and illustrated the common visions of the inhabitants - clarifying, verifying the legitimacy, and translating their wishes and desires into objectives and briefs for specialized work to be hired by the municipality. [13].

4.3 Five Living Labs for Community Engagement

The second edition of the URBOTECA Fellowship (2024) is being delineated within another research project conducted alongside eighteen academic, industry, civic, professional organisations and public institutional partners from Romania, France, Sweden, Spain, Norway, and The Netherlands. [14] This edition brings a second turn in our approach, aiming to broaden stakeholder participation compared to its predecessor. Reflecting on the evaluation of the previous iteration, several improvements have been identified: (1) The expansion of involvement opportunities for young professionals: Despite the large number of applicants in the previous round, we had to select only a limited number of fellows in order to match the availability of hosts from the civic initiative group. To avoid this constraint, the second edition has chosen several urban situations while maintaining small teams of three to six fellows per situation, where hosts are willing to facilitate field-work contributions. (2) The choice of more diverse and established hosts, as institutions and organizations that might want to become more open and relevant to the local community in which they operate. (3) Scaling up the use of the disciplinary backgrounds among potential fellows: The initial focus on integrating spatial design skills with anthropology and visual arts yielded valuable insights. However, the documentation and demonstration of both process and outcomes were not prioritized, hindering communication with other professionals, researchers, decision makers, and potential collaborators. To improve this, the second round establishes clearer interdisciplinary teams comprising students in architecture, urban design, or landscape architecture; anthropology or sociology; and visual arts, especially documentary photography and videography. This aims to create powerful storytelling materials to enhance visibility for hosts, fellows, and their individual and collaborative experiences. (4) Flexibility in the level of professional training: While the first round required enrollment in master's programs, the second edition emphasizes stronger connections across formal training programs. The curricula analysis and interviews with teaching staff from various universities revealed the need for flexibility in professional training levels, provided that fellowship activities receive recognition within formal curricula at various levels (bachelor's or master's). This tightens the alignment between the fellowship and the academic requirements, facilitating broader participation from diverse backgrounds.

Hence, the second edition is designed to create more connections: (1) among at least three professional fields within a team, and also by having at least fifteen fellows in the program, a larger variety of perspectives and possibilities of collaborations is to be expected; (2) among five teams of fellows and their hosts that are from various categories: a community of artists, an independent theater, an NGO promoting the idea of a creative neighborhood in Bucharest, a public cultural institution under the general council of Bucharest, and an URBACT project team from the planning department of the inter-community association of municipality of Bucharest and Ilfov county council; (3) among teaching staff from various universities who will have to consider the recognition of the activities not just for their students, but also of their students' contribution in interdisciplinary teams. Besides, by having several pilot areas, the hosts will also see each other and learn to appreciate that the common interest definition needs and can be a participatory process. The comparative perspective among at least five different parts of the city will also allow various shapes and indicators around the same idea of public spaces or public facilities as public interest.

5   Capacity Building in Neighborhood Communities with the Support and in Support of Architectural Education

The experiences presented in this paper are related to action research projects that are aimed at placing the connections between the communities and their spaces in the public agenda. These experiments speak of testing ideas and tools coming from the training programs of the built environment professionals together with the ones from other fields of study. In the long run, the expected result is a more democratic definition of the public interest in relation to public spaces or public facilities. The cooperation between researchers, students and neighborhood communities is helpful in creating the necessary frames for dialogue leading to these definitions.

Simultaneously, these projects correlate the three missions of the academia: (1) the research mission is about formulating the questions and defining the indicators for experimenting; (2) the professional training mission is learning by doing within processes that allow students (and not just them) to understand and deal with the complexity of real life situations; and (3) the third mission, that has became more and more important in various public policies at national and local level, is here about engaging the teaching staff, students, and researchers, with societal issues of the democratic processes of the transformations of spaces at the neighbourhood scale. Collaborating with grassroots initiatives offers a mutually beneficial opportunity for both students and community representatives and stakeholders.

For students in architecture, the conventional academic environment often focuses on discussing outcomes and analyzing design proposals as if they were destined for actual construction. However, While students may grow well-versed in the typical format of architectural proposals and urban regulations, they lack exposure to the practical processes behind these outcomes. The live project formats open spaces for engaging with diverse stakeholders, and offer valuable experience in problem-solving, teamwork, and communication, contributing to projects with societal impact. By actively participating in how civic initiatives define expectations for public spaces, students gain insight into formulating briefs and understanding underlying agendas, enhancing their ability to create informed proposals.

Conversely, students can aid civic initiatives by assisting them in articulating their expectations, by posing pertinent questions, and by leveraging professional tools to create formats improving communication between grassroots initiatives and decision-making processes. Storytelling and visualization skills can offer innovative perspectives without immediately delving into technical solutions, and theoretical knowledge can enhance local self-organization or hands-on activities. Students are in a unique position to catalyze pertinent territories for stakeholder interaction and exchange, which is essential for collaborative approaches to urban challenges.

Our neighborhoods serve as the narrative backdrop of our lives, embodying the stories and experiences of those who inhabit them. Ensuring that public funding aligns with public interest is paramount, and one approach we employ to achieve this alignment is through collaborative action research. Through this method, we engage directly with community members to understand their needs, preferences, and priorities. By involving residents in advocating for more open decision-making processes regarding the allocation of public funds, we strive to ensure that these resources are directed toward initiatives that address genuine community needs and aspirations.

References

1. Awan, N., Schneider,T., Till, J.: Spatial agency: other ways of doing architecture. 1st edition. Routledge, ebook (2011)
2. Spatial Agency Homepage, https://www.spatialagency.net/ , last accessed 2023/11/4
3. Rollot, A.M. (Dir.) L'hypothèse collaborative: conversation avec les collectifs d'architectes français. Hyperville, 978–2–9552985–7–2. Ffhal-01819337f (2018)
4. Peris-Ortiz, M., Ferreira, J., Farinha, L., Fernandes, N.: Multiple helix ecosystems for sustainable competitiveness, Springer International Publishing (2016)
5. Amaral, M.: Management and assessment of innovation environments. Triple Helix **2**, 19 (2015)
6. Carayannis, E.G., et al.: The Quintuple Helix innovation model: global warming as a challenge and driver for innovation. J. Innov. Entrepreneur. **1**, 2 (2012)
7. ATU Homepage, https://atu.org.ro/ , last accessed 2024/02/11

8. Compagnucci, L., Spigarelli, F.: The Third Mission of the university: a systematic literature review on potentials and constraints. Technol. Forecast. Soc. Chang. **161**, 120284 (2020). https://doi.org/10.1016/j.techfore.2020.120284

9. URBOTECA Homepage, http://urboteca.ro/urboteca-fellowship-programme-2018/, last accessed 2024/02/11

10. Urban Education Live Homepage, https://urbedu.live/, last accessed 2024/02/11

11. Anderson, L.W., Krathwohl, D.R.: A taxonomy for learning, teaching, and assessing: A Revision of Bloom's Taxonomy of Educational Objectives. Longman, New York (2001)

12. Raderstrong, J., Boyea-Robinson, T.: The why and how of working with communities through collective impact. Community Dev. **47**(2), 181–193 (2016). https://doi.org/10.1080/15575330.2015.1130072

13. Marin, V., Calciu, D., Bădescu, G., Dumitru, A., Mocanu, R., Community based development plan for the Bucurestii Noi Urban Area (available at http://atu.org.ro/wp-content/uploads/2021/04/ENG-Copy-of-PLAN-COMUNITAR-_compressed.pdf) (2019)

14. JPICoNECT Homepage, https://www.jpiconect.eu/ , last accessed 2024/02/11

Research by Design at the Crossroads of Architecture and Visual Arts: Exploring the Epistemological Reconfigurations

Marianna Charitonidou[(✉)] [iD]

Athens School of Fine Arts, Athens, Greece
m.charitonidou@icloud.com

Abstract. The paper aims to explore the potential of research by design in architecture and visual arts. The main objective of the paper is to analyze the different models and epistemological positions advanced in the academic milieus as far as doctoral research by design is concerned, to explore the differences and similarities between research by design in the field of architecture and in the field of visual arts. Even though the research by design in the fields of architecture and visual arts is focused on the production of knowledge through visual, associative, and experimental research, the models of research by design in visual arts differ from those prevailing in architectural programs. The paper focuses on a corpus of models of doctoral research by design carried out within different university programs in schools of architecture and visual arts, and various multi-university and supra-disciplinary programs (architecture and visual arts) which share the intention to develop their research projects through creative practice. Two questions to which this paper aims to respond are the following: Under which circumstances and conditions a work in the field of architecture (drawn, graphic, or con-structed) or its formulation (verbal or written) can be recognized as active in the constitution of knowledge? What are the modes of inquiry specific to architectural and artistic practices that make it possible to build knowledge that can be transmitted and shared?

Keywords: research by design · epistemologies of architecture · architectural pedagogies · design thinking · architecturology · visual arts · advancement of knowledge · doctoral research in architecture · art-based research · art-informed research · practice-based research · heuristics

1 Introduction

While the project is placed at the center of teaching in schools of architecture, the recognition of research by design in architecture as an academic and scientific outcome remains a controversial topic. The possibility of considering a design project a research activity at the doctoral level implies conceiving research by design as a process of exploratory, speculative, or even experimental investigation, whose object is the creation of knowledge. Two specificities of the knowledge acquired through the research by

© The Author(s) 2025
M. Barosio et al. (Eds.): EAAE AC 2023, SSDI 47, pp. 219–231, 2025.
https://doi.org/10.1007/978-3-031-71959-2_25

design in architecture are the exploration of the status of architectural drawings, on the one hand, and the reinvention of ways of establishing new relationships between the subject who produces the architectural artifacts and the subject who interprets them, on the other hand [1]. A point of departure for this paper is the recognition that these questions, which lie at the heart of any architectural practice, should be addressed by the doctoral programs of architecture schools.

Even though project-based research in the fields of architecture and visual arts share an orientation towards the production of knowledge through visual, associative, and experimental research, the practice in the visual and plastic arts is on different levels from those prevailing in architecture programs. The main objective of the paper is to question what can be the criteria and conditions for the architectural project to be accepted as a tool of thought and production of knowledge capable of contributing to the history and theory of architecture [2].

2   On the Epistemological Status of Research by Design in Architecture

The particular relationship of architecture to knowledge thanks to its positioning in the intertwining of action and knowledge is a variable that should be taken into consideration in architectural education. A series of questions that this paper aims to clarify are the following: At what point can an architectural work (drawn, graphic, constructed) or its formulation (verbal, written) be recognized as active in the constitution of knowledge? What are the modes of inquiry specific to architectural and artistic practices that make it possible to build knowledge that can be transmitted and shared? What forms can research by design should take to meet the essential criterion of the evaluation of doctoral research, namely a contribution to the advancement of knowledge? [3, 4].

The topicality of these questions is indicated by the proliferation of scientific publications and symposia around doctoral research by design. Three key moments of research through the project can help to better define the purposes related to the reflection on the epistemological status of the project in architecture. The first is the "Conference on Design Methods", which was held at Imperial College in London in London in 1962 [5]. The full title of the aforementioned conference was "The Conference on Systematic and Intuitive Methods in Engineering, Industrial Design, Architecture and Communications". It is generally is considered as the event through which the idea of design as a method of investigation was made possible. The second episode is *Exodus, or the Voluntary Prisoners of Architecture*, which was a collaborative project between Elia Zenghelis, Madelon Vriesendorp, Zoe Zenghelis and Rem Koolhaas, was presented by the latter as his thesis at the Architectural Association in London in 1972 [6, 7] (Fig. 1). The third case considered as a key moment is the "Science: Method Conference" of the Design Research Society, which took place in 1980 and revolved around the desire to overcome simplistic comparisons between science and design by problematizing their epistemological relationships [8].

A key distinction concerning the debates around research by design in architecture, is the distinction of how design is understood in academia and practice. This distinction is analyzed by Ayşe Zeynep Aydemir and Sam Jacoby, who underscore that "in academia the focus is on conceptualising a problem [, while] in practice the purpose of design tends to be more immediate in finding tangible solutions to concrete design problems" [9, p. 661]. Zeynep Aydemir and Jacoby draw a distinction between process-driven research in architecture and output-driven research in architecture, and they argue that "[p]rocess-driven research tends to have planned and cyclical and output-driven research iterative and emerging research processes" [9, p. 669].

Fig. 1. Rem Koolhaas, Zoe Zenghelis, Elia Zenghelis, Madelon Vriesendorp, Photo-collage for Exodus, or The Voluntary Prisoners of Architecture, 1972. Final project, AA School of Architecture, London, 1972. Medium: Pen, ink photo-collage in color and black and white, on silver backing. Dimensions: 295 × 418 mm. Exodus started as an answer to a competition by *Casabella* in 1972, on the theme of "the city as meaningful environment", for which the Berlin Wall is taken as model. Image courtesy of Drawing Matter. Collection No: 3151.5. Provenance: Zoe Zenghelis.

3 The State of Arts Around Doctoral Research by Design in the US, Australia, UK, Switzerland and Italy

Within the current context of Architecture Schools in Europe, there is an intensification of interest in research by design in architecture, which revolves, mainly, around the following institutional structures (networks, associations, etc.): the European Association for Architectural Education (EAAE), the Architectural Research European Network Association (ARENA), and the European League of Institutes of the Arts (ELIA). A scientific journal, that focuses on the publication of research cases by the project in architecture is *ARENA journal of architectural research*. Of pivotal importance for better grasping the epistemological debates around the status of research by design in architecture is EAAE Charter for Architectural Research, while for understanding the disciplinary questions related to research by design in the arts is the Vienna Declaration on Artistic Research. The EAAE Charter on Architectural Research is intended as a reference document for the use of universities, architecture schools, research institutions, funding agencies, professional bodies and architectural practices that are undertaking architectural research. It specifies the character and objectives of architectural research, confirms the variety of valid methodologies and supports the development of a vibrant, internationally recognized and well-funded research community. The Vienna Declaration, which was co-authored by the European Association of Conservatoires (AEC), CILECT/GEECT (the International Association of Film and Television Schools), Culture Action Europe (CAE), Cumulus, the European Association for Architectural Education (EAAE), the European League for Institutes of the Arts (ELIA), the European Platform for Artistic Research in Music (EPARM), EQ-ARTS, MusiQuE and the Society for Artistic Research (SAR), is the first outcome of a continuing collaboration between organizations and transnational networks dealing with Artistic Research at a European level and beyond [10]. Another important document for exploring the questions raised in this paper is the RIBA report *How Architects Use Research*, which was published in 2014 [11]. The Design Research Society Conference and the *Instant Journal* that brings together the papers presented at the aforementioned conference are also important catalysts for disseminating practice-based research and for researching how this mode of research can shape the relationship between different social, economic, and political actors [13].

Within the American and Australian contexts, the PhD by design in architecture is becoming more and more widespread [14]. At the same time, on an international scale, practice-oriented doctoral programs are developing in the disciplines of music, visual arts, and plastic arts, thus far exceeding those devoted to architecture [15–17]. Concerning the American context, I could mention the Doctor by Design Program (DDes) of the Graduate School of Design of Harvard University, which has already existed since 1987 and revolves around the reflection of architects on the specificity of their architectural practice [18]. Peter Rowe delivered a lecture during the DDes 30th Anniversary Program in 2017, which shed light on the endeavor of DDes to advance multi-scalar and trans-disciplinary design knowledge while addressing crucial societal issues in our increasingly complex and challenging world.

In the UK, there is the PhD Architectural Design Program of Bartlett School of Architecture in London, which was founded in 1995. Within the British context, two models can be distinguished. A first model that is characterized by the intention to propose new ways of conceiving the spatial experience and aim at the redefinition of the encounter between the architect-designer and the inhabitant-user. This model is at the core of the PhD Architectural Design Program of Bartlett School of Architecture, while the second model mentioned above was adopted by the University of Sheffield. At the center of this program is the emphasis on the creative interdependence of drawing, writing, and building in the development of innovative practices and theories of architecture. Penelope Haralambidou's PhD thesis entitled *The Blossoming of Perspective: An Investigation of Spatial Representation*, which was defended in 2003, lies between architectural design, art practice, art history, and critical theory and used drawing as a critical method [19]

(Fig. 2). Haralambidou's PhD thesis was reworked and published as *Marcel Duchamp and the Architecture of Desire* in 2013 [20]. Yeoryia Manolopoulou's PhD thesis entitled *Drawing on Chance: Indeterminacy, Perception and Design* [21], which was also defended in 2003 at UCL, "investigates how our perceptual and aesthetic habits are altered by chance events of accidents and asks whether architecture might employ them as creative devices, in particular Duchamp's *The Bride Stripped Bare by Her Bachelors, Even*, and the actual production of design work" [22]. Manolopoulou's PhD thesis was reworked and published as *Architectures of Chance* in 2013 [23].

The second model of PhD by Design in Architecture that is encountered in the UK places particular emphasis on the conception of architectural and urban practices as modes of social engagement. This model has been developed at the University of Sheffield by researchers such as Doina Petrescu and Kim Trogal [12]. The PhD by Design Conference devoted to the theme "Idea of 'Self' in Practice-based Research", which was held at the University of Sheffield between 3 and 4 April 2017, intended to engage with narratives of 'self', and to explore how the notion of self as a researcher can be assumed and embodied in research by design [13]. Moreover, the School of Architecture and Landscape Architecture (ESALA) of the University of Edinburgh offers A PhD in Architecture by Design program.

In Switzerland, the École Polytechnique Fédérale de Lausanne (EPFL) currently offers a doctoral degree explicitly based on project-based research. At EPFL, "the 'Research in/by Design' axis addresses the project both as a research topic - in its conceptual, cultural and technical foundations - and as a form of research in its whole" [24]. Between 2017 and 2021, Philip Ursprung from the Swiss Federal Institute of Technology Zurich (ETH Zürich) led a research project entitled "Design Research in Architecture"

Fig. 2. 'Illuminated Scribism', plate 19 from Haralambidou P (2003) The Blossoming of Perspective: An Investigation of Spatial Representation, PhD Thesis, UCL, London. This PhD thesis was reworked into the book *Marcel Duchamp and the Architecture of Desire* published in 2013.

[25]. This project aimed to analyze the practice of architectural research as conducted by the faculty at six selected Departments of Architecture in Scandinavia, Germany and Australia. It placed particular emphasis on how architectural design is taught in the different Departments of Architecture under study. Its main objective was to compare the methods of design research in architecture within these different contexts [26, 27]. The aforementioned project was based on a collaboration between the Department of Architecture of ETH Zurich and the Chair for History of the Modern World of ETH Zurich. In Italy, several doctoral programs are formed with explicit reference to the project. In other words, the doctorate in architectural and urban composition is quite widespread within the Italian context.

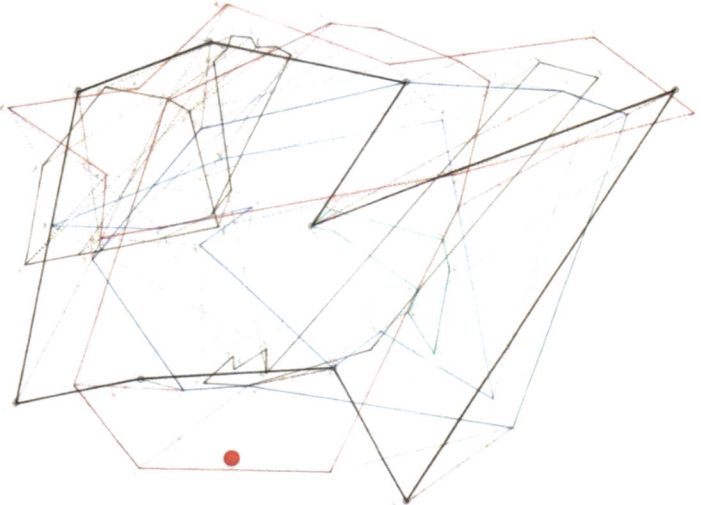

Fig. 3. Drawing from Manolopoulou Y (2003) Drawing on Chance: Indeterminacy, Perception and Design. PhD Thesis, UCL, London. This PhD thesis was reworked into the book *Architectures of Chance*, published in 2013. In this drawing Manolopoulou used 'measurable chance' to mark, multiply and connect 9 × 9 points in a drawing experiment that tries out Duchamp's concept of 'demultiplied vision'.

4   Around Different Themes, Methods and Orientations of Research by Design in Architecture

In the studies already carried out around research by design in architecture, we can distinguish various themes and orientations [28]. A first model is that of doctoral theses who are interested in the invention of new tools of architectural representation. A second model focuses on exploring the mechanisms by which architecture can function as a form of social engagement [29–31], and participate in the reinvention of the status of inhabitants [32–34]. A third model focuses on the design and development of computational and algorithmic tools [35]. A fourth model concerns doctoral theses whose

Fig. 4. Drawing from Manolopoulou Y (2003) Drawing on Chance: Indeterminacy, Perception and Design. PhD Thesis, UCL, London. This PhD thesis was reworked into the book *Architectures of Chance*, published in 2013.

reflection develops through the staging of interactive environments and experimentation with various ways of creating such environments [36]. A fifth model aims to establish innovation strategies in the field of ecological and bioclimatic design, and more broadly sustainable development at the scale of architecture and the city [37]. A final model of doctoral research by design in architecture seeks to approach urbanism and architecture in their interconnections and deals with problems related to the phenomenon of urbanization through the project [38].

An important distinction is that between research on the architectural project and that by the architectural project. To the first type belongs the book *Design Thinking* by Peter Rowe, published in 1987 [39, 40], and the research on "architecturology" by Philippe Boudon, which is exemplified in his book entitled *Introduction à l'architecturologie*, published in 1992 [41, 42], as well as Donald Schön's critique of the notion of "Design Science" in *The Reflective Practitioner: How Professionals Think in Action* published in 1983 [43]. To the second type, we could categorize Jonathan Hill's PhD thesis entitled *Creative Users, Illegal Architects* [32] which was among the first PhD theses by design at the Bartlett School of Architecture. It was defended in 2000 and was published as *Actions of Architecture: Architects and Creative Users* in 2003 [33]. Hill argues that the definition of "disegno" that presides over the practice of the project involves "both drawing… on paper and the development of an idea" [44]. If the principle of the traditional thesis is established on the dissertation and the argument, the doctorate by design raises the question of the modalities according to which the project can function as an argument, since a project is not an argument a priori [45].

The studies already carried out around research by design within the field of visual arts are much broader and more diverse than those around research by design in architecture. An important scientific journal that brings together different approaches to research by design within the field of visual arts and intends to promote a transdisciplinary approach to visual arts is *Visual Arts Research*, founded in 1982. This journal aims to provide "a forum for historical, critical, cultural, psychological, educational and conceptual research in visual arts and aesthetic education" [46]. A key moment for the research by design within the field of visual arts is "The Penn State Seminar in Art Education", held in 1965 [47]. This seminar brought together artists, art historians, critics, arts educators, curriculum specialists, psychologists, and sociologists, and aimed at exploring ways of transforming arts education. It contributed considerably to the intensification of research by design in arts of the interest in interdisciplinarity [48]. A distinction that is at the core of the research by design in visual arts is that between "art-based research", "art-informed research", and "practice-based research" [49]. These terms are related to different methods of conducting research. A core aspect of the research by design is the interaction between thought, performativity, and composition. The exchange between these three parameters is described as the triangle "thinking-performing-composing" [50].

5   Conclusion

Architecture schools should consider different types of possible interactions between doctoral research in art and architecture, and propose new modes of development for transdisciplinary doctoral programs that link fields characterized by different visual and spatial epistemological statuses. Taking as their starting point the different epistemological positions and institutional postures adopted by the different institutional and

geographical contexts, architecture schools should try to shape new methods of addressing questions related to the differences and/or affinities between the doctorate by design
in architecture and the doctorate by design in visual arts. Useful for this exploration
of new methods of research by design in architecture is the notion of heuristics, which
as Amy Kulper and Sheila Crane highly, can lay "the ground for the future possibility
of spatial discoveries" [51]. Architecture schools should also try to shed light on the
importance of the role of interdisciplinarity for research through the architecture project
and stimulate exchanges between the epistemological tools of the visual arts and those
of architecture. Both holism and interdisciplinarity lied at the heart of Constantinos A.
Doxiadis's approach to the understanding of what he called Ekistcis. Doxiadis drew
a distinction between interdisciplinary and a condisciplinary science. In "Ekistics, the
Science of Human Settlements", published in *Science* in 1970, Doxiadis highlights: "To
achieve the needed knowledge and develop the science of human settlements we must
move from an interdisciplinary to a condisciplinary science" [52].

Although doctoral projects by design in architecture are increasing thanks to the
intensification of debates around the epistemological issues of research by the architectural project, their legitimization in the context of the current university system is still
not sufficiently institutionalized in several cases. In contrast, in the field of visual arts,
project-based research is, in its current state, much better legitimized and institutionalized. It is pivotal for architecture schools to shed light on inter- and transdisciplinary
combinations between the models of the PhD by design in the field of architecture, and
the PhD by design in the field of visual arts. In order to do so it is important to explore in
a transversal way in different institutional contexts how the intention to produce research
by design at the doctoral level is treated. This would help to better grasp the ongoing
epistemological debates, and to shed light on the diversity characterizing the epistemological status of visual and spatial production [53], on the one hand, and to experiment
on the disciplinary boundaries of doctoral research in architecture, on the other hand.

An aspect that is of great significance for reflecting upon how doctoral research
in architecture can challenge the boundaries of architectural epistemology is the idea
that there are project-based research methods that are irreducible to textual language
[54]. Another direction that would be fertile for enhancing interdisciplinarity in doctoral
research by design is the exploration of multiple types of interaction between doctoral
research by design in visual arts and doctoral research by design architecture, and to
propose new modes of development for transdisciplinary doctoral programs that link
fields characterized by different visual and spatial epistemological statuses. The epistemological status of the architectural project is at once speculative, critical, and pragmatic.
To preserve the specificity of architecture without falling into disciplinary solipsism, it
is necessary to explore how doctoral research by practice is addressed in other disciplines. A characteristic of the disciplines that support the PhD through the project is the
understanding of knowledge production as intrinsically linked to a concrete efficiency,
specific to the discipline concerned. Such an approach is more dominant in disciplines
like biotechnology or materials science in which knowledge production is not characterized by such a critical dimension. Within the framework of this endeavor to explore new
directions of doctoral research by design in architecture, we should bear in mind that the

production of knowledge in architecture is intrinsically linked to an experimentation on critical disciplinary concerns.

Acknowledgements. The research project was supported by the Hellenic Foundation for Research and Innovation (H.F.R.I.) under the "3rd Call for H.F.R.I. Research Projects to support Post-Doctoral Researchers" (Project Number: 7833). I am grateful to the Hellenic Foundation for Research and Innovation (H.F.R.I.) for its support for my participation in the European Association for Architectural Education (EAAE) 2023 Annual conference meeting in Politecnico di Torino, to Drawing Matter and Niall Hobhouse (Drawing Matter) for providing authorisation to use Fig. 1 in this paper, to Prof Penelope Haralambidou (Bartlett School of Architecture, UCL) for providing authorisation to use Fig. 2 in this paper, and to Prof Yeoryia Manolopoulou (Bartlett School of Architecture, UCL) for providing authorisation to use Figs. 3 and 4 in this paper.

References

1. Charitonidou, M.: Architectural Drawings as Investigating Devices: Architecture's Changing Scope in the 20th Century. Routledge, London, New York (2023). https://doi.org/10.4324/9781003372080
2. Cohen, J.-L.: L'Architecture entre pratique et connaissance scientifique. Editions du patrimoine, Paris (2018)
3. Schatzki, T., Knorr Cetina, K., Von Savigny, E. (eds.): The Practice Turn in Contemporary Theory. Routledge, London (2001)
4. Cotte, L.: Reclaiming Artistic Research. Hatje Cantz, Berlin (2019)
5. Jones, C., Thornley, D.J. (eds.): Conference on Design Methods. Pergamon, Oxford (1963)
6. Koolhaas, R., Zenghelis. E., Vriesendorp, M., Zenghelis, Z.: Exodus, or the Voluntary Prisoners of Architecture. Architectural Association Thesis, London (1972)
7. Charitonidou, M.: Rem Koolhaas and the congestion of metropolis How the artificial would replace the reality? In: Charitonidou, M. (ed.) Architectural Drawings as Investigating Devices Architecture's Changing Scope in the 20th Century, pp. 236–250. Routledge, London, New York (2023). https://doi.org/10.4324/9781003372080
8. Jacques, R., Powell, J. (eds.): Design: Science: Method. Westbury House, Guildford (1981)
9. Zeynep Aydemir, A., Jacoby, S.: Architectural design research: drivers of practice. Des. J. **25**(4), 657–674 (2022). https://doi.org/10.1080/14606925.2022.2081303
10. ELIA (2020) The Vienna Declaration on Artistic Research. https://www.e-flux.com/announcements/338016/the-vienna-declaration-on-artistic-research/. Accessed 20 February 2024
11. RIBA (Royal Institute of British Architects). 2014. "How Architects Use Research: Case Studies from Practice." Accessed 7 December 2020. https://www.architecture.com/-/media/GatherContent/How-Architects-Use-Research/Additional-Documents/HowArchitectsUseResearch2014pdf
12. Petrescu, D., Trogal, K. (eds.) : The Social (Re)Production of Architecture Politics, Values and Actions in Contemporary Practice. Routledge, New York, London (2017)
13. Robinson, S.: Enactive interaction and performer training without a tutor: reflection from the workshop. Instant J. **1**(4), 32–33 (2017)

14. Van Schaik, L. In the medium itself: Three decades of design practice research at Royal Melbourne Institute of Technology. In: Goffi, F. (ed.) InterVIEWS Insights and Introspection on Doctoral Research in Architecture. Routledge, London, New York (2019)
15. Nilsson, F., Dunin-Woyseth, H., Janssens, N., (eds.): Perspectives on research assessment in architecture, music and the arts. Discussing Doctorateness, pp. 69–84. Routledge, London, New York (2017)
16. Sullivan, G.: Art Practice as Research: Inquiry in Visual Arts. Sage Publications, Thousand Oaks, CA (2005)
17. Frayling, C.: Research in art and design. Royal College Art Res. Papers **1**(1), 1–5 (1993)
18. Bechthold, M.: Design research at Harvard Graduate School of Design. In: Goffi, F. (ed.) InterVIEWS Insights and Introspection on Doctoral Research in Architecture. Routledge, London, New York (2019)
19. Haralambidou, P.: The blossoming of perspective: an investigation of spatial representation, PhD Thesis, UCL, London (2003)
20. Haralambidou, P.: Marcel Duchamp and the Architecture of Desire. Routledge, London, New York (2013)
21. Manolopoulou, Y.: Drawing on chance: indeterminacy, perception and design. PhD Thesis, UCL, London (2003)
22. Allen, L., Borden, I., Stevenson, R.: Bartlett School of Architecture UCL, Exhibition Catalogue 2004. Dexter Graphics, London (2004)
23. Manolopoulou, Y.: Architectures of Chance. Routledge, London, New York (2013)
24. EDAR Affiliated labs & Research expertise, https://www.epfl.ch/education/phd/edar-architect ure-and-sciences-of-the-city/edar-affiliated-labs-research-expertise/. (accessed 3 November 2023)
25. Ursprung, P.: Design Research in Architecture, https://data.snf.ch/grants/grant/172843, (accessed 3 November 2023)
26. Silberberger, J.: Architecture schools and their relationship with research: it's complicated architecture schools and their relationship with research: it's complicated. Dimensions **1**(1), 77–84 (2021)
27. Silberberger, J. (ed.): Against and for Method: Revisiting Architectural Design as Research. GTA Verlag, Zurich (2021)
28. Charitonidou, M.: Repenser la recherche doctorale par le projet architectural: les reconfigurations épistémologiques au croisement de l'architecture et des arts visuels. ETH Zurich, Institute for the History and Theory of Architecture (GTA), Zurich (2020). https://doi.org/10.3929/ethz-b-000432256
29. Awan, N., Schneider, T., Till, J. (eds.): Spatial Agency: Other Ways of Doing Architecture. Routledge, London, New York (2011)
30. Jenkins, P., Forsyth, L., Smith, H.: Research in UK architecture schools: an institutional perspective. ARQ Archit. Res. Quart. **9**(1), 33–43 (2005)
31. Till, J.: What is Architectural Research. Architectural Research: Three Myths and One Model. Royal Institute of British Architects, London (2005)
32. Hill, J.: Creative Users, Illegal Architects, PhD Thesis, UCL, London (2000)
33. Hill, J.: Actions of Architecture: Architects and Creative Users. Routledge, London, New York (2003)
34. Hill, J.: Design Research: An Eye on the Past and the Future Book. In: Goffi, F. (ed.) InterVIEWS Insights and Introspection on Doctoral Research in Architecture, Routledge, London, New York (2019)
35. Steadman, P.: An 'Artificial Science' of Architecture. In: Fraser, M. (ed.) Design Research in Architecture: An Overview, pp. 35–52. Ashgate, Farnham, Surrey (2013)
36. Green, K.E.: Architectural Robotics: Ecosystems of Bits, Bytes, and Biology. The MIT Press, Cambridge, Mass (2016)

37. Pickett, S.T., Cadenasso, M.L., Mcgrath, B.: Resilience in Ecology and Urban Design: Linking Theory and Practice for Sustainable Cities. Springer, Dordrecht (2013)
38. Aureli, P.V.: The City as a Project. Ruby Press, Berlin (2016)
39. Rowe, P.: Design Thinking. The MIT Press, Cambridge, Mass (1987)
40. Rowe, P., Chung, Y.: Design Thinking and Storytelling in Architecture. Birkhäuser, Basel (2024)
41. Boudon, P.: Introduction à l'Architecturologie. Dunod, Paris (1992)
42. Boudon, P.: The point of view of measurement in architectural conception: from the question of scale to scale as question. Nordic J. Archit. Res. **12**(1), 7–18 (1999)
43. Schön, D.: The Reflective Practitioner: How Professionals Think in Action. Basic Books, London (1983)
44. Hill, J.: Drawing forth immaterial architecture. ARQ Archit. Res. Quart. **10**(1), 51–55 (2006). https://doi.org/10.1017/S135913550600011X
45. Hanrot, S.: À la recherche de l'architecture : Essai d'épistémologie de la discipline et de la recherche architecturales. Editions L'Harmattan, Paris (2002)
46. GWH, TZ: Editorial. Visual Arts Research 8(2) (1982). http://www.jstor.org/stable/20715515 (accessed 11 November 2023)
47. Efland, A.D.: Curriculum concepts of the Penn state seminar: an evaluation in retrospect. Stud. Art Educ. **25**(4), 205–211 (1984). https://doi.org/10.2307/1320413
48. Hamblen, K.A.: Developing a theory of historical self-criticality based on the 1965 Penn State conference and subsequent events. In: Anderson, A.A., Bolin, Jr. & P.E. (eds.) History of Art Education: Proceedings of the Third Penn State International Symposium, pp. 525–532. Art Education Program, School of Visual Arts and the College of Arts and Architecture of The Pennsylvania State University, University Park, PA (1997)
49. Sullivan, G.: Research acts in art practice », studies in art education. J. Issues Res. **48**(1), 19–35 (2006). https://doi.org/10.1080/00393541.2006.11650497
50. Dyrssen, C.: Navigating in Heterogeneity: architectural thinking and art-based research. In: Biggs, M., Karlsson, H. (eds.) The Routledge Companion to Research in the Arts, pp. 223–239. Routledge, London, New York (2011)
51. Kulper, A., Sheila, C.: Designing agency: the new heuristics. J. Archit. Edu. **68**(1), 2–4 (2014). https://doi.org/10.1080/10464883.2014.870397
52. Doxiadis, C.A.: "Ekistics, the Science of Human Settlements", Science 393–404 (1970). https://www.jstor.org/stable/1729412 (1970)
53. Klinke, H. (ed.): Art Theory as Visual Epistemology. Cambridge Scholars Publishing, Newcastle upon Tyne (2014)
54. Büchler, D., Gabriela, A., Lima, G.: Drawing about images: textual and non-textual interpretation. Working Papers in Art and Design 5. https://researchprofiles.herts.ac.uk/en/public ations/drawing-about-images-textual-and-non-textual-interpretation (accessed 11 November 2023) (2008)

Topological Deformability in Architecture, or How to Learn About Differences

Maja Dragišić[(✉)]

Faculty of Architecture, University of Belgrade, Belgrade, Serbia
majchilo@gmail.com

Abstract. The main question of this paper is related to whether and in what way we understand the notion of topological deformability, which appears in numerous texts of architectural theory from the beginning of the 1990s until today. In general, the paper deals with the problem of the denotation of concepts from other scientific fields, and shows how the architectural discourse changes the meaning of a mathematical concept, determining it within its own discipline. The transition of the term topological deformability is set as an example of how architecture uses and improves its inherent interdisciplinary dimension, as well as an example of how the way we research in architecture could make a tangible social impact.

The first part of this paper will focus on introducing and analyzing the concept of topological deformability and its transition from mathematics to architectural theory.

In the second part, it will examine how, through different theoretical approaches, the term topological deformability influences the transformation of the thinking modality in architecture, and how it traces the narrow path through architectures' knowledge toward the wider audience. Therefore, architecture's dominant as an instrument of plural social reality will be emphasized. The research will show how the term topological deformability through architecture opened the way to changing the idea of otherness, leading to the essential acceptance of differences.

Keywords: architectural topology · deformability · plural society · otherness · difference

1 Topology

The analysis of topology as a contemporary mathematical discipline requires a transition from the term place to the term space because mathematics does not recognize places with their contextual particularities but examines and describes abstract mathematical spaces and everything they comprise. The relevant literature in the field of mathematical topology explains that, in general, topology studies the properties of geometric shapes that are preserved under continuous deformation, such as connectedness or compactness, i.e., mathematical topology makes no distinction between two shapes or two spaces if it is possible to shift from one to other under continuous deformation. When it comes to these spaces, it is irrelevant whether something is large or small, round or square, if it can be changed by stretching or bending, for example. The difference between the two spaces is primarily related to the components that remain unchanged under the deformation [1].

© The Author(s) 2025
M. Barosio et al. (Eds.): EAAE AC 2023, SSDI 47, pp. 232–241, 2025.
https://doi.org/10.1007/978-3-031-71959-2_26

Sergei Petrovich Novikov emphasizes that it is even intuitively clear that knowledge about the geometric properties of shapes does not end with data about their metric characteristics, such as length, height, angles, etc., i.e., that "there remains something beyond the limits of the old geometry" [2]. Regardless of the length, the line can be open, closed, tied in a knot, several lines can be chained in different ways, volumes can have holes, etc. These and similar properties of geometric shapes, but also of various mathematical objects that do not have geometric realizations, are characterized by the fact that they do not change during deformations without interruption. Some typical examples of topological spaces are the *Moebius strip, Klein's bottle, tori, various knots*, and similar objects.

During the 19th century topology was developed by several mathematicians, among others, Karl Friedrich Gauss, Bernhard Riemann, etc. but it is considered that topology as an autonomous branch of mathematics was founded by Henri Poincaré at the end of the 19th century. In the following decades, its internal problems were solved, so that in the second half of the 20th century there would be a more serious breakthrough of topological methods into modern physics and chemistry, but also a more general interpretation of topology through the discourses of the social and humanistic sciences.

It is certain that the mathematical definition of topology, when separated from the main field of research, is difficult to understand and cannot represent a basis for further analysis of the appearance of topology in the architectural discourse. Partly it can be explained by the fact that it is a scientific field that requires more complete and greater mathematical knowledge, the subject of research is far from the perceptible world, and therefore it is difficult to explore its visuality. In odred to comprehend the evolution of topology in architectural discourse, one must look deeper into the history of science, especially mathematics.

2 Toward Architectural Discourse

Morris Kline indicates that the first thoughts on topology can be found in the works of Gottfried Wilhelm Leibniz, in the book "Characteristica geometrica" from 1679, in which Leibniz introduces the term Analysis situs /position/, in order to opposite size and form, emphasizing the lack of an adequate term when talking about form [3]. Also, in the letter to Christiaan Huygens, Leibniz points out that: "We need another strictly geometrical analysis which can directly express *situm* in the way algebra expresses the Latin *magnitudem*" [4]. In order to understand his idea to differentiate the properties of geometric shapes by position and by measurement, it is important to take into consideration the fact that at the same time, Leibniz worked on the invention of calculus. It is also known that, as a branch of modern mathematics, topology initially arose from the study of geometric problems, but its methods are based on Georg Cantor's theory of sets as well as on modern algebra. The roots of topological phenomena can be found in Euler's work on seven Königsberg bridges from 1736, but the first truly modern fundamental concepts of topology were given by Henri Poincaré in 1895 in one of the most significant classical works of mathematics - "Analysis situs" [5].

As more important for this research, it is the draft overview of relevant literature of the history of mathematics that shows that the development of topology runs parallel to the achievements that will distance mathematics from the sensory world. Between 1830 and 1850, Nikolai Ivanovich Lobachevsky and János Bolyai published the first model of non-Euclidean geometry based on the understanding that Euclid's fifth postulate is independent and that there can be logically non contradictory geometries containing an opposite one [6]. In the 19th century, Bernhard Riemann developed another kind of geometry based on the generalization of Gauss's concept of "curvature". He also stated that information about points in space need not necessarily be obtained using the coordinate system, the ultimate transcendental space of the Cartesian system, but that it is possible to determine for each point its local properties contained in the space itself. Riemann thus clarified that mathematical objects can be released of the external reference system, i.e., they can be defined as fields of local information. For the broader interpretation of observed reality, a proof such as Beltrami-Klein's from 1868, which equalised two geometries, one that belong to the real world of human perception and one that does not, meant the absolute relativization of reality as people understood it so far.

Mathematics philosopher Stephen Francis Barker points out that when we talk about the curvature of space, we must not assume or imagine a visual representation of curved space. Although separated from observable reality, the consequences of discovering these geometries were fundamental to the epistemological status of mathematics and for its wider intellectual influence [7]. In general, the development of topology, along with other mathematical achievements from the same period, indicated that the prevailing philosophical platform of Immanuel Kant was being undermined, in which mathematics had a special status as the essence of all natural sciences but had to be applied. Zvonimir Šikić, in his book on the new philosophy of mathematics, discusses the problem of the relationship between the abstractness of mathematics and the reality of nature, i.e., the applicability of contemporary mathematics in reality, and emphasizes that the culmination of this concept can be found in the philosophical platform of Immanuel Kant at the end of the 18th century. It is based on the idea that abstract mathematics is always directed to the description of nature because mathematical knowledge is specific as knowledge of the *a priori* forms of space and time, which are also components of reality [8]. Despite being intuitive, mathematics was still necessary for Kant to refer to the sensible world. In this context, the new way of thinking that accompanied the discoveries did not rule out the applicability of mathematics. However, the discipline was no longer prioritizing it. As a result, over time it stopped being a priority for all sciences that rely on mathematics, and ultimately for the overall understanding and perception of the world that surrounds us.

The methodology of applying mathematical concepts to a broader range of knowledge often draws on specific knowledge in various fields, and Arkady Plotnitsky defines it with the term "quasi-mathematics"[9]. Although he does not question the philosophical influence of mathematics on the development of civilization, he states that quasi-mathematics enables the dissemination of certain mathematical concepts and principles which, although originating from it, are not exclusively defined by its tools and, as such, become possible and applicable outside its disciplinary framework. With the term quasi-mathematics, Plotnicky explains the difference in the interpretation of algebra, geometry, and topology in a general sense. He interprets algebra as the ultimate concept of formalization, be it the formalization of systems in the natural sciences, conceptual systems as in logic or philosophy, or the language system that exists in linguistics. In this sense, "algebra" means a set of specific formal elements and their relations. On the other hand, "geometry" and "topology," although both deal with questions of space, are distinguished by their mathematical origins, "geometry" arises from the measurement of space as geo-metry. In contrast, "topology" ignores quantities and deals exclusively with the structure of space (*topos*) and the essence of the form of a shape.

Such reflections have shown that different transitions of concepts from mathematics to other discourses, and therefore to architectural, where possible, whether it is about exact application or flexible appropriation of notions. With the previously presented broader image of mathematics in the field of science, it becomes clear that the path from topology to architecture has become open. During the nineties of the twentieth century, this will become particularly significant in architectural theory.

3 Topological Deformability in Architectural Theory

Even though the dominant architectural style in the most of the twentieth century – Modernism was based on the standard elements of Euclidian geometry, there were examples that architects were familiar with the more organic, freely deformed architectural form, but that was never referenced in the topological terms. However, the small number of buildings and significant research work during this period indicate that architects did not have a aspiration to include topology in the dominant movements of architecture.

At the beginning of the nineties of the twentieth century, with the appearance of adequate digital tools in the architectural design process, the conditions for more extensive research of modern mathematical theories of space arrested. Thus topology has started to become an integral part of the architectural design methodology, and therefore the architectural theory. The first attempts to record and analyze the term topological deformation in the architectural theory appear in the historical and theoretical overviews of contemporary architecture using more general term, *topological architecture.*

Mario Carpo explains the new architectural avant-garde at the beginning of the new millennium, known as topological, as an architectural response to the new digital technologies that were flourishing at the time. "Topological" architecture, as it was called then, was seen for a while "as the quintessential embodiment of the new computer age - and we all remember the excitement and exuberance that surrounded all that was digital between 1996 and 2001" [10]. Branko Kolarević uses the same term while classifying digital architecture: "This new fluidity of connectivity is manifested through folding, a design strategy that departs from Euclidean geometry of discrete volumes represented in Cartesian space, and employs topological, "rubber-sheet" geometry of continuous curves and surfaces" [11].

The similar term *topological tendencies* in architecture were introduced by Guiseppa Di Christina in her doctoral dissertation "Architecture and topology: for a theory of space in Architecture" in 1999 at the Faculty of Architecture in Rome [12], where topological tendencies were explained as "the topologizing of architectural form according to dynamic and complex configurations that lead architectural design to a renewed and often spectacular plasticity, in the wake of the baroque and of organic expressionism."

Furthermore, she started defining the appearance of topology in architectural design in the domain of creating dynamic variations of form. The focus of her research is directed towards the formal vocabulary of buildings, where topological methods are primarily used to achieve the desired dynamics of the architectural form. Di Cristina also indicates a theoretical problem related to the question of to what extent the forms obtained by the dynamic process of topologizing are dynamic in the domain of architectural work. As the main protagonists of this, for her progressive tendency, she cites Peter Eisenman, Greg Lynn, Daniel Libeskind, and Bahram Shirdel, as well as the influence of the theoretical works of Bernard Cache, Jeffrey Kipnis, Brian Massumi, and other authors, crucial for the development of topological architectural forms.

As seen from the beginning, the use of the term topological deformability in architectural discourse pointed to the problem of formulating a comprehensive definition, because the interpretations were constantly shifted between the field of architectural theory of form and the field of architectural design theories. The first half of the 1990s was evidently dedicated to the "fascination with topological objects", where for example, the project for the Guggenheim Museum in Bilbao from 1997, by architect Frank Gehry, was cited as a typical example of using the "deformation made possible by flexibility of topological geometry" with "forms that bending, twisting and folding" [13]. Moreover, the term *topological architecture* [14] is mentioned in some historical reviews, even as a strategy to create the new contemporary architectural paradigm, or a new architectural style.

As architectural criticism advanced with these tendencies, concerns about the idealization of form were raised. The majority of theoreticians and authors who influenced the development of the term topological deformability in architectural theory at the end of the nineties were directly confronted with the criticism of the idealization of geometry, that is, that placed deformability as a representative of the idea of diversity is placed exclusively under the framework of phenomenology. Referring to Di Cristina's research, Michael Speaks underlines that the topological form technique, which is based on continuity and movement, is entirely negated by the finitude of the end product [15], additionally moving the problem into the domain of the experience of the architectural space. Mario Carpo sees it as a cause-and-effect relationship between digital technologies and complex geometry. He emphasizes that generalization has led to delusion and that many projects with computer-generated formal characteristics have become inconspicuous, almost banal architectural objects, and the use of digital tools, as well as the reference to topology, did not give objects validity. Antoine Picon emphasizes an additional problem arising from the topological treatment of form, which refers to the aesthetic valorization of deformed amorphous architectural forms. He sees part of the problem in the lack of an established aesthetic evaluation system for evaluating the aesthetic characteristics of new forms and another part in the process of their creation, which he underlines with the question of what in the process of form transformation determines when it will end [16]. Similar observations are made by Michael Meredith when he says that the result of using the topological method during the nineties is reflected in isolated physical and aesthetic models, which do not have broader significance but remain within their framework [17].

However, in the end of the nineties with moving away from the theory of form more towards the theory of design, some other interpretations of the term topological deformability have been developed, which will influence the architectural projects on a much deeper and more significant level. One of the basic definitions was given by the architectural theorist Kostas Terzidis in the book *Expressive Form, a conceptual approach to computational design*, introducing the term "topological operations" which includes twisting, stretching and compression of the architectural form, excluding cutting and tearing. Any type of operation that deforms the form by hollowing out, creates two topologically different entities, which leads him to the conclusion that "topology should be used in order to achieve the unity of the form, because it preserves the integrity of the endlessly transformed geometry" [18]. He implies that certain formal properties remain unchanged, even when the geometric shape undergoes intense distortions, resulting in the loss of its metrical and projective properties. Apart from mathematical precision, the great importance of Terzidis' definition lies in the clear distancing from traditional architectural methodologies that were based on addition and substitution of forms. Similar explanations of the topological method speak of a departure from the Cartesian geometric model in architecture towards a more complex, non-linear logic of space, with which it is possible to express the flexibility and continuity of an endless number of variations.

The transition from "making form" to "finding form" occures at the moment when the question of curvature is left aside, and along with the complex network of influences mentioned above, the topological deformability has become an integral part of architectural design methodology. This phase of development implied that the topological deformability should be considered as a comprehensive spatial system, where topology is understood as a flexible structure formed by specific and clear relations, which remains unchanged as a result of transformation and deformation. To design topologically, it was implied to emphasize specific relations or certain "conditions" which are key to the logic of organization, whereby geometry is flexible when it comes to dimensions, distances, or form. By the end of the nineties of the twentieth century, the topological deformability was no longer interpreted as the geometry of architectural objects, nor its prototype, but as a demonstration of certain geometrical principles. Topological thinking implies that spaces are not about a specific form but rather about relations. The authors explain this by stating that topological principles can be manifested through various forms where "the concept of continuity is obtained only by applying algorithmic logic" [19].

Over time, these types of definitions resulted in a more diverse understanding of topological deformability, leading to a more liberated and broader interpretation of the term in various contexts. It will turn out that evident heterogeneity in use, without a clear system or unique definition, has spread the term far beyond the limits of architectural discipline to many contiguous scientific fields, showing clearly the inherent architectural interdisciplinary dimension. The more the term was used in the domain of design methodology and less in the domain of form design, the easier and faster it started to increase its social impact.

4 Transformation of the Thinking Modality

As demonstrated, the outcome of the shift in the scientific paradigm from a determined and stable to temporal and complex, which resulted in a distinct comprehension of the contemporary cultural context, brought the increasing complexity in diverse domains of architectural theory and practice. Although criticized from many aspects, its formalism, lack of relevant space logic, a fixation on digitalization etc. the topological deformability as methodology appear in the architectural discourse as a response to the more comprehensive scientific and cultural context.

On the one hand, the complexity of the spatial structures encountered by the users of the built objects undoubtedly influenced their relationship to the space, as architectural form remains inseparable from the way we experience the world, which involves our senses and perceptions. The complex relationship of people to architecture is an intimate and longstanding one, and it is strongly linked to the relationship of the human body to the wider cultural context. This relationship has evolved over time to reflect the philosophical and architectural discourses that shape both.

On the other hand, with the development of computer technologies, there has been a change in our notion of materiality. Our age can be defined as a flow of information, and architecture captures this flow, creating more complex conceptions and interactions through the space. In this regard, it cannot be ignored that aspects that influenced the occurrence of topological deformability in architectural theory were strongly supported by the development of digital technologies at the end of the twentieth century. It is clear that only with the development of technological tools has architecture become interested in these types of complex spatial relations, primarily through the research of the medium itself - software. Upon examining the chronological progression from craft to engineering and to the digital design of virtual or natural spaces, it is evident that the study of medium has consistently dominanted architectural practice. However, when discussing topological deformability in architecture, is it simply dealing with the medium, or is it something else?

The overall picture of the emergence and evolution of the concept of topological deformation from mathematical abstraction to the creation of architectural space suggests that the essence of the influence originated from a change in the thinking modality. The phenomenon of topological spatial structures where the sole relevant factor is their deformability refers to the idea that it is possible to tolerate, but initially endure, the most diverse types of deformation. This research argues that topological deformability through architecture opened the way to the essential acceptance of the different, to the changing the idea of otherness. If architecture, together with natural sciences, contributes to the creation of a specific system of world perception, then the predominant role of topological deformability in architecture is to serve as an instrument of expressing plural social realities.

5 Idea of Otherness

Otherness can be viewed as an articulation of diversity as well as a definition of differences. According to Jean-Francois Staszik, difference belongs to the realm of fact, and otherness to the realm of discourse. The notion of otherness mainly examines the idea of a criterion that allows humanity to be divided into two groups: one that embodies the norm and whose identity is valued, and another that is defined by its faults and devalued [20]. Hence, the concept of otherness is attributed less to the distinction between the other and the other person than to the perspective and discourse of the individual who perceives the other as such. Since topological spaces deal with relations and connections with a given spatial context and not a specific form, it is clear that a particular topological construction can manifest itself through numerous forms. It is more spatial relation than a spatial determinism. Furthermore, through the relation with topological spaces, one can become aware of the variability of form and, therefore, the possibility of the existence of other forms. It is possible to create an idea of a space that is subject to change, which can lead to a different enviroment for users. If such spatial structures also belong to the multifunctionality, where numerous activities are interwoven or possible, then the idea of finality and certainty of the space is changed.

Moreover, it is possible to assume that there is not only duality, the opposition between self and other, but that many spaces in many forms with the most diverse activities are possible. To put it differently, despite the inherent tendency of humans to make categorical distinctions, the categories themselves and meanings associated with them are social construction rather than natural processes. Therefore, it is possible that topologically designed spaces may open the way to the diversity of multifunctional deformable spaces rather than to construct a new architectural typology.

According to the theory, the notion of otherness originated in a spatial form, arosing from the idea of difference that is associated with the geographical nature of segregation. This approach implies that groups are divided into territories or spatial units with clearly defined boundaries that are difficult or impossible to exceed. However, topological constructions fundamentally change the relationship between the outside and the inside because, for those types of spaces, it is impossible to determine their boundaries. The only relevant characteristic of the structure is its ability to deform. In general, the notion of the boundary of space in mathematics, even when viewed chronologically, is closely associated with the notion of distance between two elements. The idea of metrical space corresponds more closely to the idea of Euclidean space, as it relies on understanding of spatial relations, such as the notion that the distance between two points is always positive. At a higher level of abstraction, the distance between two elements must be understood as a transition from one element to another, which, in the context of topology, is continuous.

As previously demonstrated, in architecture, the interpretation of purely mathematical definitions moves away from the original model. Hence, the treatment of the relationship between exterior and interior, wherein one can simultaneously be both the exterior and interior of the architectural space, requires that the users permanently change their relationship to the space. In the architectural discourse, the hierarchical treatment of the structure can be precisely discerned through the outside-inside relationship. With a model like the Möbius strip, the boundaries and images of a hierarchical structure are weakened. At the end, it becomes evident that one of the primary characteristics of topological spaces is the ability to blur the distinction between territorial boundaries, as well as to examine the traditional spatial duality between interior and exterior, employing methodologies that involve integrating the structure with the immediate environment.

6 Conclusions

This paper examined a widespread problem in architectural discourse, the denotation of concepts from other scientific fields, which, due to the interdisciplinarity of architecture, have become an integral part of the design process.

The beginning of the paper provided a brief overview of the term topological deformability within its native field - mathematics. Since topology is difficult to understand and appropriate, it has remained highly abstract to the architectural discourse. Nonetheless, taking into account its potential for expansion into other scientific fields, a multitude of interpretations appeared, along with a multitude of topological propositions that expanded the boundaries of topology beyond its native field, thereby bringing it closer to architectural theory and practice.

With an undeniable social impact, architecture traces the way of topology to a broader audience, with a specific impact on understanding the potential of the term deformation. Endless changes of form were understood as a potential to perceive transformability and to accept the differences in architecture. The significance of deformability was heightened, and it was imperative to clarify the limitations imposed on society as those that society ought to be able to tolerate. The paper examined how topological deformability in architecture, in a very subtle way, teaches us about and how to accept the differences.

References

1. Munkers, R.: Topology. Prentice Hall, New York (2000)
2. Novikov, S.: Očigledna topologija. Zavod za udžbenike i nastavna sredstva, Beograd (1984)
3. Kline, M.: Mathematical Thought from Ancient to Modern Times, vol. 1. Oxford University Press, New York (1972)
4. Kantor, J.-M.: A tale of bridges: topology and architecture. Nexus Netw. J., Birkhäuser Verlag, Basel 7(2), 13–21 (2005)
5. Poincaré, H.: Analysis situs. Journal de l'École polytechnique, an II, cahier 2. Imprimerie de la République, Paris:1–123 (1895)
6. Božić, M.: Pregled istorije i filozofije matematike. Zavod za udžbenike, Beograd (2010)
7. Barker, S.: Filozofija matematike. Nolit, Beograd (1987)
8. Šikić, Z.: Novija filozofija matematike. Nolit, Beograd (1987)
9. Plotnitsky, A.: Algebras, geometries, and topologies of the fold: Deleuze, Derrida, and quasi-mathematical thinking (with Leibniz and Mallarmé). In: Patton, P., Protevi, P. (eds.) Between Deleuze and Derrida. Continuum, London, New York (2003)
10. Carpo, M: The Alphabet and The Algorithm. MIT Press, Cambridge (2011)
11. Kolarevic, B.: Digital architecture, eternity, infinity and virtuality in architecture. In: 22nd Annual Conference of the Association for Computer-Aided Design in Architecture, ACADIA, pp. 251–256. Catholic University, Washington D. C (2000)
12. Di Christina, G.: Architettura e topologia: per una teoria spaziale dell'architettura. Librerie Dedalo, Rome (1999)
13. Lynn, G.: Architectural curvilinearity: The Folded, the Pliant and the Supple. In: Lynn, G. (ed.) Architectural Design: Folding in architecture, vol. 102, pp. 24–31. Wiley-Academy, London (2004)
14. Robinson, C.: The Material Fold: Towards a Variable Narrative of Anomalous Topologies. In: Lynn, G. (ed.) Architectural Design: Folding in architecture, vol. 102, pp. 80–81. Wiley-Academy, London (2004)

15. Speaks, M.: It's out there...the formal limits of the American avant-garde. In: Perrella, S. (ed.) AD: Hypersurface Architecture, Profile 133, vol. 68, pp. 26–31. Wiley-Academy, London (1998)
16. Picon, A: Architecture, Science, Technology and The Virtual Realm. In: Picon, A., Ponte, A. (eds.) Architecture and The Science, Exchanging Metaphors, pp. 292–313. Princeton Architectural Press, New York (2003)
17. Meredith, M.: Never Enough (transform, repeat ad nausea). In: Sakamoto, T. (ed.) From control to design: Parametric / Algorithmic Architecture, pp. 6–10. Actar, Barcelona (2008)
18. Terzidis, K.: Expressive Form. Spon Press, New York, A Conceptual Approach to Computational Design (2003)
19. Zellner, P.: Hybrid Space: New Forms in Digital Architecture. Thames&Hudson Ltd, London (1999)
20. Staszik, J-F. Other/otherness. International Encyclopedia of Human Geography: A 12 – Volume Set. Elsevier Science: 43–47 (2009)

The Complexity Conflict in Research and Practice: The Case of Public-Private Interface Configuration

Šárka Jahodová[✉]

Faculty of Architecture, Czech Technical University, Prague, Czech Republic
dolezsar@cvut.cz

Abstract. The disassociation between research and practice in architecture and urban design is a recognised issue, leading to an ongoing academic debate on the credibility of the field. Most authors agree that the irreducible nature of the practice, the complexity of the built environment itself, and its indirect and multidisciplinary shaping processes render the applicability of gained knowledge problematic, as researchers are inherently confronted with the dilemma of rigour vs. relevance. This paper explores the relevance of architectural research and constraints to the application of theoretical knowledge in practice in a particular case of urban interface. It discusses the benefits and limits of planning and design tools derived from the theoretically described causal relationship between the physical configuration of the public-private interface and its social effects on users of the adjacent public space. Through a conceptualization of the process, the article illustrates that the application of theoretical knowledge in urban design and planning can be significantly improved by considering the factors of scale and spatial context as well as the forming forces, values, and motivations of the actors involved. This theoretical dissection aims to clarify the contradiction between theoretical values and subsequent practice to help promote not universally good but, more essentially, adequate, sustainable, and equitable spaces.

Keywords: Urban Interface · Public-Private Interface · Urban Planning · Urban Design · Research Applicability · Research-Practice Gap

1 Introduction

1.1 The Divergence of Research and Practice

There is a widely recognised problem of persistent disassociation between research and practice in the field of urban studies. The most influential text shaping urban design and planning practice has limited scientific rigour and relies on a normative interpretation of subjective observations. [1–6]. This gap has been addressed in almost every research paper dealing with the urban environment in the past two decades, regardless of its main topic, indicating the gravity of the issue.

Most authors agree that the greatest challenge to the application of theoretical knowledge lies in the practice's irreducible nature, which arises from its two distinguishing

© The Author(s) 2025
M. Barosio et al. (Eds.): EAAE AC 2023, SSDI 47, pp. 242–250, 2025.
https://doi.org/10.1007/978-3-031-71959-2_27

traits. First is the complex nature of the object of inquiry—the urban environment itself. The practice of urban studies is typically multilayered and multiscalar in structure and explores the intertwined relationship between urban form and its social implications. The second challenge stems from the dynamic nature of the city, the indirect and multidisciplinary forces that form the urban environment, and the reciprocal relationship between its social and spatial conditions.

1.2 Challenges in the Study of Complex Urban Environments

The different layers and scales of urban structure, the interplay between space and human behaviour, and the complicated and indirect process of its formation and transformation make research of the urban environment extremely challenging. When confronted with the complex and dynamic urban reality, the pursuit of rigour, validity, and consistency in research design poses a challenge to the practical application of knowledge due to the inherent loss of information. Researchers often grapple with the trade-off between data quantity and granularity and the dilemma of rigour vs. relevance [7].

This article highlights the need to actively address the complexity of the urban environment and its formative processes. It relies on the author's background in both theory and practice to identify the limited applicability of theoretical claims due to the spatial and procedural complexity of today's cities. This applicability of theoretical knowledge can be significantly improved by considering the factors that relate the uncovered principles and patterns to their adequate spatial and procedural context.

2 Case of Urban Interface

2.1 The Interface as an Intersection of Different Realms

In order to illustrate the challenge that the complexity of the urban environment poses to the application of normative research in the field of urban theory, this article focuses on a particular element of the environment: its interface. "Urban interface" is a term used in this paper to describe the spatial configuration delimiting a public space from a building block. The broad theme of the urban interface appears in the theoretical literature under different terms ranging from "street edge" to "transition zone." This field of urban theory and the topic on the interface itself have become more prominent in past decades due to the increased interest in public spaces in general. The specific social and morphological role of the interface was discussed in the most influential texts in urban planning theory, as the details of its configuration, for example, its permeability, articulation or setback, can have a major impact on the liveability of the space on both sides of the border [8–10]. The interface is one of the elements that co-shapes the image of the city [9, 11–14] and can be interpreted as a reflection of society.

Interface is a place where opposing interests tend to collide—a physical manifestation of a dividing line—but also a vibrant zone shared by the two realms in between which it stands. The public/private interface can be viewed as a symptomatic space for many current urban issues, as it is conceptually a space between public and private interests characterized by shifts of values and powers. Transversing different spatial scales from architectural detail to urban macro-morphology and extending across different realms of influence, the public/private interface illustrates both abovementioned challenges: 1/ the process of balancing interests of individual actors and the public interest, and 2/ the consideration of the local and global context.

2.2 The Causal Relationship Between Interface Configuration and Human Perception and Behavior

The majority of available literature on urban interface focuses on the causal relationship between the physical parameters and spatial configuration of the interface and its social effects on users of the space—both their perception of the space and their behaviour within it.

Two examples of influential theoretical claims concerning interface attributes and configurations were analysed for their practical applicability using urban planning and design tools. These claims were selected based on previous literature reviews and represent different spatial characteristics and perceived values.

– "Ground floor visual permeability and the frequency of openings reinforce safety." These features allow for natural surveillance and deter potential criminality through a concept described by Jane Jacobs as the "eyes on the street" [15, 16]. Transparency also increases the capacity for visual exchange between the public and private realms, which promotes social interaction and activity and creates a sense of openness [10].
– "Rhythm and articulation of facades can increase diversity and significantly influence a pedestrian's perception of a space." People naturally prefer streets with relatively shorter facades, as they conform to the "human scale" in relation to their walking speed [10, 17, 18].

2.3 Conflict and Contradiction

Despite the progress in planning theory, it seems that the practical application of such knowledge is rather underdeveloped. The aforementioned theoretical understanding of the impact of the public-private interface on urban social life has led to the development of specific design and planning tools. These tools span from the overall spatial and functional regulations promoting the legibility of a space, such as a "building line" and setback from the street line, to the detailed parametrical design codes promoting the human scale, which prescribe, for example, the maximal length of a façade or the number of entrances. Yet there are still many cases, even in very recent city development, where the interface configuration of new buildings does not positively contribute to the creation of a high-quality environment.

The problem is often attributed to a lack of knowledge or care on behalf of the planning authority, investor, or architect. Despite a general advancement in knowledge, the prevalence of such issues indicates that there might be inherent conflicts and misconceptions that can be avoided or at least mitigated when taken into consideration. Through analysing such cases, we aim to outline the most important factors that might have been omitted in the process.

Our investigation of unfavourable urban interfaces is divided into two main lines of inquiry that correspond to the two previously outlined problems, namely, the complexity of the urban environment and its indirect formative processes. The first question is connected to the issue of information loss in the description of a complex urban environment: If the theoretically defined principle was applied in urban development, why did the spatial configuration not have the intended effect? The second question concerns the complex spatial development process: Why do valuable characteristics not occur naturally within the new development?

3 Disregarded Factors, Limits, Moderators and Mediators

3.1 The Indirect Process of Translating Theoretical Knowledge into Practice

The research and theory that are so closely allied to the applied fields of urban design and architecture have a strong tendency toward a normative approach that is directly implemented in practice. Due to the inevitable information loss in scientific analyses of complex urban environments, it is crucial to avoid oversimplification. A typical pitfall of strictly morphological analysis is the omission of the role of non-spatial factors in the actual formation of an urban environment.

A declared objective of many research projects is "informing policy-making" or "developing a knowledge base for urban designers and planners," with the overall goal of ensuring vitality, safety, comfort, etc., establishing best practices, or developing toolboxes to help design a "successful" or "good" space. The idea is that gaining a better understanding of the particular effects of specific spatial configurations on positively interpreted human behaviour (for example, intensities of movement or visual engagement) by replicating or avoiding that configuration will have a positive effect on the urban environment. This process of implementation is based on a prevalent but incorrect view of the design profession as applied planning theory. It assumes a similar approach to knowledge dissemination through the "translation model" that is typical in the natural sciences, where theory can be directly applied to new technology, as opposed to the more indirect "enlightenment model," wherein knowledge is disseminated to multiple relevant audiences [1].

The specifics, overlaps and ambiguities of a rather new theoretical field [3] that is, at least partially, in the process of development of its general theoretical framework and vocabulary [2]. The specific knowledge from focused inquiries lacks a rigid theoretical anchor and is vulnerable to seemingly opposing interpretations, misconceptions and improper uses. The aim of theory should be to develop a more robust and inclusive method for building up a shared theoretical framework by synthesising particular pieces of knowledge. It is also imperative for the design of normative research to understand the formation and adaptation processes shaping the urban environment and the limited role of a practising architect or planner within these processes. The following simplified diagram outlines the complexities of the mechanism linking the research outcomes with their impact in reality (Fig. 1).

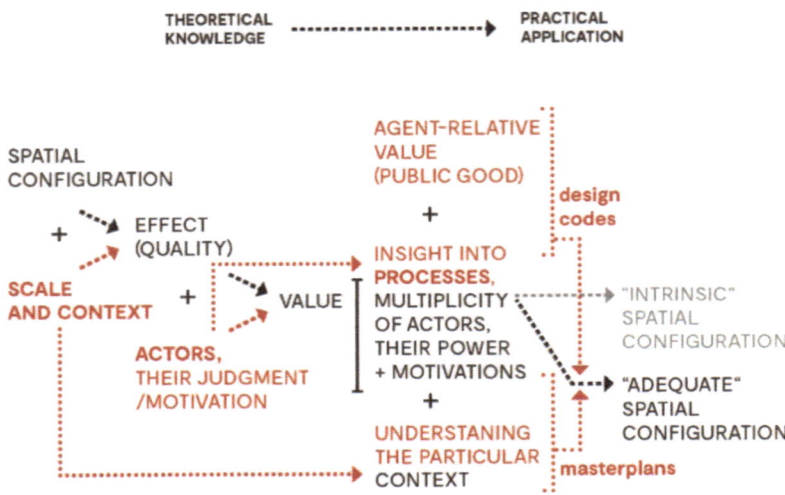

Fig. 1. A conceptual diagram of the theoretical knowledge application process in urban planning and design. By author.

3.2 Scale and Context

The answer to the first of the previously formulated questions, " *If the theoretically defined principle was applied in urban development, why did the spatial configuration not have the intended effect?"* is shown on the left side of the conceptual diagram. There is a possible omission of scale and context moderator in the relationship between space and effect, as the configuration might have been applied in the wrong scale or context.

The instinctive tendency to use design codes and effectively apply specific knowledge as a universal problem-solver leads to the overuse of specific regulations or designs in larger areas without considering specific spatial or functional settings, morphologies or land-use regulations. For example, there are cases of planning documentation prescribing a specific setback across entire municipalities. It is also not unusual to encounter such suggestions within the research papers or theoretical literature. For example, the usage of active frontage is recommended *"whenever possible"*, while other studies suggest that *"the more windows and doors into the public realm the better"*[19]. Such recommendations disregard the actual urban context or land use of a ground floor. The application of too-specific or too-universal rules might prevent planners and designers from finding the right solution at the right scale.

3.3 Actors and Processes

The second question connected to the process of spatial development, *"Why do the valuable characteristics not appear naturally?"* is slightly more complex and touches on the role of the urban planning and urban design professions among the conflicting interests of specific actors. The answer lies in a confusion of effect and value—notions often used interchangeably despite the relativity of value according to different actors, judgments and motivations. The motivations and intentions of different actors naturally do not perfectly align at the borderline. Given the multiplicity of roles and meanings of an interface, it is only natural that numerous contradictions and paradoxes arise. The diagram illustrates the necessity of professional insight into the formative processes and the multiplicity of actors, their power and motivation. These elements are crucial to balancing the interests in a particular context, safeguarding the public good, steering the architectural process, and mediating the discussion to achieve a balanced, adequate and sustainable spatial and functional configuration.

4   Illustrative Case Studies

To further explain the contradiction originating in a misconception of the role of factors such as "scale and context" and "actors and processes", the claims about urban interface configuration derived from the theoretical literature, as outlined in chapter 2.2, will be examined. It is, however, important to mention that the cases were not selected to disprove these statements but are extreme exceptions to the rule, outlining the limits of simplified fragments of theoretical knowledge taken out of context. By studying limits and outliers, we can better understand the intricacies of the research–practice interaction and help formulate a corrective mechanism. The following cases also show the importance of "relational" research.

4.1   Case Study 01: Limits of Visual Permeability

Fig. 2. Case 01: Ground floor visual permeability and the frequency of openings reinforce safety. Left: Amsterdam, 2018; right: Prague, 2020 (Photographs by Šárka Jahodová)

Scale and Context: The macro-morphological setting is critical to many safety-improving strategies. Permeable façades do not achieve the intended impact in quiet, low-density residential areas or service side streets (Fig. 2 right) that lack sufficient through-movement [15, 20] and tend to reinforce the preceding concept of "defensible space" [21]. More extensive studies also indicated that critical micro-morphological aspects such as inter-visibility, entrance density and orientation, topological distance and ground floor use have a substantial impact above and beyond the transparent façade [16].

Actors and Processes: This case also shows that there is no singular "client" in urban planning and design. The interface can be interpreted in two directions: on the one hand, it protects the public space from private interference, and on the other hand, it protects the private, intimate space from outside views [13], so the people occupying the different sides may have opposing perceptions of privacy, usually resulting in the adaptation of the physical form by blinding and covering the openings (as seen in the right part of Fig. 2). The balance at the borderline also varies based on a particular cultural context.

4.2 Case Study 02: Rhythm and Scale

Fig. 3. Case 02:Rhythm and articulation of facades can increase diversity. Left: Malmö, 2016; right: Prague, 2020 (Photographs by Šárka Jahodová)

The excessive pursuit of architectural articulation and formal detail without understanding the mechanisms of human perception or the underlying economics of large-scale development can result in paradoxically mismatched urban scenes (Fig. 3).

Scale and Context: Visually, there is a tipping point where the rhythmic articulation of a façade passes from a structural property to an indefinite decorative pattern uniformly coating a building. That threshold is relative to the observer's distance and speed of movement. Recent research points out the need for a shift in focus from formal façade design and detailing towards the structural aspects behind the subdivision of a street edge [18]. In this approach, the segmentation of a ground floor façade is linked to distinct territorial units [22].

Actors and Processes: The aesthetic motivation for a short façade usually does not align with the current economic and operational reality of most prevalent typologies. The concern with human scale is a rather recent movement within urban theory, originating as a reaction to the increase in large-scale development and the economic aggregation and densification that go hand-in-hand with a major shift in ownership structure. A wide range of design codes and policies regulate the maximal length of a façade and support its detailed articulation, often disregarding the building structure and volume and its typological and economic reality. It is important to choose the right tool that conforms to the economic and typological context, as any fragmentation could increase the cost of a project, complicate the usability of a building or lead to a decorative pastiche resulting in a "Potemkin village" effect.

5   Conclusion and Summary

The implicit pursuit of rigour, validity and consistency in research design poses a challenge to its subsequent applicability when the confronted with a complex urban reality. The examples presented above indicate that the information loss inevitable in scientific analyses of complex urban environments can be mitigated by considering scale and context through relational, multiscalar, or multidisciplinary research. Complementing the investigation of urban detail with the impact of its wider context and enriching macro-morphological research with information about configuration at a micro-morphological level can help to indicate their interrelationships. Such an approach can be beneficial in consolidating the theoretical framework by linking specifically focused inquiries and thereby increasing the scientific integrity of the broad field of architectural and urban design theory. Moreover, a prerequisite for the successful application of acquired knowledge is consideration of the actors and processes affecting the urban form. A conceptual understanding of the tools and methods used in practice and an acknowledgement of the different actors, motivations and priorities within the development process increase the relevance of the applied knowledge.

The aforementioned systematic, incremental enhancement of both theoretical and practical perspectives can further clarify the apparent contradictions between theoretically defined values and subsequent practices. The presented case studies indicate that the implementation of theoretical knowledge in urban design and planning can be significantly improved by considering factors relating the discovered principles to their spatial and procedural contexts. Such act of refinement can help in practice to create not just vaguely good but more thoughtfully adequate and sustainable spaces.

References

1. Alexander, E.R.: Introduction: does planning theory affect practice, and if so, how? Plan. Theory **9**(2), 99–107 (2010). https://doi.org/10.1177/1473095209357862
2. Marshall, S.: Science, pseudo-science and urban design. Urban Des. Int. **17**(4), 257–271 (2012). https://doi.org/10.1057/udi.2012.22
3. Carmona, M.: Explorations in Urban Design: An Urban Design Research Primer. Ashgate. J. (2014)
4. Palermo, P.C.: What ever is happening to urban planning and urban design? Musings on the current gap between theory and practice. City, T. Archit. **1**(1), 1–9 (2014). https://doi.org/10.1186/2195-2701-1-7
5. Scott, A.J., Storper, M.: The nature of cities: the scope and limits of urban theory. Int. J. Urban Reg. Res. **39**(1), 1–15 (2015). https://doi.org/10.1111/1468-2427.12134

6. Dovey, K., Pafka, E.: The science of urban design? Urban Des. Int. **21**(1), 1 (2016). https://doi.org/10.1057/udi.2015.28
7. Schön, D.A.: The reflective practitioner: how professionals think in action. Basic Books (1984)
8. Jacobs, J.: The death and life of great American cities (pp. 1–458) (1961)
9. Alexander, C., Ishikawa, S., Silverstein, M.: A pattern language. In Ch. Alexander (pp. 1–1218) (1977)
10. Gehl, J., Rogers, R.: Cities for People. Island Press (2013)
11. Lynch, K.: Image of the City. MIT Press (1964)
12. Habraken, N.J. (1998) The structure of the ordinary form and control in the built environment (J. Teicher (ed.)). MIT Press
13. Madanipour, A.: Public and private spaces of the city. Public and Private Spaces of the City, October, 1–237 (2003) https://doi.org/10.4324/9780203402856
14. Carmona, M., Heath, T., Taner Oc, S.T.: Public Places - Urban Spaces. Routledge (2010)
15. Hillier, B., Sahbaz ,O.: High resolution analysis of crime patterns in urban street networks: an initial statistical sketch from an ongoing study of a London borough. In: Proceedings of the 5th Space Syntax Symposium Delf, I, 451–478 (2005)
16. Rønneberg Nordhov, N.A., Van Nes, A. (2019) Proceedings of the 12 th Space Syntax Symposium The Role of Building Entrances Towards Streets and the Perception of Safety in Six Neighbourhoods in Bergen
17. Simpson, J., Freeth, M., Simpson, K.J., Thwaites, K.: Visual engagement with urban street edges: insights using mobile eye-tracking. J. Urban. **12**(3), 259–278 (2019). https://doi.org/10.1080/17549175.2018.1552884
18. Gatti, M., Nollert, M., Pibernik, E.: Regulating Façade length for streetscapes of human scale. Land **11**(12), 2308 (2022). https://doi.org/10.3390/land11122308
19. Llewelyn-Davies: Urban design compendium. In: Design. English Partnerships (2000)
20. Hillier, B., Sahbaz, O.: An evidence based approach to crime and urban design or, can we have vitality, sustainability and security all at once? Designing sustainable cities: decision-making tools and resources for design, pp. 163–186. Wiley Blackwell, March, Oxford (2008)
21. Newman, O.: Defensible Space. Macmillan, Crime Prevention Through Urban Design (1972)
22. Simpson, J., Freeth, M., Simpson, K.J., Thwaites, K.: Street edge subdivision: structuring ground floor interfaces to stimulate pedestrian visual engagement. Environ. Plan. B: Urban Anal. City Sci. **49**(6), 1775–1791 (2022). https://doi.org/10.1177/23998083211068050

Regenerating Public Housing in Italy with the Support of the Next Generation EU Fund. Lessons Learned from a Research by Design Experience

Fabio Lepratto[1] and Giuliana Miglierina[2]([✉])

[1] DAStU, Politecnico di Milano, Milan, Italy
fabio.lepratto@polimi.it
[2] MAS in Housing, ETH Zurich, Zürich, Switzerland
gmiglierina@student.ethz.ch

Abstract. In Italy, the Next Generation EU instrument has made it possible to launch what could be optimistically defined as a new season for public housing after at least three decades of decreasing financial support. The PINQuA (National Innovative Program for Housing Quality) was supported with 2.8 billion euros, prompting a race among eligible public bodies to submit a proposal. Thanks to the funding, 159 projects of 271 presented were selected and are currently in the implementation phase. Quality and innovation, two concepts prominently featured in the programme's title, encourage a sense of optimism. However, they also require critical examination to understand how the issues related to these concepts have been interpreted in terms of proposal development, evaluation, subsequent selection, and implementation. If we focus on aspects related to the culture of design, can we consider this programme an opportunity for architecture as a discipline? The research approach undertaken involved a direct participation in one of the projects: the urban regeneration of the Piazzale Visconti housing complex in Bergamo. The article reports the outcomes of a "reflection in action" [1] gained from a privileged observation point. Although referring to a single case, the research-by-design activity was an opportunity to get to the core of general issues, triggering a reflection on two complementary dimensions (1) the verification of project potential, expressed by the programme, (2) the understanding of the opportunities and certain problems intrinsic to the process.

Keywords: Public Housing · Regeneration · Quality of architecture · Process dimension

1 Research Perspective

1.1 Introduction

Through a design experience this contribution integrates a set of research activities conducted at the Politecnico di Milano regarding the quality of contemporary design in public housing. The proposed reflection adds a voice to the emerging debate on the results

M. Barosio et al. (Eds.): EAAE AC 2023, SSDI 47, pp. 251–262, 2025.
https://doi.org/10.1007/978-3-031-71959-2_28

of the National Innovation Programme for Housing Quality (PINQuA). The text presents considerations that emerged during a design action, addressing both architectural and procedural issues.

1.2 A New Season for Public Housing

In Italy, the Next Generation EU instrument has made it possible to launch what could be optimistically defined as a new (short) season for public housing, focusing on the regeneration of existing stock, and with an intense experimental orientation, after at least three decades of decreasing financial support [2]. The PINQuA (programme acronym) was launched in 2020 and supported with 2.8 billion euros, prompting a race among eligible public bodies to submit a proposal. Thanks to the funding, 159 projects of the 271 presented were selected and are currently in the implementation phase across national territory (the ranking list of projects accepted for funding was published in October 2021). The projects are scheduled for completion by March 2026, which is the extremely limited timeframe imposed by the conditions for accessing the funding – a key aspect with a profound impact on the entire experience. This occurred when the program – initially set up with fewer funds – was expanded and absorbed into National Recovery and Resilience Plan resources. This compression of design time, which exceeds any rational logic, puts significant pressure on the involved stakeholders and announces considerable fallout on the quality of the proposals, which are often critically deprived of the time for reasoned development and for exploring concepts that differ from routine expectations. As a further consequence, this condensed timeline hinders the possibility of engaging the local community in meaningful participatory processes.

According to the numbers released by the ministry, the program will involve renovation, replacement, and new construction of 16,500 housing units [3] – approximately 2% of all Italian public housing stock estimated at 806,146 units [4]. Quality and innovation, two concepts prominently featured in the program title, encourage a sense of optimism. However, they also require critical examination to understand how the issues related to these concepts have been interpreted in terms of proposal development, evaluation, subsequent selection, and implementation.

In other words, if we focus on aspects related to the culture of design, can we see this programme as an opportunity that has also been designed for architecture as a discipline and not solely for architecture as a technical service? It would thus provide space for experimentation beyond business as usual. In the absence of actual results, it is too early to formulate a definitive answer at this stage. Although most of the design decisions are already taken, the outcomes will only be visible after the projects have run their entire course. Only then we will have a comprehensive vision and be able to start a meticulous evaluation phase. We will then understand whether this major investment has enabled Italy to contribute to the international debate on social housing design, bridging the current gap compared to major European countries – a gap confirmed by the almost complete absence of projects carried out in Italy in major specialist European literature.

Today, however, by taking part in one of the interventions, we can anticipate some of the themes and questions that may fuel the debate in the future. In parallel, even in the absence of a systematic collection of official detailed material from other interventions, a brief report published by Mims (Ministry of Infrastructure and Sustainable Mobility) allows us to draw up a shortlist of the 159 projects, similar in scale and program, that will be worth comparing in the future when in search of successful solutions and possible failures – some of these projects are: Ri-Abito qui in Potenza; ex SAIRO area in Udine; ViviBbusto 2030 in Busto Arsizio; Ponte San Giovanni in Perugia; San Giovanni: un quartiere verde, inclusivo e smart in Trieste; Pinqua Vallette in Torino; Nuove Ca.Se. in Calenzano e Sesto Fiorentino; Librino Città Moderna in Catania; Terra in Andria; Contrada Torregiana-Fontescodella in Macerata.

1.3 A Reflection in Action

Having refined the goal of the investigation, there are several possible research approaches that can be more suitable for acquiring an understanding of the change potential and challenges of this tool, as well as for testing its ability to guide outcomes toward the best possible results. The approach undertaken within the framework of research-by-design [5] involved direct participation in one of the 159 PINQuA projects. Thus, it was possible to conduct – and produce – a "reflection in action" [1] from a privileged observation point. Although the experience relates to a single case, it was an opportunity to get to the core of general issues, triggering a reflection on both the design and process dimensions: two complementary aspects. The study of the first dimension makes it possible to test the program's design potential, also in response to the current challenges in the implementation of low-budget housing solutions. The study of the second dimension examines the main opportunities and problems intrinsic to the PINQuA process that impact the physical results, as they prepare the ground. This contribution seeks to bring together these two interdependent levels that influence architectural results.

1.4 A Point of Observation: The Case Study of Piazzale Visconti in Bergamo

The project that prompted this reflection focuses on the urban regeneration of an area facing various challenges: the Piazzale Visconti housing complex, located on the southwest outskirts of Bergamo and separated by the consolidated part of the city. The area, commonly known as *Villaggio degli Sposi*, traces its roots back to 1955 when the local parish initiated a self-managed housing project financed by subsidised mortgages aiming to provide housing for young couples – hence the name of the quarter, which means Newly-weds Village. The outcome was a low-density neighbourhood comprising single-family homes, rowhouses, and small apartment buildings, each comprising two to four storeys. The site being regenerated lies 500 m north of the neighbourhood centre, where a few local services and micro-commercial businesses can be found around the church. The five public buildings occupying this area are in a state of decay. In particular, two have reached the end of their life cycle and are currently empty. The central public space, characterized by a monotonous functionalist design, serves mainly as a parking lot, with the only positive quality provided by the presence of a few trees. Size-wise, the plot is relatively compact, approximately 5,000 sqm, bordered by public roads. Of the five buildings, two are linear blocks perpendicular to each other (owned by ALER, the regional public housing authority) and three are standalone buildings (owned by the municipality). The aim of the intervention, in line with the program, is to reverse the decay in progress and define a new focal point for the entire neighbourhood through the creation of a pedestrian square, with addition of public functions on the ground floors; replacement and increase of social housing stock (a 60% increase in volume is planned); and the energy and seismic upgrading of the remaining buildings.

Our contribution – a scientific consultancy for the Technical-Economic Feasibility Project provided on direct commission of the company awarded the design services contract, Progettisti Associati Tecnarc s.r.l., in partnership with Mythos S.C.AR.L. – falls midway in the overall process, bridging the gap between the administrative phase, (started with the publication of the call for proposals and concluded with the allocation of funding), and the competition for the assignment of design services (which, in turn, precedes contracting of execution of works), providing an overview of the entire process. In this specific case, it should be noted that the transfer of responsibilities from the public authority to the design holder was facilitated through a document of intent. The public authority, during the drafting of the proposal and for the subsequent development phases, engaged in a scientific consultancy provided by Politecnico di Milano (responsible for the consultancy: O. E. Bellini, Department of Architecture, Built Environment and Construction Engineering) with the aim of creating a Design Guideline Document (Documento di Indirizzo Progettuale – DIP) containing directives for objectives, themes, innovative aspects, and quantitative requirements to be adhered strictly in the design phases. This procedure, a virtuous attempt of the commissioner to protect the design scope, does not represent a mandatory step but arises from the initiative and civil responsibility of the single decision-maker.

2   Two Complementary Research Dimensions

2.1   The Project Dimension

The first reflections on the actions carried out have made it possible to verify the real transformative capacity of the PINQuA programme, which is partly determined by an adequate match between the initial ambitions (as declared in the call and included in the goals set by the DIP) and the financial resources allocated. While this might seem obvious, past national experiences with the so-called *Contratti di Quartiere* (Neighbourhood Contracts) programme, for example, revealed a misalignment in this sense, resulting in a downsizing of the originally envisioned scope during the implementation phase [6]. Secondly, we had the chance to highlight the main issues arising from the specific context and to relate them to problems common to other fragile areas targeted by the program, together with a range of possible responses. In this regard, the case study allowed us to observe how the program offered designers flexibility to develop adequate solutions, in line with the most recent experiments in housing design.

These findings, grouped below into six themes, are then an example of what the implementation of the PINQuA can enable, allowing for some elements of design experimentation while respecting strict constraints. The following thematic paragraphs identify problems common to any similar intervention and provide a starting point for addressing equivalent challenges [7].

Advantages and Consequences of Densification (1). Given the acute shortage of social housing, when involving such estates, regeneration tends to include the issue of volume increase: a more intensive use of already available and urbanized land is absolutely advantageous for increasing housing stock and/or expanding the user base attracting higher-income households [8]. This phenomenon is common in other European contexts, like the practices of housing associations in the Netherlands. Here, regeneration serves not only to increase the supply of social housing but also integrate a proportion of open-market housing, thus fostering social diversity [9].

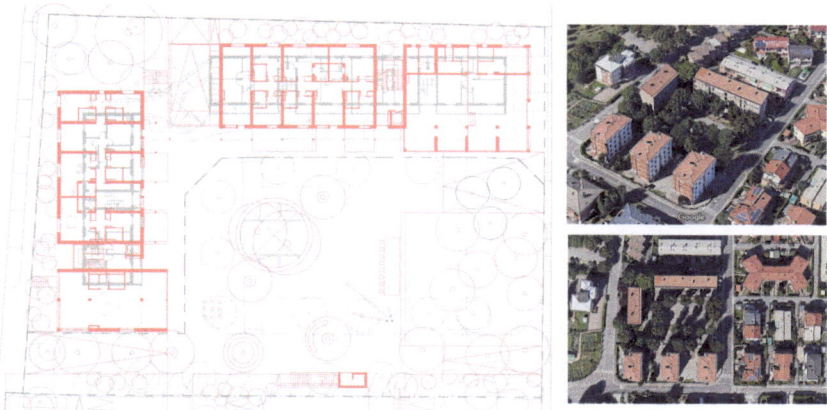

Fig. 1. Preliminary hypothesis of the ground floor plan for the feasibility study (left), comparing the building to be demolished with the new construction (pink), accompanied by images of the neighbourhood (right), 2023. (Drawings reworked by the authors of the paper. Images ©Google.

Beyond its primary objective, a net increase in density can, however, become challenging for settlement and typology, especially in small, highly constrained areas like that under consideration. These difficulties are often accentuated by the conservative attitude that permeates many urban planning regulations, requiring compliance with existing settlement criteria or restricting maximum building heights. An initial hypothesis was to consolidate additional volumes within a tall building located at the end of a low L-shaped structure. However, this option was dismissed since a derogation from local regulations was required, despite its typological advantages. So, to preserve the original building heights and adhere to land constraints, the only viable course of action for increasing density was to construct thicker buildings. This led to the development of residential structures with deep typologies, achieved through the design of three parallel bands. The central band contains bathroom and kitchen, while the two lateral bands house the main rooms – as will be explained later. (see Fig. 1 and Fig. 5).

Opportunities for Reactivating Urban Quality (2). While addressing the issue of monofunctionality and underutilization of open spaces, a common problem in most mass-housing contexts, urban spaces can be transformed into welcoming and inclusive places by rethinking the ground floor. This can be done by increasing types of use (not solely residential) and strengthening the permeability of the outward-facing frontage – this principle is well expressed in the words of Aldo van Eyck: "Forty doors make a good street" [10], referring to the inseparable relationship between home and street rooted in the Flemish tradition (Fig. 2).

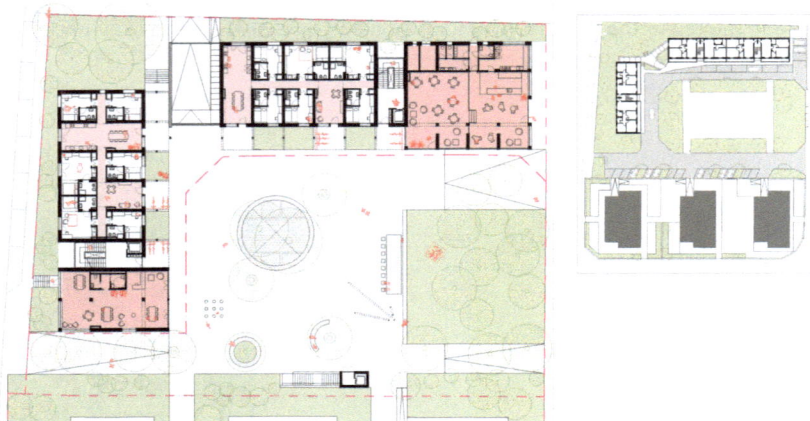

Fig. 2. Preliminary hypothesis of the ground floor plan for the feasibility study, with communal areas highlighted in pink (left), and ground floor plan of the building before intervention (right), 2023 (Drawings reworked by the authors of the paper).

In line with these principles, the design strategies to make Piazzale Visconti a vibrant place included concentrating commercial or community spaces in the most accessible corners of the ground floor; placing only specific shared residential types at the urban level (such as temporary co-living or student accommodation); and providing direct access to all ground floor spaces from the square. In particular, in the case of Bergamo, the project included ground floor housing intended for collective living, organized as Cluster-Wohnung [11], a typology better suited to extending the outdoor part of shared domestic life (see Fig. 1).

Impacts of Regulations on Parking Spaces (3). The current national regulation requires the provision of parking spaces for new residential constructions in proportion to the volume of the building, without considering potential future decrease in private mobility. This regulation contrasts with recent European initiatives where incentives for shared transport have led to the minimization of parking space provision. For example, in the Mehr als Wohnen cooperative neighbourhood in Zurich, residents agree to give up private car ownership to become members of the cooperative. Other projects, such as the Sonnwendviertel in Wien and the Résidence Rosalind Franklin in Paris-Saclay, tackle the issue with above-ground, removable, or reconfigurable structures that envision a car-free future. In the case at hand, there is no viable alternative to designing a conventional underground garage to meet code requirements. However, fulfilling parking needs on such a small site demands significant compromises, which unfortunately have a deep impact on environmental aspects.

These compromises include reduced soil permeability, removal of existing trees, and challenges in finding suitable locations for new plantings. Such conditions have

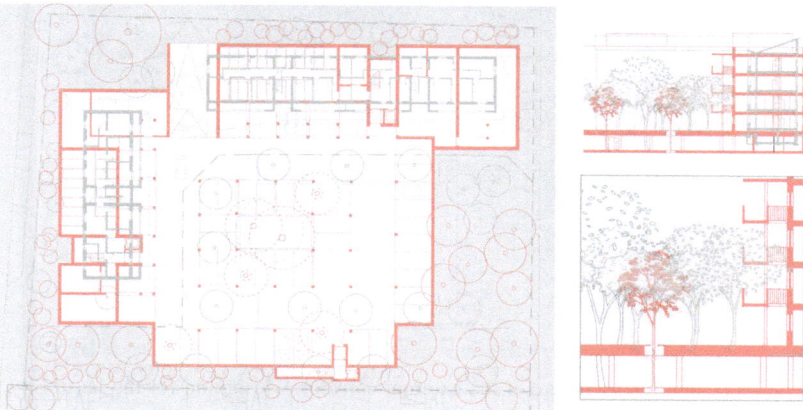

Fig. 3. Preliminary hypothesis of the underground floor plan (left) and section (right) for the feasibility study, comparing the building to be demolished and the new construction (pink), 2023 (Drawings reworked by the authors of the paper).

led to some compensatory solutions that allow some trees, planted at the underground level, to emerge on the square after crossing the garage in section. Two prominent trees emerge from a central patio, as in the CasaNova district in Bolzano, designed by Fritz van Dongen. Additionally, others are individually scattered between the parking spaces, corresponding to an equal number of holes drilled in the ceiling that allow the trunk to overhang the public space above – an innovative approach previously tested in the Het Kastel residential complex in Amsterdam. The combination of these options allows the square to be shaded and to contain heat islands, although it is a compromise solution. (see Fig. 3).

Solutions for Efficient and Welcoming Distribution (4). The choice of how to implement the internal distribution of a social housing building can contribute strategically to controlling the amount of construction and operating costs (a burdened for tenants), allowing an efficient use of the limited resources available without compromising the spatial results. An example of efficient distribution is the gallery solution as it provides access to a large floor area per level without multiplying staircases and lifts, and without limiting the double exposure of the dwellings. Although this option is considered taboo by many because of the well-known window privacy issues, its use can be strategic in terms of efficiency and spatial opportunities – even in a renovated version, as proposed in the case of Bergamo. In addition to increasing costs, a solution with two staircases per building would have meant missing the opportunity to characterize the access to home as a place for fostering social relations and to consider circulation spaces as urban spaces, reinterpreting the Smithsons' idea of a 'street in the sky'.

Numerous contemporary examples of social housing, moreover, have effectively experimented with the inclusion of welcoming and generously proportioned circulation spaces to improve the overall living environment, while also cutting down window privacy issues through specific design solutions. Of the cases analysed during the design process, the Briesestraße housing project in Berlin (EM2N studio) and the Fratelli di Dio

Fig. 4. Initial design render image of the external corridor distribution for the feasibility study, 2023 (Drawing reworked by the authors of the paper).

complex in Sesto San Giovanni (Giancalo De Carlo) shared several similarities with the project. Like these two examples, in Bergamo the gallery depth is bigger than usual and makes it possible to distance walkways from windows to respect privacy; it also makes the spaces in front of the individual entrances habitable, transforming the system into a large loggia. (see Fig. 4).

Advantages of a Modular and Flexible Floor Plan (5). Ongoing social changes that affect domestic life encourage the concept of buildings capable of adapting to possible changes over time. Adaptability becomes a way to sustainability. The lines of research active in this field respond with multipurpose or flexible layouts, the former understood as spaces that can adapt to different uses because of their geometrical conformation [12]; the latter understood as structures composed of a permanent part completed by versatile elements – a theory initially developed in the 1960s with the term "support" by Habraken [13] and the SAR group. Among the various recent experiences, particular interest has been directed on interventions in Catalunya. For example, the Peris + Toral social housing projects in Cornellà were built with a strict structural grid that gave origin to same-size rooms that could be combined in different ways and rearranged in the future; or the project by Lacol for La Balma cooperative housing, which features a longitudinal tripartite block, where a central band, accommodating bathrooms and kitchens, separates two bands that host the main rooms of the house.

Similarly to the second case, the Bergamo project develops a modular plan that accommodates a wide variety of housing types and, at the same time, is easily reorganizable in the medium to long term. The regular cadence of the wooden structure, in addition to guaranteeing simplicity of construction, divides the building into regular spans, determining a basic module useful for current and future aggregations. The depth of the building, which is dictated by settlement factors, is resolved by a central strip of bathrooms and kitchens – analogous to the reference – that subdivides each span into two equivalent spaces, drawing a pattern of juxtaposed rooms that can be easily reassembled (see Fig. 5).

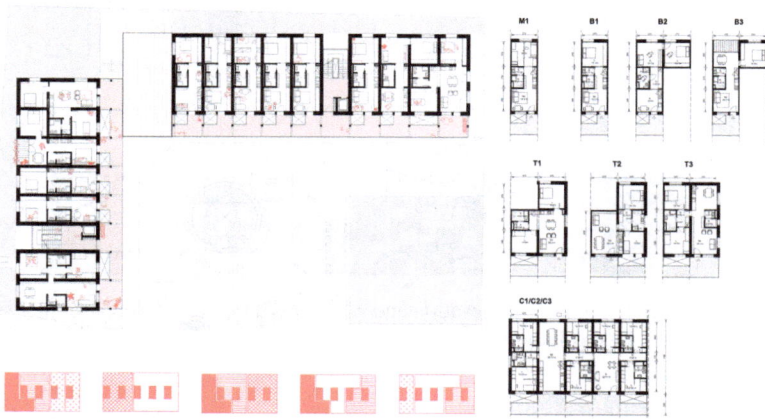

Fig. 5. Preliminary hypothesis of the standard floor plan for the feasibility study, accompanied by typologies layout and diagram of possible future configurations, 2023 (Drawing reworked by the authors of the paper).

Language Responding to Criteria of Necessity and Durability (6). When it comes to figurative and material choices, designing with limited economic resources favours solutions that respond to principles of necessity and durability, foregoing any non-essential elements, sophisticated materials, or fragile fixtures. Also, designing with limited economic resources implies a strategy which relies heavily on geometric choices, eliminating the superfluous. This is probably the foundation for an aesthetic appropriate to the theme of public residential construction. According to these principles, the figurative character of the Piazzale Visconti complex relies on the meticulous composition of a limited number of standard components, and not by adopting customised, sophisticated elements or by using noble expensive materials. The façades are of two types: hermetic toward the exterior of the plot; porous toward the public space. The former defines an apparently random pattern, with four types of windows arranged on a plastered background (see Fig. 6). The latter – the gallery front – is a direct expression of the structural steel elements: cantilevered beams and pillars set back from the outer edge, joined by metal railings (see Fig. 4). Most interventions, however, share that they were carried out after the 2008 crisis, a time that prompted many designers to look towards more elementary solutions, which coincided with an increasing focus on envelope design to meet specific energy performance requirements.

2.2 The Process Dimension

As for the reflections on the potential, criticality and risks of the process, the feeling is similar to what has already been noted by other experts: that we have focused on "how much", without finding indirect measures that could control the "how" [14]. Despite the transformative potential and the centrality given to physical makeover, at least three weaknesses emerge, all related to the preservation of the quality of architectural and urban design. Firstly: the time available to prepare proposals, which required a rather

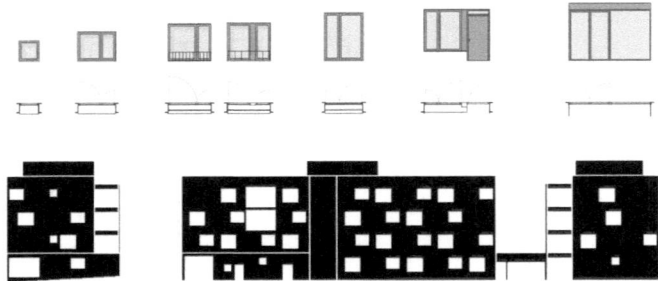

Fig. 6. Preliminary hypothesis of elevation and window typologies for the feasibility study, 2023 (Drawing reworked by the authors of the paper).

advanced level of complexity, was extremely limited, especially when considering the ambitious objectives to be achieved, particularly for a sector that has been under-funded for years and is no longer accustomed to addressing the theme of new design [15]. This key factor could have a negative impact on the quality of proposals, which may result sometimes from the revival of projects developed in the past and then abandoned for various reasons, sometimes from new projects stripped of the necessary breathing space to innovate. A side effect that cannot be reversed, even if the timeframe for the completion of the proposals is reconsidered. Secondly, the evaluation of the preliminary proposals, entrusted to a ministry-appointed High Commission, is centred on measurable criteria and quantitative parameters. This reflects an attempt to make the proposals comparable, based on an unquestionable system, but at the risk of losing effectiveness when faced with predominantly qualitative issues. None of the seven criteria, moreover, places architectural and urban design value at the centre of the evaluation (here is the list of criteria: Quality of the proposal and consistency with the purposes of the program; Scale of the interventions with respect to ERP properties; Recovery and promotion of cultural, environmental and landscape resources; "Zero balance" of soil consumption; Activation of public and private resources; Involvement of private operators; Application of BIM methodology, and of innovative management models). The third issue concerns the selection of professionals. This is done through a public tender based on the proven ability to complete the assignment, expressed through a technical offer influenced by an economic offer. There is no explicit evaluation of the architectural and urban quality of the project to be developed, implicitly emphasizing architecture as a technical service. This does not necessarily lead to a negative outcome but rather places the quality of the result in the hands of the awareness, culture, and capability of the tasked professional (or team). This selection is based on entirely different criteria, however, as if it were a secondary aspect, overshadowed by the correctness of execution.

3 Final Considerations

3.1 The Project Dimension

The experience conducted allowed us to test the potential and understand some structural weaknesses of the entire PINQuA program, which can be pinpointed in the seemingly secondary position afforded to the quality of architectural and urban design. The successful results – as one would expect given the number of proposals – will owe their achievement to fortunate coincidences, rather than to the will and/or ability of the process to direct the outcomes towards the best possible results. Thus, beyond the words that accompany the call, there is a significant risk that exceptional opportunities, such as those made possible by PINQuA, may result in ordinary responses that fall short of their full potential.

In this, as in other circumstances, the absence of a law for architecture in Italy is keenly felt – a law capable of defining, for example, the ways in which projects must be evaluated and professionals must be selected. Such a law, widely discussed but without tangible outcomes, would aim to create regulatory conditions for architectural culture to assume a decisive role in the processes guiding the transformation of the built environment (The MAXXI National Museum of 21st Century Arts has recently added fuel to the debate about this law). This would allow Italy to regain an active role in the contemporary architectural debate, proposing solutions that seek interaction with current international best practices. A law should help to ensure that this does not happen as a result of positive coincidences (as the PINQuA experience will hopefully be able to demonstrate), but as a systematic result of a political will, properly regulated and included in the tools and mechanisms behind every public work. This would require, upstream, a greater recognition of the role of the discipline of architecture in proposing innovative solutions that can respond more effectively to initial objectives, understanding current and anticipating future needs.

Acknowledgment. The contribution is the result of a joint reflection by the authors. Sections 1 and 3, as well as Subsect. 2.2, are attributed to both authors. Subsection 2.1 is attributed to Fabio Lepratto.

References

1. Schön, D.A.: Il professionista riflessivo: per una nuova epistemologia della pratica professionale. Dedalo, Bari (1993)
2. Storto, G.: La casa abbandonata. Il racconto delle politiche abitative dal piano decennale ai programmi per le periferie. Officina edizioni, Roma (2018)
3. MIMS (2022) PINQUA Programma Innovativo Nazionale per La Qualità Dell'Abitare. Progetti e Prime Evidenze. Unità di Missione PNRR del Ministero delle Infrastrutture e della Mobilità Sostenibili in collaborazione con la DIGES
4. Federcasa (2014) Abitazioni Sociali. Motore Di Sviluppo - Fattore Di Coesione
5. European Association for Architectural Education (2022) EAAE Charter on Architectural Research. Retrieved from https://www.eaae.be/wp-content/uploads/2022/08/EAAE-Charter-on-Architectural-Research-2022-update-version-130722.pdf

6. Fiacchini, M.: La meteora dei Contratti di Quartiere nell'esperienza del San Siro a Milano. Techne **4**, 189–197 (2012)
7. Kuhn, T.S.: The Structure of Scientific Revolutions. University of Chicago Press, Chicago (1962)
8. Talluri, L.: La sfida dell'abitare sociale in Italia. Aumentare il numero di alloggi rigenerando le città e rinnovando la gestione. In: Delera, A., Ginelli, E. (eds.) Storie di quartieri pubblici. Progetti e sperimentazioni per valorizzare l'abitare, Mimesis, Milano (2022)
9. Priemus, H.: Regeneration of Dutch post-war urban districts: the role of housing associations. J. Housing Built Environ. **21**(4), 365–375 (2006). https://doi.org/10.1007/s10901-006-9055-4
10. Smithson, A.: Team 10 Primer. Studio Vista, London (1968)
11. Guidarini, S.: New urban housing. Biblioteca universitaria Skira, Milano (2018)
12. Abaigar, A. et al.: Geometrías Habitables. Una introducción al proyecto de arquitectura desde el 9SG Problem. Recolectores Urbanos Editorial, Málaga (2021)
13. Habraken, N.J.: Supports: An Alternative to Mass Housing. The Architectural Press, London (1972)
14. Dattomo, N., Rizzica, C.: Tutti Pazzi per Pinqua. La Qualità Dell'abitare Alla Prova Della Innovazione. In: Gli Stati Generali, www.glistatigenerali.com/architettura-urbanistica/tutti-pazzi-per-pinqua-la-qualita-dellabitare-alla-prova-della-innovazione/. Last accessed 27/04/2023 (2021)
15. Lepratto, F.: Experimenting with mass-housing regeneration in Italy: two pioneer actions in Bolzano as part of the SINFONIA project. Plan J. **7**(2), 529–552 (2021)

Interscalar and Interdisciplinary Approaches for a Valley Community. The Case of Sappada

Alessandro Massarente[1]([✉]), Alessandro Tessari[1], and Elena Guidetti[2]

[1] Department of Architecture, University of Ferrara, Ferrara, Italy
alessandro.massarente@unife.it
[2] Department of Architecture and Design, The Future Urban Legacy Lab, Polytechnic of Turin, Torino, Italy

Abstract. Our research aimed to explore an inter-scalar field in which architectural design tools are intertwined with urban and territorial scales. The case is an enclave between different limits: a valley near the border with Austria, located on the edge of north of Veneto region with an ethnic-linguistic heritage of Germanic matrix, administratively migrated in 2017 to Friuli Venezia Giulia region. Due to this specific geographical identity, local communities and municipal administration needed to define new sustainable development models of their territory. This gave us the chance to fine-tune a design exploration process holding territorial plan, with its analytical and programmatic categories. Additionally, it allowed us to address prefigurations, of a dispositive and dimensional nature, that enable the definition of thematic aspects through urban and landscape tools in various locations. Through discussion tables with local communities, various stakeholders, and interdisciplinary experts, as well as cross-referencing data provided by the administration itself, we elaborated open web-based GIS tools. These tools allow for the interrogation of data at different levels. From this analytical phase, five main themes emerged: naturalistic system, tourism resources, local economies, territorial infrastructures, and historical identity heritage. These themes represent the inter-scalar field within which different design explorations at urban and landscape scales intend to interweave possible relationships with territorial and urban plans.

Keywords: Inter-scalar field · Enclave · Spatial Prefigurations · Geographical identity · Local communities · Territorial plan

1 Architecture Beyond the Plan

The research work within which this project was developed aimed to explore an inter-scalar field in which tools of architectural composition intertwine with those of the project on an urban and territorial scale.

The Alpine landscape constitutively includes a spatial complexity that can only be explored through the interaction of different cartographic and representation tools, which often only partially allow to reflect this complexity.

© The Author(s) 2025
M. Barosio et al. (Eds.): EAAE AC 2023, SSDI 47, pp. 263–272, 2025.
https://doi.org/10.1007/978-3-031-71959-2_29

The territory of Sappada is already in itself an enclave between different borders: terminal valley whose northern mountain crests mark the border with Austria, located on the edge of the Comelico belonging to the high Veneto Cadore in which an ethnic and linguistic heritage of Germanic matrix since 2017 administratively migrated to the territory of the Friuli Venezia Giulia region.

These specific identity traits have determined on the one hand the need to redefine the models of sustainable development of the territory, on the other hand the opportunity to develop a design exploration process. This process is capable of intertwining different tools: cartographic instruments of the Plan, with its analytical and programmatic categories; prefigurations, by arrangement and size, that the urban and landscape project has allowed to define in the various places and themes identified; definitions of individual architectural elements serving mobility networks and public spaces.

This analytical phase unfolded in parallel and in constant interaction with a participatory discussion process, comprising various open discussion tables for the local population, stakeholders, and municipal administration, delineating a series of fundamental shared themes and strategies. From the elaboration of these themes emerged five thematic axes enabling the intertwining of potential relationships with the Plan.

The Plan, therefore, besides conventionally synthesizing analyses and summarizing urban regulations and operational tools, in this instance, selects and indicates short- and medium-term strategies through spatial prefigurations, which the Municipality of Sappada intends to translate into planning phases at different levels. In the wake of a tradition intertwining the Plan and studies on the city with urban design [1], this research experience potentially broadens the scope from Alpine landscapes to other territories, experimenting with a theoretical dimension transcending the Plan, through the exploration of the form as a precise sign that is placed in the reality and meanwhile is the measure of a transformation process. [2].

2 Themes and Methodologies

These distinct identity traits mentioned before have led to two significant outcomes. First, they have explored a redefinition of sustainable development models for the region. Second, they have provided an opportunity to start a design exploration process that integrates the cartographic tools of the Plan (Figs. 1, 2, 3).

These tools include analytical and programmatic categories [7], incorporating the prefiguration of arrangements and scales enabled by urban and landscape projects. [3, 5] They are defined across various locations and themes. Moreover, the endeavor involves crafting individual architectural elements impacting mobility networks and public spaces. Following this analytical phase, conducted with participatory discussion panels addressing issues such as natural resource management, culture and education, tourism, employment development, and social inclusion and safety, five thematic axes emerged. These axes, encompassing the naturalistic system, tourist resources, local economies, territorial infrastructures, and historical identity heritage, facilitated the structuring of a specific proposal for a strategic plan.

The naturalistic system is formed by three main areas, the first one is Piave river park. The second one is Sesis Valley, that represents the higher course of Piave river

Fig. 1. Orthophoto of the Sappada territory, Sappada, 2023 (courtesy of the Municipality of Sappada)

until its springs. Thus, the Plan not only organizes conventional analyses, compiles urban regulations, and outlines practical tools, but also goes a step further [6]. In this instance, it specifically identifies and recommends short- and medium-term strategies by means of spatial prefiguration.

The Municipality of Sappada aims to convert these strategies into various planning stages across different levels. In case of the lodges located in this mountainous area, that provide shelter and basic amenities to hikers, climbers, skiers. And this wider network of infrastructure will serve plateau of Sappada 2000, that show many potentials for these users. In summary, the key themes that shape this area of study are the Piave river branch, the Val Sesis Naturalistic Oasis, and the naturalistic system of Sappada 2000. These three areas significantly contribute to establishing the prominent "environmental axes" of Sappada. The potential for the community's future lies within these three natural systems: the envisioned path involves strategically linking them with the town center and its associated services, along with the municipal and territorial infrastructure hubs.

Within a broader perspective, the Plan operates on two fronts: firstly, intervening within the three natural systems by structuring a mobility framework, with a specific focus on sustainable transportation and public transit services, and enhancing their utility through the addition of new service infrastructures. Secondly, efforts are directed towards fostering connections between the three overarching systems and the town center.

The first phase of the research allowed an analytical reading of the territory from multiple disciplinary viewpoints: from the geographical to the naturalistic-environmental dimension, from the historical-cultural to the socio-economic aspects. This initial phase was conducted parallel to the development process of the participatory Strategic Plan "Sappada/Plodn 2050," undertaken by the administration in collaboration with ComPA FVG (a public body focused on strengthening the institutional, planning, organizational and management capacity of local authorities, bodies and of public companies in Friuli Venezia Giulia region), aimed at gathering input through open work tables involving citizens, local stakeholders, and external references. [4].

Fig. 2. Map of the Sappada territory with its three naturalistic macro-systems, Sappada, 2023 (image created by the research team and owned by the authors)

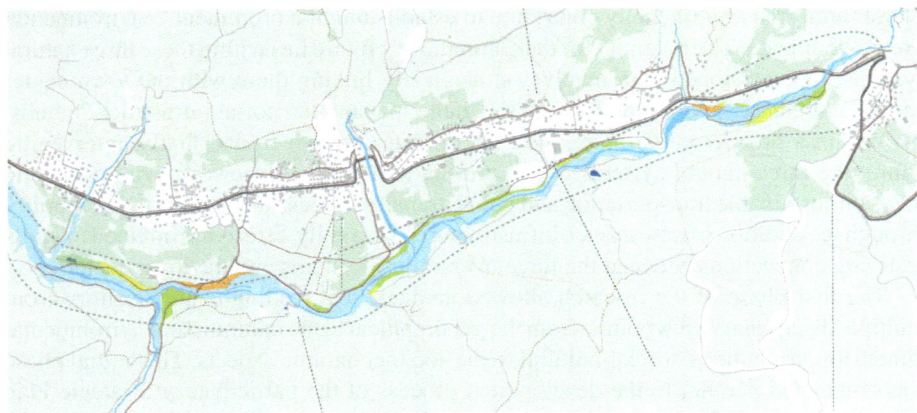

Fig. 3. The urban organization of Sappada along the east-west axis, parallel to the river Piave, Sappada, 2023 (image created by the research team and owned by the authors)

The elaboration of the general map of the river landscape, illustrating the overall state of existing elements, reaches a basic definition that includes, superimposed in layers, all the systems relevant to the park project: morphological, hydrogeological, and fluvial, both natural and anthropic. Alongside the research process, five thematic tables were organized in response to the aforementioned issues, conducted by the Municipality of Sappada with the University of Padua, enabling a better understanding of Sappada's critical issues and potentials through dialogue between citizens and experts, aiming for a more calibrated solution to the emerging problems. The role of the administration has been to participate without guiding the choices of local participants, but rather by introducing the issues discussed from time to time, contextualizing them within the social and political landscape of the community.

The thematic tables are integral to the understanding process of Sappada's territory and have been divided according to the different areas of interest indicated above. They also conclude the elaboration of the programmatic framework, aiming to obtain a concrete response to the critical issues raised for each theme, addressable and achievable over the next 10 to 20 years through properly planned interventions. Once the addresses and research topics have been collected and specifically identified, and subjected to content synthesis, the main macro-themes have been further explored, covering various territorial problems and potentials of Sappada on several fronts, to achieve their full recognition and development, leveraging their resources and opportunities.

For each macro-theme, cartographic elaborations have been identified to allow direct comparisons, seeking the broadest spatial understanding of the local reality, and to realize proposals that adequately meet the needs of the community and citizens of Sappada. Georeferenced files, photos, and images from municipal archives, including existing data in initiated and ongoing plans and projects over the past 20 years, have been collected for the construction of this diagnosis. To enable a direct comparison with research conducted in parallel, each macro-theme has been juxtaposed with analytical data on project initiatives emerging from discussion tables (Figs. 4, 5, 6).

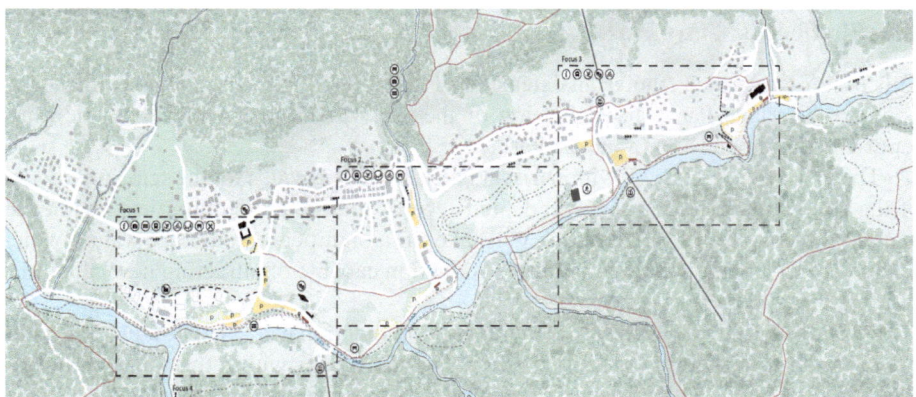

Fig. 4. General map of the river Park of Piave with the three project focuses, Sappada, 2023 (image created by the research team and owned by the authors)

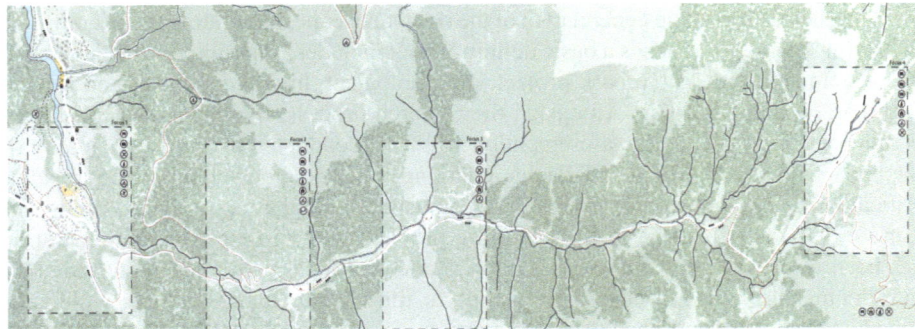

Fig. 5. General map of the Val Sesis Oasis with the four project focuses, Sappada, 2023 (image created by the research team and owned by the authors)

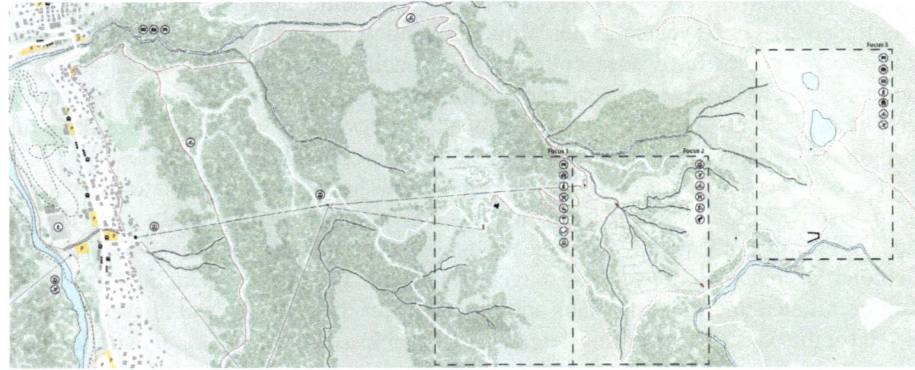

Fig. 6. General map of Sappada 2000 with the three project focuses, Sappada, 2023 (as above)

2.1   First Focus: Piave Park

This first focus, which frames an extension of about one kilometer, includes the access point to the Piave Park, the expansion of the craft area and the first thematic place represented by the vision of an Art Park.

The vision of an Art Park within a river basin, which provides for the semi-immersion of the stalls, guarantees an unprecedented type of tourist and cultural use. Within an aquatic dimension, the installations will take on a significant landscape value, which will create an authentic pole of cultural production capable of attracting new categories of users throughout the year.

The development of a linear park with specific functional areas, distributed throughout its extension, requires specific modular service blocks within the route, as indicators of a change in the landscape, capable of reverberating the material-environmental characteristics of the river landscape.

A second focus is dedicated to an eastern area along the Piave river, where will be possible to reach directly the river for bathing and walking.

A third focus is centered on food products coming from local agricultural activities, in connection with a previous dismissed military area.

This area was involved in a teaching activity in University of Ferrara since September 2022 to February 2023: students of 3rd year attending Master degree course on Architecture have developed architectural design exercises for the reuse of this area and its surroundings to get different scenarios, as outlined in a further chapter (Fig. 7).

Fig. 7. Art Park along Piave river. Axonometric view, Sappada, 2023 (image created by the research team and owned by the authors)

2.2 Second Focus: Val Sesis

The main element that characterizes the Nature Oasis of Val Sesis are the springs of the Piave River, as well as the river itself in its first stretch that descends with a torrential character until the height of Cima Sappada. It is precisely from this first village that a paved path starts, parallel to the riverbed that connects the town center to the springs, located at the foot of Mount Peralba.

The entire naturalistic compartment has an extraordinary landscape value that nevertheless presents few elements of tourist, recreational, receptive and cultural. The Piave springs themselves appear to be choked by parking lots, are barely visible, small in size and marked by a precarious arrangement of elements commemorative, protective, urban and landscaping totally inadequate to a monument with a very strong historical, cultural, symbolic and identity value for the entire nation.

The general project strategy begins with the intervention in the village of Cima Sappada, which is to be decongested from car traffic through a new parking area and the activation of a public transport service by bus-shuttle that solves the connection with the springs, maintaining the stops in the intermediate shelters. A specific stop module is designed at these points, also equipped with support equipment for cyclists. Then there is the provision of some parking lots for all the periods in which the service of public transport is not provided. The intervention on the springs of the Piave River involves the complete redesign of the area to give a wider scope to the springs themselves, attempting to simultaneously restore dignity and sacredness to the site (Fig. 8).

Fig. 8. The new access and reorganization of the Piave springs. Axonometric view, Sappada, 2023 (image created by the research team and owned by the authors)

2.3   Third Focus: Sappada 2000

The naturalistic territory of Sappada 2000 limited to the peaks arranged on the northern side of Sappada, develops only at high altitudes, bordering to the north, along the crests of Monte Lastroni, with Austria, to the east with the first virgin stretch of the Piave river, to the west by the Monte Ferro line and to the south by the town of Sappada. The intervention envisaged in this naturalistic context must develop in a scenario of amplification of the offers inherent to the high altitude, admitted only by the elaboration and implementation of the strategy for connecting the historic center to Sappada 2000.

The general objective is to structure a program no longer based only on sport users, but extended and aimed at family groups, at children, therefore, at all those activities generated in the recreational-sports-excursion field (Fig. 9).

The first focus of the intervention envisages the restructuring of the Sappada 2000 refuge, through some interventions: a system of naturalistic residences, built on the same

Fig. 9. The new sports and recreational infrastructures of Sappada 2000. Axonometric view, Sappada, 2023 (image created by the research team and owned by the authors)

ridge, together with a covered observatory; new playground built on the natural slopes of the land, near the refuge itself; a zip-line with departure at high altitude and arrival at the valley station; a paragliding starting point, with dedicated platforms, built just east of the Chapel dedicated to the fallen soldiers in Russia.

The design of the new panoramic service module for the entire territory of Sappada 2000 is the result of a sophisticated thematic development, supported by a severe need for adaptability to the context, from a geological, structural, meteorological and, above all, identity point of view. It is a module that can be positioned in different points of the high-altitude region, a decision resulting from the desire to restore the extraordinary visual observation potential in some of these places, thus enhancing those naturally generated panoramic points.

3   Didactic Experience

The didactic experience conducted in the Design Studio 3 of the University of Ferrara, within the framework of the Master Degree Course in Architecture, focused on the adaptive reuse of the site of the former Alpine Barracks Fasil in Sappada (Fig. 10).

The Studio, structured into three ateliers, engaged 133 third-year Architecture students, organized into 39 design teams. These teams grappled with formulating a design approach that meticulously considered the urban and landscape configuration of the Sappada region. Under the guidance of 7 instructors, the projects were geared towards three primary functional scenarios: innovative tourist accommodations; educational facilities rooted in the region's excellence; pioneering residential and coworking spaces. The projects developed challenged the conventional approach to existing heritage, particularly the abandoned barracks building. These proposals presented design solutions where the interplay between the existing structure and the new intervention was encapsulated

Fig. 10. Exercise on adaptive reuse of former Alpine Barracks Fasil in Sappada. Model, Ferrara, 2023 (photo by Alessandro Tessari, owned by the author)

within two overarching morphological strategies: "inside" which focused on the barracks building itself and involved processes of addition and subtraction, and "outside" which focused on extensions and clusters of new buildings.

This didactic experience underscored the complexity in intervening within a multifaceted landscape, rich in identity, urban and architectural significance and memory. The multifarious nature of the solutions presented underscored the complex features of the site, offering a glimpse into a multitude of potential adaptive reuse scenarios.

References

1. Secchi, B.: Edifici-mondo. Progetti per il centro storico di Salerno, Casabella, 667 (1999)
2. Rossi, A.: Architettura per i musei. In: Locatelli, A. (ed.) Teoria della progettazione architettonica, pp. 132. Dedalo, Bari (1968)
3. Smith, J., Jones, A., Brown, K.: Urban and landscape projects: analytical and programmatic approaches. J. Urb. Plan. **20**(3) (2020)
4. Healey, P.: Collaborative planning in perspective. Plan. Theory **2**(2), 101–123 (2003)
5. Lynch, K.: What Time Is This Place? The MIT Press, Cambridge (1972)
6. Paez, R.: Operative Mapping: Maps as Design Tools. Actar, Barcelona (2019)
7. Cavalieri, C., Viganò, P. (eds.): The Horizontal Metropolis: A Radical Projects. Park Books, Zurich (2020)

Architectural Design Studio: Embracing a Transdisciplinary Approach

Christina Panayi[✉], Effrosyni Roussou, and Nadia Charalambous

University of Cyprus, Nicosia, Cyprus
panayichrist@gmail.com

Abstract. The co-creation design studio at the Department of Architecture, University of Cyprus, has been acting as a meeting place for students, educators, researchers, citizens, and external stakeholders since 2021, aiming to bridge gaps between architectural research, pedagogy, civil society, and local governance bodies through a transdisciplinary pedagogical framework. Acknowledging that many societal challenges are complex and multifaceted and cannot be adequately addressed by any single discipline or sector alone, the studio focuses on the co-production of knowledge with stakeholders outside of academia. This entails involving them from the outset of the project as well as co-designing design activities and proposals that are relevant to their needs and interests, ensuring a grounded process in real-world challenges. The studio's pedagogical framework and methodology have been designed, implemented, and evaluated three times until now through participatory action research methodology, investigating the impact on the design result, the development of skills for the students, and their attitude towards their role as future professionals. The paper highlights the findings of these three years of research in a reflective way, suggesting future steps for improvement. Its long-term repetition will gradually build a knowledge base, aiming to revisit existing educational methods to respond to current and future challenges in an efficient and inclusive way.

Keywords: Architectural Design Studio · Transdisciplinarity · Community Engagement · Architectural Pedagogy · Architectural Research · Participatory Action Research

1 Introduction

1.1 Current Approaches to the Planning of Cities and Neighbourhoods

As cities around the world have been radically and continuously changing in response to diverse challenges, urban planning and city development need to evolve into complex and multidimensional processes that place urban resilience, social sustainability, and a sense of 'belonging' at the forefront. Understanding urban environments as places with different actors and stakeholders requires the collaborative involvement of many and diverse stakeholders in their design. Transdisciplinary cooperation in decision-making, co-creation approaches in the design process, as well as innovative initiatives regarding urban commons, city and neighborhood planning, appear to be more important than ever.

© The Author(s) 2025
M. Barosio et al. (Eds.): EAAE AC 2023, SSDI 47, pp. 273–282, 2025.
https://doi.org/10.1007/978-3-031-71959-2_30

Urban resilience has become an important goal for cities, particularly in the face of contemporary challenges [1]; it has to do with recovering from crises and is directly related to the concept of sustainability, as the two concepts are recognized as interrelated, highlighting the potential to achieve sustainable urban development [2]. Social sustainability is acknowledged as an important part of sustainable development as a process that improves a community's quality of life [3], resulting in their happiness, security, health, and quality of life [4]. Social relationships and a sense of "belonging" are important factors in a community's social sustainability and resilience [5, 6], contributing to shared emotional connection through interaction with others, strong shared values, shared norms and codes [7, 8]. Educators, scholars, practitioners and activists among many others, have been highlighting inclusion and participation in the decision making and design processes regarding urban space, as a vital element towards achieving and sustaining urban resilience [9, 10]. This argument illustrates the importance of transdisciplinary approaches in knowledge production and of co-creation practices in the design of urban spaces.

Transdisciplinary and transdisciplinarity are relatively recent terms, meant to highlight the need to transcend disciplinary boundaries and create holistic problem-solving approaches for city planning through the combination knowledge, tools and methods stemming from both the academic and non-academic world [11]. A transdisciplinary approach supports bottom-up collaborations, and creates a nurturing environment for mutual learning and cross-pollination between different participants [12]. Such a setting is a prerequisite for a co-creation process to flourish. In urban planning, co-creation can be described as the last steps on Arnstein's ladder of participation, "delegated power" and "citizen control" [13]: co-creation is defined as a process where participants can have active and prominent decision-making roles, exhibiting traces of self-organisation, increased commitment and a sense of ownership of the process [14].

Recent research is exploring new methods of urban planning in which citizens and municipalities co-create new planning rules or collectively shape and prioritise actions related to urban space [15], as is the case of the Co-Cities initiative in Bologna, a co-governance scheme between citizens and municipal authorities that provides a transdisciplinary, co-creation framework for joint decision making in matters regarding the city's services, policies and spaces [16]. Such practices highlight a continuity in engagement and a level of delegation of power that are crucial in fostering a sense of ownership and belonging among citizens and community members, towards both the urban spaces, and the processes that shape them. Therefore, the urban commons paradigm becomes an increasingly relevant model for urban governance. The term urban commons refer to urban resources (commons) that are managed by their users collaboratively (commoners), through collectively agreed-upon rules in a non-profit-oriented way (commoning) [17]. As a co-governance model, the urban commons paradigm highlights the importance of redefining urban values in contemporary cities while, in parallel, helping to build creative and open processes that foster experimentation, collaborative knowledge production, and trust among those involved.

1.2   Architectural Research and Pedagogy Having a Social Impact

The need for continuous, inclusive, and pro-social practices in urban design, development, and governance should also be reflected in the way future design professionals, specifically architects, are trained. Consequently, the need for a more responsive architectural pedagogy raises questions about the relevance of knowledge and research in architecture as an educational discipline and as a profession [18]. The traditional role of the architect, as well as the knowledge and skills required of a future architect, are being called into question. According to academic Ashraf M. Salama [18], there should be a continuous focus on the skills required for successful practice by incorporating scenarios that involve real interaction with the everyday environment into architectural pedagogy programs to promote critical thinking. Various architectural educational methods focused on community engagement, situated learning, and transdisciplinarity are now at the center of discussion in this framework, supporting the student's active role in the process of learning [19]. Since the beginning of the 21st century, a wide range of different pedagogical architectural approaches and methodologies that propose the architectural design studio to be embedded in the real environment through a co-creation framework have been widely discussed, applied, and criticized, such as community design [20], "live projects" [21, 22], and "design-build projects" [23].

Moreover, universities have a civic responsibility to the cities of which they are a part, both in terms of sharing the expertise and knowledge they produce [24] and in terms of their role as "significant institutional actors" in urban development processes with great power and visibility [19]. Communities and universities, according to Sara and Jones [24], can develop a twoway collaboration between citizens, architects, students, and academics by co-creating new knowledge. This knowledge co-production makes the university "an urban agent with transformative potential for co-creating more sustainable, resilient communities" and has the potential to be transformative as the city and its residents become change agents for the institution.

In this framework, the co-creation studio at the Department of Architecture at the University of Cyprus (https://www.instagram.com/cocreationstudio.ucy/) was established and implemented in 2021 in an attempt to put a higher priority on social sustainability issues during the architectural design studio and systematize research and its relationship with pedagogy. It embraces a transdisciplinary framework and incorporates community-engaged approaches and situated learning. Simultaneously, it examines the impact of the approach through participatory action research and knowledge co-production with stakeholders outside of academia, linking research with teaching and learning. Since 2021, the studio has been implementing co-creation activities in three different Nicosia neighborhoods (Fig. 1).

Fig. 1. The design of a small public space in Mylou neighborhood in Latsia Municipality, (Work by Co-creation Studio UCY, 2022).

2   Methodology

In the execution of this research and the attainment of its objectives, Participatory Action Research (PAR) is employed. Its participatory nature allows all involved stakeholders, to actively contribute to knowledge production, and its cyclical development, characterized by iterative phases of planning, action, observation, and reflection that bridge theory with practice, fosters continuous improvement and change [25]. Educators-researchers engage in a form of "self-reflective teaching" with the overarching goal of fostering improvement and change [26]. The reflective framework and cyclical nature of continuous improvement and review establish meaningful connections between research findings and practical educational applications. The implementation of this research unfolds in three consecutive phases-cycles of exploration (2021–2023). Each phase-cycle is grounded in the theoretical framework and methodology of Urban Living Labs [27, 28], as well as in the iterative cycle of action research proposed by Kemmis and McTaggart [29], consisting of four main stages: framework design, implementation, co-evaluation, and reflection and re-design (Fig. 2). Anything occurring during the framework implementation is fluid, non-linear, and susceptible to changes, primarily following a non-linear three-stage process without clear boundaries: a) Co-identification stage: Co-assessing and understanding the existing situation, identifying issues or needs; b) Co-creation stage: Co-development and co-selection of solutions and scenarios; c) Co-design and implementation stage: Implementation of the optimal solutions are implemented in the agreed form, scale, and degree of implementation.

The first phase of research is embedded in the 2nd-year architectural design studio at the University of Cyprus during the Spring Semester of 2021, involving 28 s-year architecture students. Due to the COVID-19 pandemic, the studio is exclusively conducted online. The thematic focus centers on contemporary living concepts, including "collective housing," "cohabitation," and "sharing," explored in Pallouriotissa, Nicosia. The second phase unfolds in the Spring Semester of 2022, with 30 s-year students and a summer workshop in June 2022 involving 32 participants. The emphasis remains on "collective housing" and "sharing" in residential and public spaces. Students are tasked with designing collective housing forms in specific plots within a Latsia Municipality neighborhood, Nicosia, near a small public space. Detailed planning of the public space occurs during the summer workshop, alongside the implementation and evaluation of specific constructions with the support of the University of Cyprus Fab Lab. In the third phase, during the Spring Semester of 2023, 29 s-year students participate. Concurrently, a Spring course engages 20 students (third-year and fourth-year). This phase critically explores "collective housing" and "sharing" within the neighborhood, examining the school's role and its relationship with housing in Latsia Municipality, Nicosia, adjacent to the Latsia Lyceum. Parallel groups in the studio develop ideas for collective housing, while the students of the parallel course are tasked to design a structure in the school-yard as a connecting link with the neighborhood. Both courses are followed by a summer workshop in June 2023, involving 21 participants, implementing one of the proposals for construction in the schoolyard with the assistance of the University of Cyprus Fab Lab.

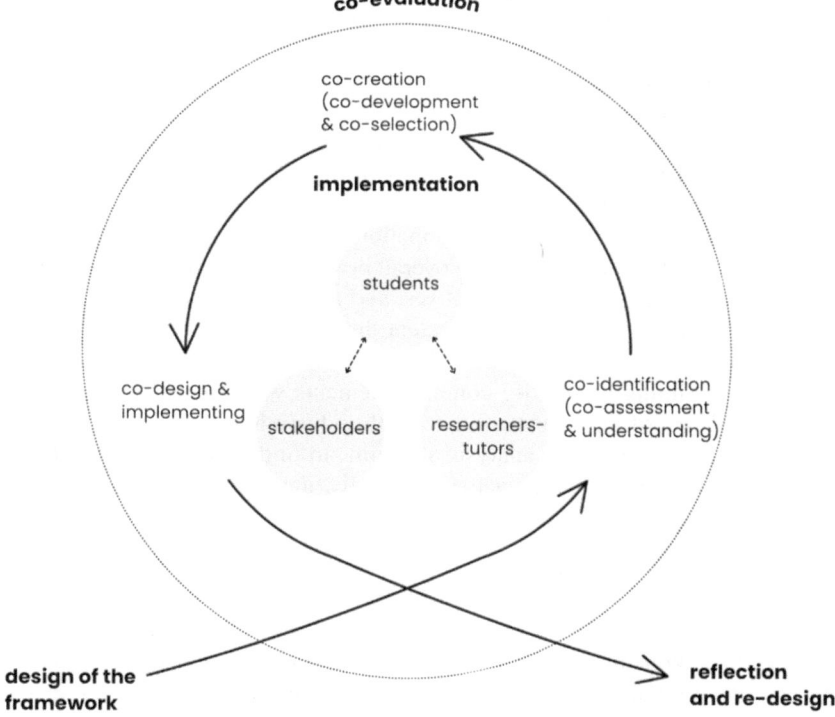

Fig. 2. The circular methodology of the Co-creation Studio, consisting of four phases: design, implementation, co-evaluation, reflection, and re-design (© Panayi Christina, UCY Library, 2023).

Co-evaluation and reflection process takes place simultaneously with the implementation and is structured around three key axes: a) evaluation of the structure and methodology of the studio as a teaching and learning process and a co-creation process; b) evaluation of the impact of the pedagogical approach on the students (skill development, learning outcomes, motivations, experiences, attitude towards their future professional role) and the design outcome; c) evaluation of participatory action research as a methodology for co-creation of knowledge. Data collection and analysis aims at yielding insights into the entire co-creation process, its effects on the various stakeholders involved, and the overall effectiveness of the pedagogical framework. The co-evaluation process involves data collection from the educators-researchers, students, and other stakeholders (e.g., citizens). Students participate in focused group discussions held at specific moments during the process to capture emotions, thoughts, and informal opinions. They are also required to respond to specific questionnaires, typically before and after each course, focusing on their motivations, experiences, roles in the process, self-reflection of the design outcome, skill development, and achievement of learning outcomes. Students also take part in semi-structured interviews, primarily consisting of open-ended questions, aiming to record emotions and spontaneous reactions. During the interviews, they are asked to comment on their overall experience, the design outcome, their gains, recommendations for future improvements regarding the learning process, and their general thoughts and ideas. The educators-researchers continually monitor the process to document students' spontaneous behaviors and reactions, collecting data through observation and recording them in a reflective diary. Additionally, continuous evaluation of the design outcomes delivered by students occurs through individual tasks and presentations at regular intervals. Lastly, the participating stakeholders are interviewed, expressing their opinions on their experience and gains and evaluating their involvement in the process. They also play an active role in assessing the design outcome through interviews and their participation in intermediate or final presentations.

Both the data collection tools and the overall process of co-evaluation are adapted to each semester's program, learning objectives, and the students' potential. The validity of the data is ensured to a considerable extent through triangulation, both in terms of using different tools and collecting data from different sources. The data are qualitatively analyzing, aiming to identify common elements within each of the three axes. After the second year of implementation, to confirm hypotheses and identify any recurring patterns over the years, the analysis also aims to find horizontal connections and correlations between the years in each axis. Finally, the analysis method also aims to identify correlations between the three axes.

3 Co-creating Knowledge Over These Three years of Research

Several outcomes about the co-creation framework as a participation, teaching, and learning process, its impact on students, and the design result(s) have emerged incrementally over the three years of research, with the process being updated through improvement actions from year to year. The outcomes of 2023 are still ongoing and have not been completed. Some of the most important ones are summarized below, concerning students' attitudes toward their role as future architects, their perspectives on urban commons and architecture's social responsibility, their gains and motivation, as well as the researcher's/tutor's role.

Students' exposure to real-life settings and engagement with other people enabled them to explore their role in a diverse team and view negotiation and conflict as opportunities for productive and fruitful discussion. Throughout the process, students often changed their roles, from passive listeners to discussants, leaders, and negotiators, learning to use arguments to defend their point of view. They also started to explore the multiplicity of the roles of the professional (architect, urban planner), understanding the complexity of urban environments and urban actors. Interestingly, when students were asked about how they perceive the role of the architect in contemporary urban processes and consequently their own role as future professionals, the answers varied considerably. Several students viewed the architect as needing to have a central role in the design process(es), leading the discussions and developing the design vision based on the input from prospective users. Others perceive the architect's role as that of a mediator, someone who facilitates the design and co-creation processes and is in continuous negotiation with the involved stakeholders. Few saw the role of the architect as that of an ally to local communities, emphasizing the importance of enabling community-engaged or community-initiated processes and mildly highlighting the political aspect of a co-created urban development (Fig. 3).

Fig. 3. Designing and implementing a meeting place in Latsia High School yard (Photos by Effrosyni Roussou, 2023).

Regarding urban commons, the majority of the participating students stated that they recognise their value as a way to counter the high individualisation of contemporary societies. They regard positively the potential of the urban commons to foster social and environmental values as well as create mutual learning hubs for communities. Their own role, as many of the students highlighted, is to be active both as citizens and architects and act as "a unifying force" between the various stakeholders.

From 2022 onward, additional (optional) design and build activities were introduced as intensive summer workshops, the built outcomes of which would then be handed over to the communities that participated in their conceptualisation during the semester.

Through these activities, students were exposed to the complicated world of architecture's interdependencies; from material costs and delivery times to navigating design for 1:1 scale, they were able to realize the complex and volatile chain of actions necessary to transition from paper to reality. Through hands-on, building activities, they were able to empathize with and acknowledge the contribution of those involved in a construction process (construction workers, engineers, material suppliers, etc.), have increased commitment in the process, and exhibit aspects of increased collaboration, communication, and self-organization, anchored in the belief that they are working on something meaningful but also in the joy stemming from playful creation.

Regarding the researcher's/ tutor's role, the inherent reflexivity of the action research process and, in this case, the double role of educators as researchers led to personal changes, re-evaluations of the individual pedagogical practices, and positioning towards the students, as well as -ultimately- to personal growth. The tutor's role is thus under question in terms of how intrusive it can be, the amount of advice or support it can provide, and the ways that reviewing and assessment can take place in such a co-creative approach.

4 Discussion

Live pedagogies offer the opportunity to transcend the boundaries of the classrooms and consequently reconfigure the predefined hierarchies of teacher-student. In processes of co-creation and design and build, the clear boundaries between learner and teacher, as well as the clearly defined "roles" and behavioral "etiquette", are more easily blurred. This allows for instances of informality and sharing to permeate the process, which create bonds of trust and cooperation and ultimately engage both students and tutors as whole-people in a nurturing and caring learning environment. The disparity in the students' responses in regards to their perception of the role of the architect in contemporary urban development processes is indicative of the limits of "isolated" pedagogical efforts. In order for meaningful and profound change to occur in students' attitudes and perceptions, one course or studio will likely not be enough. What is needed is broader collaboration between the teaching staff, aligning agendas and complementing topics and methodologies, as well as collectively trying to expand beyond disciplinary boundaries both academic and non-academic partners.

Finally, architecture schools need to become active stakeholders and actors in urban processes as hubs of knowledge and experimentation that open up new possibilities and visions. Both research and pedagogy can and should have a tangible social impact that is substantial for communities and people. The social and political dimensions of architecture should be at the forefront, highlighting the need for a socially sustainable, community-engaged and politically aware design practice that could counteract the deregulatory dominance of the global market. In this manner, the participation of communities and actors beyond academia in architectural research and pedagogy is significant, involving civil society in conceptualizing and structuring the research process and knowledge co-production. Establishing and maintaining long-term relationships between academic institutions, governing organizations, and civil society for urban decision-making can be transformative for all participants and the city, cultivating a sense of responsibility for urban living environments. Thus, the city may be viewed as a change agent for the university (research and education), while the university can be viewed as a change agent for the city [24].

References

1. Bautista-Puig, N., Benayas, J., Mañana-Rodríguez, J., Suárez, M., Sanz-Casado, E.: The role of urban resilience in research and its contribution to sustainability. Cities **126**, 103715 (2022)
2. Zeng, X., Yu, Y., Yang, S., Lv, Y., Sarker, M.N.I.: Urban resilience for urban sustainability: concepts, dimensions, and perspectives. Sustainability **14**(5), 2481 (2022)
3. Colantonio, A.: Urban social sustainability themes and assessment methods. Proc. Inst. Civil Eng.-Urb. Des. Plan. **163**(2), 79–88 (2010)
4. Grum, B., Kobal Grum, D.: Concepts of social sustainability based on social infrastructure and quality of life. Facilities **38**(11/12), 783–800 (2020)
5. Dempsey, N., Bramley, G., Power, S., Brown, C.: The social dimension of sustainable development: defining urban social sustainability. Sustain. Dev. **19**(5), 289–300 (2011)
6. Rapaport, C., Hornik-Lurie, T., Cohen, O., Lahad, M., Leykin, D., Aharonson-Daniel, L.: The relationship between community type and community resilience. Int. J. Disaster Risk Reduct **31**, 470–477 (2018)
7. Kearns, A., Forrest, R.: Social cohesion and multilevel urban governance. Urb. Stud. **37**(5–6), 995–1017 (2000)
8. Talen, E.: Sense of community and neighbourhood form: an assessment of the social doctrine of new urbanism. Urb. Stud. **36**(8), 1361–1379 (1999)
9. LopezDeAsiain, M., Díaz-García, V.: The importance of the participatory dimension in urban resilience improvement processes. Sustainability (Switzerland) **12**(18) (2020). https://doi.org/10.3390/SU12187305
10. Esteban, T.A.O.: Building resilience through collective engagement. Architecture_MPS. **17**(1) (2020). https://doi.org/10.14324/111.444.amps.2020v17i1.001
11. Doucet, I., Janssens, N.: Transdisciplinary knowledge production: towards hybrid modes of inquiry in architecture and urbanism (2011). https://doi.org/10.1007/978-94-007-0104-5
12. Klein, J.T., Grossenbacher-Mansuy, W., Häberli, R., Bill, A., Scholz, R., Welti, M.: Transdisciplinarity: joint problem solving among science, technology, and society an effective way for managing complexity (2001). https://doi.org/10.1007/978-3-0348-8419-8_2
13. Arnstein, S.R.: A ladder of citizen participation. J. Am. Plann. Assoc. **35**(4), 216–224 (1969). https://doi.org/10.1080/01944366908977225
14. Puerari, E., de Koning, J.I.J.C., von Wirth, T., Karré, P.M., Mulder, I.J., Loorbach, D.A.: Co-creation dynamics in urban living labs. Sustainability (Switzerland), **10**(6). (2018). https://doi.org/10.3390/su10061893
15. Bisschops, S., Beunen, R.: A new role for citizens' initiatives: the difficulties in co-creating institutional change in urban planning. J. Environ. Planning Manage. **62**(1), 72–87 (2019). https://doi.org/10.1080/09640568.2018.1436532
16. Foster, S.R., Iaione, C.: Co-cities : innovative transitions toward just and self-sustaining communities. MIT Press (2022)
17. Urban Commons Research Collective: Urban Commons Handbook, 1st ed. University of Sheffield (2022)
18. Salama, A.M.: Skill-based/knowledge-based architectural pedagogies: an argument for creating humane environments. In: 7th International Conference on Humane Habitat-ICHH (2005)

19. Natarajan, L., Short, M., (eds.): Engaged Urban Pedagogy: Participatory practices in planning and place-making. UCL Press (2023)
20. Salama, A.M.: Seeking responsive forms of pedagogy in architectural education. Field **5**(1), 9–30 (2013)
21. Anderson, J., Priest, C.: Developing a live projects network and flexible methodology for live projects. In: Live Projects Pedagogy International Symposium (2012)
22. Harriss, H., Widder, L., (eds.): Architecture live projects: Pedagogy into practice. Routledge (2014)
23. Stonorov, T., (ed.): The design-build studio: crafting meaningful work in architecture education. Routledge (2017)
24. Sara, R., Jones, M.: The university as agent of change in the city: co-creation of live community architecture. Int. J. Archit. Res. **12**(1), 326–337 (2018)
25. Walter, M.: Participatory Action Research. Social Research Methods (2009)
26. Tran, T.T.H.: Why is action research suitable for education?. VNU J. Foreign Stud. **25**(2) (2009)
27. Bulkeley, H., et al.: Urban living labs: governing urban sustainability transitions. Curr. Opin. Environ. Sustain. **22**, 13–21 (2016)
28. Menny, M., Palgan, Y.V., McCormick, K.: Urban living labs and the role of users in co-creation. GAIA-Ecol. Perspect. Sci. Soc. **27**(1), 68–77 (2018)
29. Kemmis, S., McTaggart, R.: The Action Research Planner. Deakin University, Australia (1990)

Mountains in Motion, Visions in Nutshells. The Alpine Way for Common Living

Alisia Tognon$^{(\boxtimes)}$

Department of Architecture and Urban Studies, Politecnico di Milano, Milan, Italy
`alisia.tognon@polimi.it`

Abstract. Architectural research is defined as community practice and for communities. Architecture has always been constituted as an action of territorial settlement and domestication for the community liveability of places.

This article, referring to the concepts of resilience and transformability of mountain systems, proposes to reflect on how rural mountain areas have always been able to adapt to changes thanks to the ability of communities to conceive common and shared design directions.

The paper reflects on the conceptualization of commons and how they can be a model for resilient development in socially inclusive contexts. For this reason, an extensive review of historical palimpsests was conducted to understand how to start from the roots of the operating models.

The article includes a field approach in Trentino's valleys. The results are exploratory, as the process cannot be concluded and requires a long time for validation. From it, some working hypotheses are derived to be explored in further studies, such as reintroducing the commons into projects and societies in transformation.

Architectural research in these contexts contributes to orienting knowledge through fundamental collaboration with other disciplines, which is helpful in defining regenerative processes consistent with specific environmental contexts.

Keywords: Regeneration · Commons · Community engagement · Alpine region · Trentino cultural landscape

1 Domesticated Mountains

The mountains result from a lengthy process of human domestication aimed at rendering them inhabitable.

This adaptation is closely tied to the various forms of exploitation that have emerged from the study of the knowledge of these spaces, with their inherent natural and environmental characteristics. In this regard, collective practices have fostered a transformative relationship between architecture, landscape, and community actions within these mountainous territories. In this line, the value of the mountains lies in their capacity for self-regeneration, and today, the existing connection with the local community remains a foundational practice that imbues these places with meaning, transforming them through new relationships (Fig. 1).

M. Barosio et al. (Eds.): EAAE AC 2023, SSDI 47, pp. 283–293, 2025.
https://doi.org/10.1007/978-3-031-71959-2_31

Fig. 1. Abandoned Masi in Celado, Tesino Valley (TN) (©Tognon 2023)

The paper focuses on the ongoing research[1] which deals with understanding how new regenerative possibilities on the mountain's territories emerge from the interaction of ecological, cultural, and social criteria that underlie mountain civilization and appear to be fundamental to living and producing in the mountains and, therefore, non-negotiable. Envisioning a future for mountains means basing it on the reestablishment of a strong relationship between communities and the land by integrating what Pazzagli refers to as the three "P's": "landscape," "villages," and "products," which are endogenous elements for the long-term care of the area. It is necessary to direct the narrative towards a conscious appreciation of the territory, investigating and emphasizing latent knowledge capable of including the individual, uniting the community, generating sustainable economies, and reconstructing a history while containing the future one. In this perspective, the concept of "fragility" [1] as a lens for examining the territory seeks to explore the specificities that can be related not only from a natural perspective but also from a social, economic, and architectural standpoint.

The research focuses on the alpine region, a "unique region at the center of Europe" [2], to which special attention must be paid. Operating in a design logic on these territories, it is not necessary to limit to a national perspective, but it is essential to think of the territory in a transnational key [3] since the Alpine territories belong to the same mountain range, although with their regional peculiarities. Moreover, the Alps geographically range from high snowy peaks to the sea, resulting in a rich complexity of interconnected variables: different topographies or orographic masses, variations in the shapes and environmental conditions, and unique conditions imposed by life [4]. Overlapping with these geographical components is the immaterial aspect defined by their close relationship with humanity.

2   Alps, Common Ground

As Ferrari [5] states, defining the Alps could even lead to tautology: the mountain is what is perceived as such. In other words, the mountain is perception and representation and what we perceive in our minds through our eyes. How to view the Alps is a complex theme that has evolved over the years because it encapsulates the ability to create a mental geography that transcends the physical aspect. The changes in perceptual direction have portrayed them in muted ways over the years: from inhospitable places inhabited by wilderness communities to valleys as romantic expressions of a new Arcadia for regenerating city dwellers, from John Ruskin's "cathedrals of the earth" for explorers, to playful places with ski total resorts [6]. Even today, the challenge lies in the multiplicity of perspectives from which we view the Alps. Nevertheless, it is essential to understand the implications these interpretations have on the physical landscapes and the future. In this context, the Alps should be recognized as having an intrinsic value that goes beyond economic considerations [5] while avoiding falling into an ideologically driven environmentalism that disregards the necessary balances for human life in a region that has been historically influenced by human activity.

Humans in the Alps should be seen as active participants in the natural balance, historically shaping and preserving this equilibrium. The Alps could be described as an anthropogeographical landscape, the result of millennia of gradual transformations. Initially, the Alps were covered with dense and almost impenetrable forests up to high altitudes, while the extensive and flat valley bottoms were often marshy, threatened by floods, and only passable and usable during dry seasons [7]. Only the alpine meadows above the tree line were suitable for immediate exploitation, albeit to a limited extent.

To establish permanent settlements in this region, humans had to intervene in the ecosystems and transform them according to their needs, shaping the alpine region as an architecture.

2.1 The Vertical Dimension

The alpine territory, analyzed systemically for design purposes, should be scanned and considered not only in the horizontal connection of valley systems but mainly in the vertical relationship between bottom valleys and peaks. The vertical component leads to reading the valley system as a tomographic series of cross-sections. This analytical, interpretive key stems from the vertical structure of the valleys, which required the inhabitants to cope with natural catastrophic events that threatened settlements both downstream and upstream. This made settling the land and its use for people's livelihoods challenging. Nonetheless, human interventions also took the form of modifying plant species and individual ecosystems, as well as landscape structuring to mitigate the discontinuous dynamics of natural risks into more manageable and controllable events.

In a general sense, according to Bätzing [2], the transformation of the Alpine ecosystem can be categorized into three levels of interventions: (a) creation of the Alpine pastureland - expanding pastures through clearing and transforming vegetation cover via deforestation, and other related activities; (b) creation of valley farmland - clearing and cultivating land in the valley areas; (c) reclamation and clearing of large valley bottoms. This involved drainage and clearance of extensive low-lying valley areas. These actions led to the development of an architectural landscape, the signs and features of which can still be observed today [2: 111–126].

This type of landscape is an ideal vertical section, which reflects the construction of an artificial space divided into three parts to make it habitable. This scenery is referred to a cultural (or anthropized) environment and is situated alongside the natural land in dynamic and intermediate stages, and it is often described as a "semi-natural" or a "semi-cultural" landscape. Aligning with Bätzing's definition, the "cultural landscape" is understood as an area in which human activities have shaped the natural ecosystem in such a way that an ecosystem has replaced it with a cultural imprint. The stability of this new ecosystem is now the responsibility of humans, as the new ecosystem is not inherently stable [2: 117].

This landscape was also shaped by a collective management model, which has been fundamental in caring for the territory in the Alps by land (agricultural, pastures, forests) and maintaining the built environment associated with them. As already highlighted, the regulation of the mountain territories is an ancient process that began around the XII Century in the Alpine region, where ecological and social systems have historically co-evolved and become deeply rooted [8, 9]. Traditionally, highland areas have been regulated by "civic use", and these models of collective management [10] were developed to address local populations' need to manage resource scarcity autonomously and collectively. They created specific norms of land use to ensure their livelihoods while maintaining a balance between productive activities and environmental preservation.

From a spatial perspective, it is interesting to note how these management models have been physically translated into spatial contexts and, consequently, into environmental and social palimpsests deeply ingrained in local traditions [11]. We highlight that preserving these practices has facilitated the care of the material and immaterial cultural heritage over time.

The model continues to serve as an essential historical and cultural framework for local development in mountain regions [12]. Indeed, within the context of mountain societies and ecosystems' changes and management challenges today, commons represent unique habitats. They warrant not only in preserving traditional knowledge but also in upholding ecological resources and in their ability to anticipate future scenarios.

2.2 Rural Commons, the Palimpsest of the Alps

The cultural relevance of this approach to territorial management for the Alps also reverberates in contemporary principles of local administration. In fact, within the evolving concept of economic development, "commons" have established themselves as foundational areas of interest, defined as "sustainable local economies" [13].

In the ongoing research we consider commons not as a mere subsistence economy but rather as a rational mode of resource management and utilization [14], where the use of local natural resources, combined with a system of mutual management, enables the creation and survival of communities even in challenging and isolated territories [15]. The "Regole" [16] created a connection between the collective of inhabitants, known as "vicini", and the territory on which they lived. This regulation established a link between people and territory (understood as a natural reality), the assembly of household heads, and individual local entities within the broader valley context. The purpose of the administration under these regulations was to control the standardized exploitation of common lands and safeguard small private holdings, referred to as "divisi".

Over the years, the methods of utilizing natural resources, based on communal use of common assets, have remained essentially unchanged. Cultivable plots of land belonged to families, although even privately defined areas of cultivated fields could be subject to limitations in favor of the community. According to Nequirito[16], analyzing the community's territory, four distinct levels are structured. (a) The village represents the private side of each family unit and worked as core of the housing area. (b) Cultivated lands, typically located near the inhabited centers, consisted of fields and meadows defined by hedges, fences, or dry-stone walls. The route is allowed during unproductive seasons for all the community. (c) Forests: broadleaf forests were situated in areas bordering the village settlements and were used for firewood. Meanwhile, coniferous forests at higher altitudes provided timber for trade. (d) Pasturelands were collectively exploited at various altitudes and utilized according to the seasons. They were overseen by common huts ("malghe"), where those appointed by the community, including herdsmen and shepherds, produced milk from flocks or herds brought to pasture. (Fig. 2).

In addition, the infrastructure of the territory was structured, and the water system was planned by the community, which were involved maintaining canals and embankments, monitoring the purity of water sources, regulating water flow, and managing its use through mills and sawmills, among other facilities. The road system was also a matter of mandatory contributions from the community or those with property along the roads. The "regola" defined specific control regulations and imposed strict penalties for those who damaged public roads and places. Additionally, essential activities such as butchery or baking, as well as specific aspects related to religious life regulations, were governed and managed by the "Carte di Regola" (trad. Regulation Charters).

The long life of these "regole" has left visible traces in various Alpine areas. The forms of land use for silvo-pastoral assets are reflected today in the toponymic heritage of places, a reminder of ancient customs and methods of using Alpine resources. This historical legacy is also visible in the landscape's configuration and the enduring institutions.

The Trentino case

The two autonomous Italian provinces of Trentino and South Tyrol are a peculiar case in the Italian Alps, since historically encompass the highest number of collectively managed lands, covering a significant percentage of their territory [14]. Together the provinces create a region where the concept of "common good" has been expanded not only to material assets, but also to a substrate of traditions and practices of intangible heritage.

Looking specifically to Trentino, the commons as management model are still strongly present in several valleys (e.g., Magnifica Comunità of Val di Fiemme, Regola of Spinale and Manez, etc.), defining a model of virtuous land management, but also identifying a community closely linked by a strong sense of identity with the land in which they live over generations [17].

As profit motives did not drive these communities, all proceeds were reinvested for the benefit of the territory, with fair resource distribution. Due to their ability to self-regulate, these local communities have achieved a degree of autonomy, which still reflects in local autonomy of Trentino South Tyrol. Over the past 60 years, both the Autonomous Provinces have invested in policies to preserve the social and cultural fabric and safeguard natural resources [18]. They recognize the intangible value of collective institutions, which play a fundamental role in shaping the landscape [19]. In accordance with this, the legacy of the collective management model is still prominently evident in a participatory planning approach involving the local community, predominantly indigenous, who can identify with the region's agricultural and production systems. An illustrative example is the presence of cooperatives for community production of goods (e.g., Melinda, Mezzacorona, Famiglia Cooperativa, etc.), which are a manifestation of this cultural heritage.

Fig. 2. The sketch translates into imagery the analyses that Nequirito has read regarding the subdivision of the community's territory into four levels, elements, and infrastructure (©Tognon 2023)

3   The Community as the Driving Force of Territorial Development

As illustrated in the previous chapters, communities have long been the driving force behind the domestication of complex regions such as mountainous areas. Therefore, even today, when architects and researchers investigate how these territories can be resilient and capable of meeting the challenges related to climate change and ecological transition through innovative approaches, it is essential to understand how communities traditionally found internal solutions to ensure their adaptability.

Over the past three years of research on mountainous areas, particularly in the Trentino province, we highlighted how architecture, in its dual form of discipline and utility, becomes a conspicuous part of the "solid memory", which becomes a tool for understanding how to operate in a regenerative manner. From this perspective, architectural research, thanks to the multiplicity of its meanings, works in the present, establishing a systemic relationship between the territory and the community, which is seen as a driving element. This becomes an essential combination for transmitting values from one generation to the next and is crucial in regeneration processes.

As discussed in the previous chapters, through the analysis of the historical significance of collective properties in Trentino, the Separate Administrations of Civic Use (ASUC) are today considered the heirs of the ancient rights exercised over the territory by rural or village communities since their origins. However, it is difficult, except in a few cases, to reconstruct their history philologically due to various political and administrative upheavals. Indeed, apart from well-known cases such as the Magnifica comunità della Val di Fiemme, Regola Feudale di Predazzo, and Regola di Spinale e Manez, which have survived over time, most collective properties have been lost along with their archives too. Their gradual disappearance is linked to political changes from the enlightened absolutism of the XIX Century through the despotisms of the fascist period. Subsequently, the first ASUCs were established within the municipalities, and over time, they multiplied and are now unevenly distributed across the provincial territory (numerous in Val di Sole, almost absent in the Valsugana land).

As reflected in the temporal analyses carried out on certain valleys[2], the survival or demise of collective properties also reflects the care they exercised in maintaining the territory. Signs such as spontaneous reforestation, the loss of pastures, the thickening of forests, and the disappearance of profitable trees such as chestnut groves, are indicative of the broader trend of abandoning productive mountain areas. Conversely, it is evident that the presence of collective properties serves as an irreducible stronghold that cares for and preserves the land. However, we can assert that, although the model of collective resource management has been deeply rooted in the Alpine region, current socioeconomic changes are challenging how resources are conceived, utilized, and managed by communities, as well as the very concept of community as a reference for a collective resource. As Della Torre et al. ii [20] point out, few studies have focused on the transformation and adaptation to ongoing changes, such as the gradual penetration of global economic and demographic megatrends at the local level, to decode the new tension between community needs and societal demands.

However, among the initial outcomes of the studies conducted on the territory, it is essential to acknowledge that there is a particularly fervent vitality within valley communities to implement "bottom-up" actions toward regeneration processes, as stands in chapter 2.2. These initiatives are promoted by the commitment of administrations, decision-makers, and stakeholders (valley communities, tourism boards, etc.), who actively involve the community in defining future visions.

The methodologies adopted during the community engagement processes since 2022[3] range from "visioning" to "capacity building". This alternative depends on the objectives set by the project leaders and the collaboration approaches with varied communities from different backgrounds. The translation of ideas into tangible, spatial, or directive results is always entrusted to specialized external mediators (research centers, universities, specialized companies). Using various tools such as brainstorming sessions, workshops, debates, interviews, drawings, and keywords, whose outcomes have been recorded and translated into mappings, the bottom-up process requires a varying degree of community engagement. Additionally, the translation of stimuli from fragments into a coherent system of actions and spaces requires several months of evolution, while a series of follow-up meetings with the community integrates the visions for the territory.

Indeed, the aim of the engagements is to establish a conscious connectivity between the community and the territory through a guided questioning of the current state, with the goals of viewing the territory from innovative and holistic perspective. This involves identifying the needs of the community, as well as preserving biodiversity and safeguarding the cultural heritage and the identity value of the places.

Within this experimental framework, ongoing research considers the model of "commons" as collective practices and strategies of care that can aid in the revitalization and enhancement of the territory and its resources, deemed important by a community of people. The fundamentals are the "community" that recognizes certain "common goods", both tangible and intangible, which are collectively understood through actions and agreements of "ommuning," where rules are established by the community for the collective management of the good [21]. In this way, especially in contexts often considered marginal, the commons promote the enhancement of tangible and intangible heritage, triggering transformative capacities through social innovation actions, space regeneration projects, and measures to counter abandonment at various scales, as well as fostering diversified economies.

3.1 Two Experiments in the Trentino Context

The application to the Trentino case has allowed work over the past two years on a territory characterized by a completely mountainous landscape[4]. The urban areas of Trento are matched by peri-urban areas, regions with significant infrastructure (Valsugana), areas with intensive tourism (Val di Sole), and areas impacted by intensive monoculture (Piana Rotaliana Königsberg). Despite their diverse characteristics, these zones share the feature of being in an Italian Alpine region, peripheral to main urban centers and interconnective nodes, and characterized by small-scale planning.

A consideration of the participatory design activities underway highlights the significant European interest in creating resilient communities through bottom-up actions.

This paper selects the pan-European project SATURN [22] which aimed to explore the value of the landscape in the context of three European regions (Italy, United Kingdom, Sweden) and to create tools and methods to support places pursuing climate resilience. In the Italian context, the case studies are developed in three areas of Trentino (Pergine, Valle dell'Adige, Arco). Fostering on the success of the SATURN approach, several other experiments have been conducted in various areas of the province of Trento, including the "Participatory Pathway for Creating a 2050 Vision for Borgo Valsugana" (February 2022). The team consisted mainly of experts who had participated in the SATURN project, using the "vision" method as a participatory and co-creation tool. Through a multi-phase participatory process, the goal was to develop solutions to improve specific areas of the Borgo in Valsugana territory, reducing negative impacts and increasing the overall well-being of residents and visitors (considering the impact of Arte Sella). The "vision" tool proved effective in involving local stakeholders in the development of policies and actions for territorial or strategic planning. However, at the time of writing, it is still unclear what the concrete impacts on pragmatic administrative planning have been.

A second case study is currently being tested in the Piana Rotaliana Königsberg area, where the tourism consortium began a series of workshops in 2021 (with various monthly follow-ups), coordinated by an external consulting agency. Specialists from different sectors (agricultural, tourism, entrepreneurs,...) and associations were invited to work together. The goal is to build coordination among the various individual needs, envision a sustainable and "beautiful" future for the "wine garden of Europe" through the development of innovative ideas. For this reason, with the association of Politecnico di Milano (DAStU), the project was presented in Brussels and partaken in the NEB Festival [23] fully embracing the three pillars: "sustainability," from climate goals to circularity, zero pollution, and biodiversity; "aesthetics," quality of experience and style beyond functionality; "inclusion," from valuing diversity to securing accessibility and affordability. In this latter case, the process is ongoing, and it is not yet possible to provide an initial assessment. However, the multidisciplinary nature of the three partners (Politecnico di Milano – DAStU, Fondazione E. Mach CTT, PRK Tourism Consortium) allows for constant consideration on the objectives set for the next three years.

4 Consideration

As previously noted, the limited time available for leading community engagement experiences does not currently allow for testing their efficiency in terms of their decline rate. However, when compared with the historical model of commons, it is already possible to recognize how the activation of a shared and inclusive design process contributes to the construction or reinforcement of identities, reinterpreting territorial values. By starting from the (even latent) resources already present in the territory, possible alternative futures can be imagined, protecting the collective interest. Furthermore, in the creation of "communities of communities", it is also possible to reflect on the complex interdependencies between rural and urban areas and to negotiate forms of hybrid coexistence between humans and nature that respond more effectively and resiliently to current challenges.

Moreover, the motivation to include these projects within European research and the interest shown by Europe ensure the broad scope that alliances and network building can create through fruitful collaborations with territories facing similar challenges and finding common solutions. By recognizing these interdependencies, communities and commons can become catalysts for a concrete and feasible alternative for territorial management and governance.

Notes

1. This paper presents some experiences that were conducted by the authors in the field as research fellow for the Territorial Fragilities Project [2018–2022] at DAStU "Departments of Excellence" (L. 232/2016), Ministry of University and Research (MIUR); as Assistant Professor (RTDa), DAStU, Department of Excellence 2023–2027 (MUR); as Associate Researcher [2022–2026] at LabiSAlp USI Mendrisio (CH), with a research project titled "Rural Commons, Spatial Relationships, and Memory in the Contemporary Alpine Landscape: The Trentino Case".
2. The analyses conducted are part of the research at LabiSAlp and the Politecnico di Milano, specifically focusing on two areas (Adige Valley and Valsugana) between 2022–2023. These analyses involved comparing historical archive maps with the evolution of orthophotos between 1972 and 2015, aiming to assess the permanence and mutations in the territory over time.

3. Projects with communities have been carried out in different valleys and municipalities located at altitudes with peculiar environmental characteristics and economies. These include Val di Sole, Valsugana, Valle dell'Adige, Piana Rotaliana Königsberg.
4. The classification by degree of mountainousness, which provides for the subdivision of municipalities into "totally mountainous," "partially mountainous," and "non-mountainous," is the outcome of the application of Article 1 of Law 991/1952 - Determination of mountain territories. This classification was transmitted to Istat by UNCEM.

References

1. Dezio, C., et al.: Territorial fragilities in Italy. Defin. Common Lex. Territorio **9**, 22–54 (2019). https://doi.org/10.3280/TR2019-091003
2. Bätzing, W.: Le Alpi una regione al centro dell'Europa. Bollati Boringhieri, Torino (2005)
3. Interreg – Alpine Space. https://www.alpine-space.eu. Accessed 21 Feb 2024
4. Varotto, M.: Montagne di mezzo. Una nuova geografia, Einaudi, Torino (2020)
5. Ferrari, M.A.: Assalto alle Alpi. Einaudi, Torino (2023)
6. De Rossi, A.: La costruzione delle Alpi. Donzelli, Roma (2016)
7. Guichonnet, P.: Storia e civiltà delle Alpi. vol. 1: destino storico. 5, Jaca Book, Milano (1986)
8. Pace, D.: Amministrazioni separate di uso civico. In: Nervi, P. (ed.) Il ruolo economico e sociale dei demani civici e delle proprietà collettive. Cedam, Padova (1999)
9. Ostrom, E.: Governing the Commons: The Evolutions of Institutions for Collective Actions. Cambridge University Press, New York (1990)
10. Greco, M.: Le statistiche sulle Common Land nell'Unione Europea e in Italia. Agriregionieuropa, **36** (2014). https://agriregionieuropa.univpm.it/it/content/article/31/36/le-statistiche-sulle-common-land-nellunione-europea-e-italia. Accessed 28 Oct 2023
11. Gretter, A., Rizzi, C., Favargiotti, S., Betta, A., Ulrici, G.: Trento social commons. coinvolgimento comunitario come modalità per una nuova relazione fisica e culturale tra spazi urbani periferici e rurali. J. Alp. Res. Rev. de Géogr. Alp. **106**(2) (2018). https://doi.org/10.4000/rga.4499. Accessed 28 Oct 2023
12. Bender, O., Haller, A.: The cultural embeddedness of population mobility in the Alps: consequences for sustainable development. Nor. Geogr. Tidsskr. – Nor. J. Geogr. **71**(3), 132–145 (2017). https://doi.org/10.1080/00291951.2017.1317661. Accessed 28 Oct 2023
13. Finco, A., Valentini, S.: Economia delle risorse e proprietà collettiva. La riscoperta delle comunanze agrarie nella Regione Marche. Archivio Scialoja-Bolla. Annali di Studi sulla Proprietà Collettiva, **1**, 162–182 (2008)
14. Gretter, A.: Dare valore al bene comune. Proprietà collettive, diritti d'uso e servizi ecosistemici: spunti per una comparazione tra Trentino, Lake District e Highlands scozzesi. Archivio Scialoja-Bolla. Annali di Studi sulla Proprietà Collettiva **1**, 285–293 (2008)
15. Granet-Abisset, A.: Natural territories, cultural territories – tensions and conflicting challenges surrounding French high Alpine real estate since the nineteenth century. In: Grüne, N., Hübner, J., Siegl, G. (eds.) Rural Commons - Collective use of resources in the European Agrarian Economy, pp. 116–127. StudienVerlag, Innsbruck Wien (2015)
16. Nequirito, M.: A norma di regola. Le comunità di villaggio trentine dal medioevo alla fine del '700 Provincia autonoma di Trento: Beni librari e archivistici del Trentino (2002)
17. Usi Civici, beni comuni, proprietà collettive e diritto demaniale http://www.usicivici.it/wp-content/uploads/2014/02/2-taa-d-p-r-1952-norme-attuazione-statuto.pdf. Accessed 16 Feb 2024

18. Giovannini, G.: Studio della filiera foresta. Legno per la valorizzazione delle risorse locali nella provincia autonoma di Trento. Tesi di dottorato in Territorio, Ambiente, Risorse e Salute, Indirizzo: Tecnologie Meccaniche e dei Processi Agricoli e Forestali, Ciclo XXI, Università degli Studi di Padova, Dipartimento Territorio e Sistemi Agro-Forestali. (2009). http://tesi.cab.unipd.it/25036/ Accessed 28 Oct 2023
19. Cole, J.W., Wolf, E.R.: The hidden frontier: ecology and ethnicity in an Alpine valley - with a new introduction. California University Press, Berkeley & Los Angeles (1974)
20. Dalla Torre, C., Ravazzoli, E., Omizzolo, A., Gretter, A., Membretti, A.: Questioning mountain rural commons in changing alpine regions. an exploratory study in Trentino, Italy. J. Alp. Res. I Rev. de Géogr. Alp. **109**(1) (2021). https://doi.org/10.4000/rga.8589
21. I commons rurali: un'alternativa concreta di gestione del territorio, https://www.researchgate.net/publication/357656914_I_commons_rurali_un'alternativa_concreta_di_gestione_del_territorio. Accessed 26 Jan 2024
22. Saturn, https://eventi.fmach.it/Saturn, supported by the EIT Climate-KIC program https://www.climate-kic.org. Accessed 29 Nov 2023
23. Polimi NEB. https://www.neb.polimi.it/between-mountains-the-wine-garden-of-europe/. Accessed 20 Mar 2024

Research on Environmental Perception and Preferences of Traditional Villages from the Perspective of Local Gaze: A Chinese Case Study

Wei Xintong[1]([⊠]) and Zhou Haoming[1,2]

[1] Academy of Arts and Design, Tsinghua University, Beijing 100084, China
xintong_wei@foxmail.com
[2] Department of Architecture, National University of Singapore, Singapore 117566, Singapore

Abstract. The advent of mass tourism has endowed traditional villages with multiple identities as heritage sites, communities, and tourist attractions. As hosts, local villagers have begun to introspect and reevaluate the village environment that was once a part of their everyday landscape, giving rise to new spatial perceptions. As the primary stakeholders in village communities, indigenous inhabitants' perceptions, preferences, and identifications with village spaces hold significant significance in preserving rural characteristics and sustaining village vitality. Using Hongkeng Village in southwest Fujian Province, China, as a case study, this study investigated local villagers' perception of traditional village daily life space and activity paths from three dimensions: cultural cognition, emotional preference, and behavioral activities through observation, questionnaire surveys, and cognitive maps. The results show that villagers generally have high cognitive and low emotional perceptions of the exhibition space and the former staging space; they have high emotional perceptions of the neighborhood interaction space and the collective memory place; the activity path spreads from home to the surrounding area, and there are gender differences in the scope of activities. The study suggests that the development of the tourism industry has often overlooked historical context and the spirit of places. It emphasizes the need to rekindle the identity of "home" within the Tulou clusters of Hongkeng Village while maintaining a balanced distribution of public facilities and enriching residents' leisure lives. This study is expected to provide insights for improving the living environment of tourism-oriented villages in China.

Keywords: Traditional Villages · Local Gaze · Environmental Perception · Villagers · Cognitive Mapping

1 Introduction

In recent years, an increasing number of traditional Chinese villages have adopted a coexistence model of heritage conservation and tourism development, gradually evolving into popular tourist destinations. However, behind this tourism boom, it is essential

M. Barosio et al. (Eds.): EAAE AC 2023, SSDI 47, pp. 294–305, 2025.
https://doi.org/10.1007/978-3-031-71959-2_32

to recognize that traditional villages, despite assuming the roles of tourist attractions and heritage sites, fundamentally remain crucial communities for local residents' residence, production, and daily life, and the agency of indigenous inhabitants should be emphasized. Prior to the influx of tourists, the landscapes of traditional villages were often overlooked by villagers due to their everyday familiarity. Nevertheless, with the arrival of mass tourists, villagers have begun to retrospect and reevaluate the once-overlooked village landscapes, giving rise to new spatial perceptions. How to start from the perspective of the villagers' main body to evaluate the current use of the village environment and improve the quality of the village habitat is an issue worth studying at this stage. Given this context, this paper selects Hongkeng Village in southwestern Fujian, China, as the research subject. Established approximately 500 years ago and home to numerous well-preserved Tulou buildings, the village was inscribed as a World Cultural Heritage Site in 2008 [1]. It has since adopted a development model that simultaneously emphasizes heritage conservation and tourism development, serving as a quintessential example of tourism-oriented villages in China. This study combines survey questionnaires and cognitive mapping to explore local residents' perceptions and preferences of the village space.

2   Theoretical Foundation

2.1   The Theory of Local Gaze

In 2006, Israeli scholar Darya Maoz introduced the concept of the "local gaze", building upon the theory of the "tourist gaze" proposed by British sociologist John Urry [2], which shifted the focus from tourists to local residents [3] 2. It can be understood as follows: in the process of tourism development, local inhabitants typically consider the natural and cultural landscapes of their region as ordinary facets of daily life. However, with the continuous influx of tourists who identify with these local landscapes, local residents often begin to reevaluate their own environment [4]. Under the tourist gaze, local hosts present their culture, including their own identities, as commodities showcased to tourists, leading to a certain degree of "Staged Authenticity" in the social living spaces of the local hosts. The "front stage" of a tourist destination refers to the space for local exhibitions and performances to promote local economic development, while "the back stage" refers to the cultural reserve, i.e. the space for indigenous culture [5, 6]. On this basis, Yang proposed a transitional "curtain zone" as a cultural buffer between the front stage and the back stage [7]. Today, the "gaze theory" has gradually evolved into a prominent analytical tool that finds widespread application in various research domains [8–10].

2.2   The Theory of Environmental Perception

Environmental perception refers to individuals' recognition and understanding of their surroundings, encompassing their inner representations or cognitions of structures, entities, and spatial relationships within an environment [11]. As people enter a traditional village, they will establish subjective judgments and evaluations about the environment and spatial features of the village through cognitive information and decide their own behavior in the space. The primary research methods for studying environmental perception include the semantic differential method and cognitive mapping, with the latter first applied by American urban planning expert Kevin Lynch [12, 13]. Research on environmental perception covers a diverse range of subjects, from macro-level urban conglomerates to micro-level individual village spaces [14]. Research subjects have primarily focused on tourists [15], while studies on the spatial perception of local residents have often revolved around spatial transformations and identity identification [16, 17]. In conclusion, there remains a need to further explore research related to the environmental perception of traditional village residents, taking the perspective of the local gaze, and placing villagers as the primary subjects of investigation.

3   Selection of the Case Study Site

Situated in the southwestern part of Fujian Province, China, Hongkeng Village boasts a substantial number of Tulou buildings, characterized by their historical significance and well-preserved state. In 2008, this village was inscribed as a UNESCO World Heritage Site. Subsequently, it adopted a development model that combines heritage conservation with tourism development. As a result, the tourism industry in the village has experienced rapid growth, leading to its recognition as a 5A-level national tourist attraction in China in 2011 [1]. However, this development has also brought about challenges, including the transformation of living space into consumption space and the difficulty in maintaining the original authenticity of the village, and the villagers' feelings towards the village have also undergone subtle changes. Therefore, using this village as a research case on the environmental perception problem of traditional villages is highly representative.

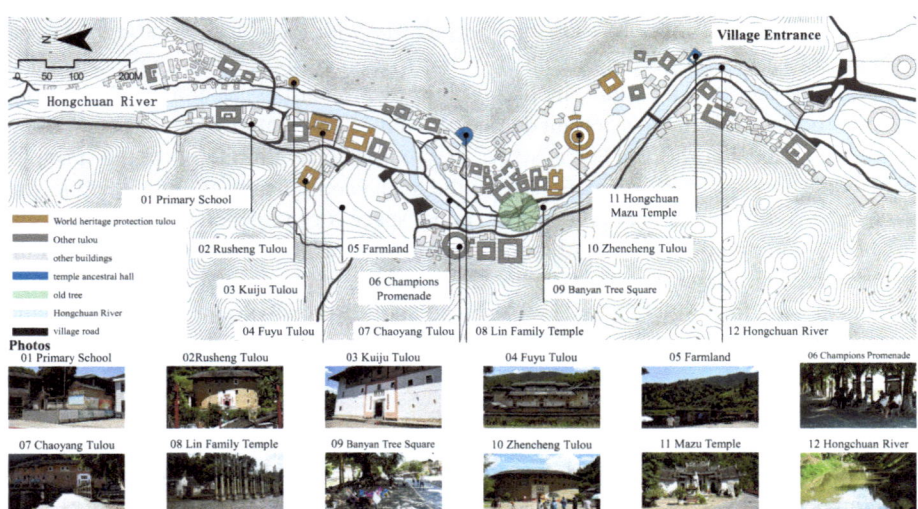

Fig. 1. Site Plan and Key Nodal Spaces of Hongkeng Village (Source: photos by the authors)

Hongkeng Village stretches along a river, featuring a narrow and elongated layout. The villagers used to take the Lin Family Temple as the center point, dividing the village into two parts, the upper village and the lower village (Fig. 1). The most famous one in the village is the cluster of Tulou buildings represented by Zhencheng Tulou, known as the "Prince of the Tulou buildings". To cater to tourism development, the ground floor of Zhencheng Tulou has been converted from residential space into retail shops. Additionally, daily performances are conducted outside, rendering it a "Cultural stage space" in the eyes of the villagers. Furthermore, Fuyu Tulou has been transformed into a hotel and restaurant, while Kuiju Building still maintains its residential function. The entrance to the village is marked by the Mazu Temple, and the village center hosts the Lin Family Temple, both serving as essential places of worship and repositories of collective memory for the villagers. The most frequented location for social interactions among village residents is the banyan tree square, situated near the central dividing point of the village and adjacent to the waterfront. This square serves as a gathering place for villagers to socialize and seek respite. Lastly, the century-old "Rixin School (Hongchuan Primary School)" within the village bears witness to the enduring influence of cultural and educational ideals in the village's history.

4 Research Design

4.1 Questionnaire and Cognitive Mapping Design

Based on the local gaze theory, this study classifies the spaces in Hongkeng Village into "front stage", "back stage", "curtain space" and daily life space. In order to better describe the local villagers' perception of different spaces, the survey design draws on the research method of Place Attachment theory from three dimensions: cultural cognition, emotional preference, and behavioral activities [18]. Since the cultural cognition dimension shows the degree of individual cognitive understanding of the place, the emotional preference dimension shows the degree of individual emotional connection or integration to the place, and the behavioral activity dimension shows the degree of familiarity and dependence to the place.

The research employed a combination of survey questionnaires and cognitive mapping. The survey questionnaire covered respondents' basic information and inquired about their perceptions of the village environment in terms of cultural cognition (which spaces within the village are most representative) and emotional preferences (which spaces within the village they like the most). The dimension of behavioral activities was assessed through cognitive mapping. Villagers were provided with proportionally scaled printed village maps and were asked to mark the paths of their daily activities and the locations where they frequently engage in these activities.

Table 1. Statistics on basic information of villagers

Category	Variable	Sort	Quantity	Percentage
Respondents' background	Gender	Male	38	46.3%
		Female	44	53.7%
	Age	Under 17	4	4.9%
		18–45	33	40.2%
		46–60	25	30.5%
		61 and above	20	24.4%
	Living in Tulou	Yes	55	67.1%
		No	27	32.9%
	Local resident	Yes	63	76.8%
		No	19	23.2%
Behavioral characteristics	Engaged in tourism-related work	Yes	26	31.7%
		No	50	61.0%
		Both	6	7.3%
In total			82	100.0%

4.2 Data Collection and Processing

Table 2. Objects of high-frequency perception in village space.

Village places	PFCC	PFEP	Place type	Place function
Zhencheng Tulou	40	14	Front stage	External display and performance space
The banyan tree square	24	36	Daily life space	Villagers' daily life space
Kuiju Tulou	18	7	Front stage	External display, villagers' living space
Fuyu Tulou	18	10	Front stage	External display and commercial space
Tulou buildings cluster	12	–	Front stage	External display and performance space
Mazu temple	10	4	Back stage	Villagers' prayer space
Hongchuan primary school	9	9	Back stage	collective memory space
Lin family temple	7	5	Back stage	Villagers' ancestor worship space
Rusheng Tulou	6	–	Front stage	External display, villagers' living space
Fuxing Tulou	6	–	Front stage	External display, villagers' living space
Hongchuan river	4	16	Daily life space	Villagers' daily life space
Yue'e bridge	4	–	Daily life space	Villagers' daily life and farming space
Home	–	19	Daily life space	personal domain

(*continued*)

Table 2. (*continued*)

Village places	PFCC	PFEP	Place type	Place function
Chaoyang Tulou	–	6	Daily life space	External display and commercial space
Champions promenade	–	5	Curtain space	External display, villagers' leisure space
Observation Deck	–	4	Curtain space	External display, villagers' leisure space

(PFCC = Perceived Frequency of Cultural Cognition, PFEP = Perceived Frequency of Emotional Preferences)

The field research work was carried out from September 14th to 18th, 2022. Respondents within Hongkeng Village were randomly selected, and questionnaires were distributed. A total of 82 valid questionnaires were recovered. The statistics of the basic information of the respondents' data are shown in Table 1. The survey questionnaire results were processed by organizing the data and ranking them based on word frequency. This process led to the identification of perceived frequencies for different locations. Locations perceived to have frequencies equal to or greater than 4 were defined as high-frequency perception targets. In this context, 12 places were identified as such for both cultural cognition and emotional preference, as shown in Table 2. Regarding the behavioral activity aspect, information was processed by plotting activity trajectories and marking stopping points based on the activity paths of different groups within the village. A total of 28 valid cognitive maps were collected.

5 Analysis of Villagers' Environmental Perception

5.1 Analysis of Perceptions in the Cultural Cognitive Dimension

Tulou cluster is regarded as the most representative and well-known space by local villagers. Among them, Zhencheng Tulou is perceived most frequently, with a cumulative total of 40 times, and the building is also most favored by tourists due to better landscape maintenance, good publicity, and regular folklore performance activities. Additionally, historical buildings like Fuyu Tulou, designated as cultural heritage preservation units, are considered representative spaces by the villagers. The Banyan Tree Square was perceived 24 times in total due to its good location with the compound functions of sitting and chatting, as well as leisure, vending and praying for blessings. The Hongchuan River, closely related to villagers' daily lives, is also deemed representative. Temples such as the Mazu Temple and the Lin Family Temple, which carry the collective memories of the villagers, have received high levels of recognition.

1. Zhencheng Tulou
2. Banyan Tree Square
3. Kuiju Tulou
4. Fuyu Tulou
5. Mazu Temple
6. Hongchuan Primary School
7. Lin Family Temple
8. Rusheng Tulou
9. Fuxing Tulou
10. Hongchuan River
11. Yue'e Bridge

Fig. 2. Distribution of high-frequency perceived places in cultural cognitive dimension (Source: produced by the authors)

Based on frequency statistics, the Hongkeng Village map was imported into ArcGIS software, and the heat map analysis tool was used to create spatial distribution maps of perception levels for different locations (Fig. 2). According to the distribution map, the high-frequency perceived places in Hongkeng Village have obvious agglomeration and form two major grouping areas: the grouping space in series with Zhencheng Tulou and the Banyan Tree Square and the grouping space in series with Rusheng Tulou and Kuiju Tulou. The former has a high degree of functional complexity, including both the "front stage" for tourists' cultural and entertainment performances, sales and displays, as well as the daily life space for residents' living and leisure; while the latter mainly provides the "front stage" for cultural displays and consumption.

5.2 Analysis of Perceptions in the Emotional Preference Dimension

In terms of emotional preferences, villagers exhibit a greater fondness for daily living spaces that facilitate neighborly interactions and leisure activities. The Banyan Tree Square garnered the highest cumulative perception frequency, with 36 mentions. The Hongchuan River was mentioned 16 times. In addition, tree-lined corridors and Observation Decks with a wide view are the places favored by villagers. The concept of "home", as an individual's personal domain, was mentioned a total of 19 times, evoking sentiments of warmth, familiarity, and attachment among the villagers. As cultural exhibition spaces, the cluster of Tulou buildings hosts regular entertainment performances and light shows. Apart from serving as "Cultural stage space" showcased to tourists, they have also enriched the lives of the villagers to some extent. Spaces like the Lin Family Temple and the Mazu Temple bear certain symbolic significance, supporting the local and identity affiliations of the residents.

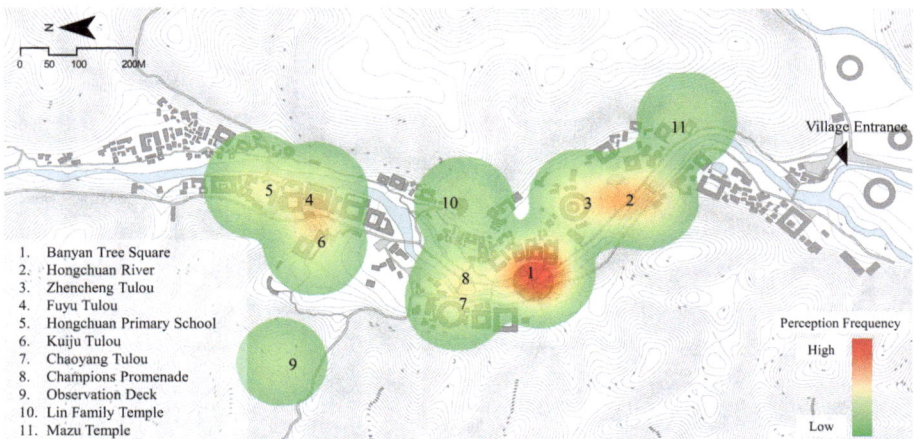

1. Banyan Tree Square
2. Hongchuan River
3. Zhencheng Tulou
4. Fuyu Tulou
5. Hongchuan Primary School
6. Kuiju Tulou
7. Chaoyang Tulou
8. Champions Promenade
9. Observation Deck
10. Lin Family Temple
11. Mazu Temple

Fig. 3. Distribution of high-frequency perceived places in emotional preference dimension (Source: produced by the authors)

Using the heat map analysis tool of ArcGIS once again, it is evident that preferred locations are predominantly distributed in the southern part of Hongkeng Village (the lower village), forming a patchy pattern with a degree of clustering (Fig. 3). Outdoor activity areas, such as those interlinked by shaded banyan trees and Zhencheng Tulou due to their convenient locations, are associated with the diverse scattered daily activities of the villagers, carrying the function for leisure and interaction of the villagers. Groupings of spaces like Chaoyang Tulou and Lin Family Temple constitute a combination of linear passages and point-like nodes, attracting the gathering of neighboring villagers. While Fuyu Tulou and Kuiju Tulou have distinctive architectural features, and Hongchuan Primary School has a long history, the three are representative of the village's external display space.

5.3   Analysis of Perceptions in the Behavior Activity Dimension

Based on villagers' recollections of their daily behavioral paths and stopping points, a total of 28 valid path information records were obtained, comprising 15 female respondents and 13 male respondents. Among these, 12 individuals reside in the upper village of Hongkeng, while 16 reside in the lower village. On a broader scale, the segment of the route from the village entrance arch to Chaoyang Tulou, situated on the eastern side of Hongchuan River, witnessed the highest pedestrian flow. The segment adjacent to Hongchuan Primary School on the south side of the river bridge, known as Xibei Bridge, experienced the second-highest pedestrian flow. Simultaneously, this route serves as the main thoroughfare within the village. The area around the Banyan Tree Square witnessed the highest number of people stopping, with a predominance of females. The fields near Yue'e Bridge, which serve as agricultural spaces, and the banks of Hongchuan River utilized for leisurely fishing activities, also frequently attracted villagers' stopovers (Fig. 4 & Fig. 5).

Fig. 4. Overall activity paths and stop locations of villagers (Source: produced by the authors)

Performance in front of Zhencheng Tulou   Crowds under the Banyan Tree Square

Fig. 5. On-site photos (Source: photos by the authors)

Regarding activity ranges, residents from both the upper and lower villages tended to center their activities around their homes. Despite residing in the same village, an intangible boundary seemed to exist, demarcated by the route from the Lin Family Temple to Yue'e Bridge and Xibei Bridge, which divided villagers' activity paths (Fig. 6 & Fig. 7). Upper village residents primarily focused their activities around the earthen buildings in the upper village and the vicinity of Yue'e Bridge, with a noticeable difference in the activity range between females and males. Lower village residents similarly seldom ventured to the upper village, with their primary activity path being the main thoroughfare on the eastern side of the Hongchuan River. However, the activity range of lower village female residents was comparable to that of male residents, with both genders inclined to take walks toward the village entrance arch. Additionally, upper village residents had fewer stopping points, often resting in the vicinity of their homes.

In contrast, lower village residents had a greater number of stopping points, which to some extent reflects the inadequacy of public facilities in the upper village.

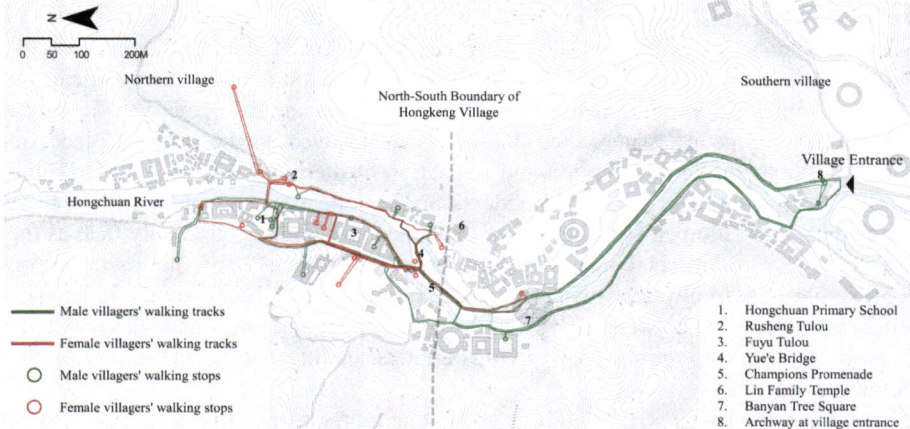

Fig. 6. Tracks and stops of upper village residents (Source: produced by the authors)

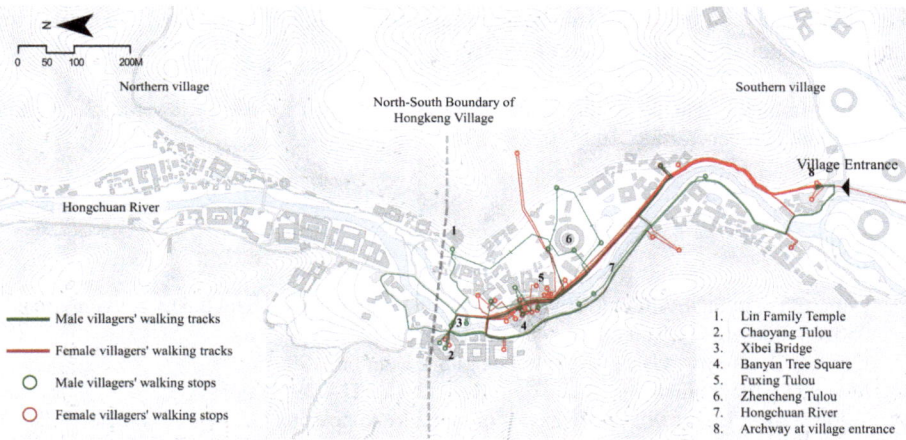

Fig. 7. Tracks and stops of lower village residents (Source: produced by the authors)

6 Conclusion

For traditional villages, tourism development provides economic support for heritage conservation but also triggers the neglect of heritage originality as well as villagers' subjectivity. This paper, using Hongkeng Village as an example, explores the environmental perceptions of villagers in a tourism-oriented village under the influence of the "local gaze" and discusses these perceptions across three dimensions: cultural cognition, emotional preferences, and behavioral activities. The research reveals that villagers generally exhibit high levels of cultural cognition in the theme display space, or named "Cultural stage space" within the village, but their emotional preferences for these spaces are low. Conversely, they show a strong preference for the everyday living spaces within the village. In terms of behavioral activities, villagers' activity area generally extends outward from the center of their home. And, the activity area of female villagers is smaller than that of males, and the space of neighborhood interaction is of great significance.

The unique Tulou buildings in Hongkeng Village is an advantage, yet the current tourism development has neglected the historical lineage and the spirit of place, and is more of a symbolic interception and utilisation of the local culture [19]. Originally serving as living spaces for local residents and a bond for connecting with relatives and friends, the Tulou clusters have seen the concept of "home" gradually fade as they transform into commercial spaces. "Front stage" and "back stage", terms of cultural context, have led to physical spatial segregation in tourist destinations: the imbalanced tourism development between Hongkeng Village's upper and lower parts has resulted in disparities in the distribution of daily living facilities, leading to a split in the development of village spaces.

Based on the findings, this study suggests that tourism-oriented villages should pay close attention to the spatial perception of villagers. It recommends establishing a village environment that harmoniously integrates tourism development, cultural preservation, and the daily life of residents. This can be achieved by focusing on perpetuating the collective cultural memory of the inhabitants, enriching public life facilities, and enhancing interactions between villagers and tourists.

Acknowledgments. Thanks to the China Scholarship Council for a grant awarded to the first author (grant number 202206210097).

References

1. Chinese Traditional Village Museum: Hongkeng Village. https://main.dmctv.com.cn/villages/35080310601/Index.html. Accessed 10 July 2023
2. Urry, J.: The Tourist Gaze. SAGE Publications, London (1990)
3. Maoz, D.: The mutual gaze. Ann. Tour. Res. **33**(1), 221–239 (2006)
4. Wang, T., Lu, L., Lu, X.: The behavior of tourists and hosts by Gazing the Huizhou villages. Tour. Trib. **30**(4), 23–32 (2015). (In Chinese)
5. MacCannell, D.: Staged authenticity: arrangements of social space in tourist settings. Am. J. Sociol. **79**(3), 589–603 (1973)
6. Zhou, Y., Wu, M., Zhou, Y., Zhu, Y.: Theory of "Authenticity" and its comparison in tourism study. Tour. Trib. **6**, 42–47 (2007). (In Chinese)
7. Yang, Z.: The front stage, curtain and back stage. Ethno-Natl. Stud. **2006**(02), 39–46 (2006). (In Chinese)
8. Zeng, S., Xie, Y., Shi, Y.: The space-time axis of tourist experience: the representation and landscape gaze of collective memory regarding everyday life in historical streets. Tour. Trib. **36**(2), 70–79 (2021). (In Chinese)
9. Liu, Y., Yin, S.: Study on the family hotel space construction in heritage site based on the host gaze: take Hongcun as an example. J. Huangshan Univ. **21**(6), 26–31 (2019). (In Chinese)
10. Fan, Y., Xie, Y.: Memory, display and gaze: a research on the synergy of protection & utilization about rural cultural heritages. Tour. Sci. **29**(1), 11–24 (2015). (In Chinese)
11. Hu, Z., Lin, Y.: Environmental psychology, 4th edn. China Architecture & Building Press, Beijing, Beijing (2018). (In Chinese)

12. Tan, C., Li, J.: Innovative research on traditional village protection methods by cognitive map. Dev. Small Cities & Towns **37**(9), 77–83 (2019). (In Chinese)
13. Lynch, K.: The Image of The City. MIT press, Cambridge (1964)
14. Zhang, Z., Chen, W., Shen, M., Shang, S.: Research on resident perception and inheritance of spatial gene of traditional villages in suzhou: a case of luxiang village. Urb. Dev. Stud. **27**(12), 1–6 (2020). (In Chinese)
15. Xu, G., Wan, C., Gan, M.: Research on tourists' image space perception differences in Sanfangqixiang historical district of Fuzhou. J. Chongqing Norm. Univ. (Natural Science) **29**(2), 94–98 (2012). (In Chinese)
16. Zhao, P., Yao, Z.: From spatial perception to spatial identity: the conservation and renewal of the grand canal historic and cultural neighborhoods. Dong Yue Trib. **43**(8), 154–160 (2022). (In Chinese)
17. Zhu, H., Qian, J., Lv, X.: Place identity and sense of place in the context of urban spatial transformation: a case study of Xiaozhou village in Guangzhou, China. Sci. Geogr. Sinica **32**(1), 18–24 (2012). (In Chinese)
18. Li, C., Zeng, W.: The connection between people and place: the place attachment. Adv. Psychol. **8**(4), 585–599 (2018). (In Chinese)
19. Liao, W.: Spatial Production of Urban Heritage Tourism Landscapes from A Consumerist Perspective. Science Press, Beijing (2014). (In Chinese)

Rethinking Architecture in the Digital Age: From Parametric Design Thinking to Philosophical Perspectives

Hongye Wu[✉]

Department of Architecture and Design (DAD), Politecnico di Torino, Viale Mattioli 39, 10125 Turin, Italy
hongye.wu@polito.it

Abstract. We are currently living in an age where the advancement of digital technologies has been rapidly progressing. Simultaneously, the digital avant-garde has been flourishing in the field of architecture for decades. The emergence of design thinking has been significant in this era, especially with the rise of parametric design thinking in contemporary discourse due to the advancements in digital technology. In this context, particular attention is given to parametric design thinking in this paper, with a specific focus on the current trend and the concept of "social performativity". Moreover, the current tasks in architecture are more challenging in dealing with its relationship with social diversity and even conflicts. Hence, this contribution questions the positionality of architecture in multiple interactions and explores the state and impact of architecture in the digital age, considering both material and immaterial aspects. To investigate this, the study unfolds philosophical perspectives on the relationship between humans and space by expounding on several fundamental ideas from Taoism and the philosophy of Martin Heidegger. The study's results contribute to a deeper comprehension of the relationship between humans and space, shedding light on both the potential and the complexities inherent in architecture within the realm of parametric design in the digital age.

Keywords: Architecture · Parametric Design Thinking · Philosophy · Space

1 Introduction

We are living in the digital age where the digital technology is playing a much more important role. This is an age where the digital avant-garde has been blooming for almost three decades in the practice of the architectural discipline. This is an age when the inevitable march of digital technologies has increasingly gained momentum, whereas it is also even more challenging for architects and researchers to tackle their roles in this tendency. Nowadays, the tasks in architecture are more demanding on dealing the relationship with social diversity and even conflicts, which are becoming more and more complex. In responding to the call of EAAE Annual Conference 2023 "not just Architecture but Architecture(s)" in terms of the interdisciplinary dimension of architecture,

M. Barosio et al. (Eds.): EAAE AC 2023, SSDI 47, pp. 306–313, 2025.
https://doi.org/10.1007/978-3-031-71959-2_33

meaning that it is time to reflect the positionality of today's architecture, the paper comes up with the question "what is the state and impact of 'architecture' in our digital age?".

In parallel with the development of digital technology, the model of design thinking has transitioned from cognitive models to computational models to the model of parametric design thinking [1]. In this light, parametric design thinking has been given particular attention in this paper. However, can parametric design approach respond to the demands of contemporary social challenges? Consequently, the current work tends to revive the self-consciousness and self-reflection of architecture, by drawing on Patrik Schumacher's idea of architecture's "social performativity" as well as the well-known "parametricism".

Furthermore, from sociological aspect, architecture is not only about inhabiting people physically but also inhabiting the full sense of people whoever involved in. Wherever social activities take place, there is a sort of architectural and urban spaces, that is to say, space is always connected with society and people. Hence, another purpose is to reflect on the relationship between human experiences and the physical environment, by turning back to the fundamental ideas in the field of philosophy. This involves seeking philosophical engagement in architecture, focusing not only on the materiality but also on the immateriality.

2 Parametric Design Thinking

2.1 State of the Art: Parametric Design Thinking

Parametric design thinking, the earliest could be traced to 1st century BC and the medieval age. Mario Carpo emphasizes that there was a class of objects defined by generative rules mathematically, because some of Vitruvius' rules were like "procedural algorithm" what we would call today [2]. Parametric design thinking can be found in the work of Antoni Gaudí. Gaudí used to design by his exacting use of geometry and had a revolution on the fusion of intersecting hyperbolic paraboloids with hyperboloids. In this sense, his design was parametrically variable, flexible architectural design [3]. For Patrik Schumacher, Frei Otto is the only true precursor of parametricism, who applied physical processes as a design engine and simulations to find out forms rather than to draw invented forms [4]. Until then, parametric thinking is a logical approach on design regardless of the aid of digital techniques. The Italian architect Luigi Moretti was probably the first who conducted the approach of parametric thinking in architectural design and created three-dimensional architectural form using a complex set of parametric relationships resolved by digital computation. Moretti also coined the term "Architettura Parametrica (Parametric Architecture)" in the 1940s [5].

Nowadays, the explosion of parametric design trend in the field of architectural design is highly related to the aid of the widespread use of parametric modelling techniques and software. As a matter of fact, in the digital age the advanced computational techniques not only empower designers' abilities to realize but also arouse their ideas by simulating endless potential solutions and overcoming the constrains in the simulation of complex forms and patterns in the earlier age. In the recent fifteen to twenty years, parametric design approach has become a more attractive topic in architectural design and research, especially when Patrik Schumacher, architect, and architectural theorist, unleashed "parametricism" to the world, at the Dark Side Club, 11th Architecture Biennale, Venice (2008), which can be seen as an avant-garde architecture and design movement [6]. Since then, parametric design became more and more fashionable.

2.2 The Social Performativity of Parametricism

The concept of social performativity in parametric design emphasizes the importance of considering the social functionality and performance of architectural designs. Patrik Schumacher, the proponent of parametricism, argues that the movement should shift its focus from foregrounding formal principles and design processes to the foregrounding of design research that is more strategic, applied, and performance-oriented, with a strong emphasis on social performativity [7]. This involves simulating the social functionality in design as a key aspect of architectural design. The performative design strategies in architecture also highlight the importance of considering social, environmental, and economic performances in the design process [2]. Parametric modeling is seen as a tool that allows for the streamlined testing of different values for variables, which can contribute to the consideration of social performance in design [8].

Patrik Schumacher calls Parametricism as a new style, which is "architecture's answer to contemporary, computationally empowered civilisation, and is the only architectural style that can take full advantage of the computational revolution that now drives all domains of society" [9]. In 2016, "Parametricism 2.0" as a "self-critical redirection" was relaunched, aiming to reverse the increasing marginalisation of Parametricism. Schumacher argues about "anti-icon polemic" and "neo-rationalism" in his article. For him, the anti-icon polemic misunderstands that "parametricism" as an architectural new style that is rigorously developed on the basis of radically new, innovative principles becomes conspicuous by default rather than by intention. "Both the anti-icon camp and the neo-rationalist camp fail to understand that urban and architectural complexity are called for by the new societal complexity" [9]. Schumacher states that "Parametricism 2.0" not only covers up various infrastructural, economic and environmental parameters, but offers the promises of social functionality [7, 9, 10]. He points out parametriscism's obvious superiority, in terms of the build environment, is "social functionality" rather than technical functionality. "The built forms are not speaking their structural performance (which is of no interest to users) but about their social purposes, and this communication facilitates these very purposes" [7].

To face the increasing societal complexity, Schumacher emphasizes the importance of the supporting from new methodology and tools. With the concern of this, the design approach has shifted from foregrounding formal principle and design processes to "more strategic, applied and performance oriented, with a strong emphasis on social performativity [7, 10]". Moreover, Schumacher and his group worked on this enhanced capacity—social performativity by semiologically empowered life-process simulations [7]. Therefore, the computational capacity of parametric modelling today would suggest a more potent opportunity for parametric design thinking to integrate with the societal complexity by ongoing research and practice.

2.3 Possible Social Impact Through Parametric Design Approach

In terms of social impact, parametric design can influence architecture by enabling the creation of innovative and sustainable buildings that reflect the values and culture of a city and its inhabitants [11]. Schumacher introduces the concept of design processes using evolutionary algorithms and agent-based life-process simulations with social interaction frequencies as success measures to optimize social functionality [7, 12]. This research, conducted at the University of Applied Arts and Zaha Hadid Architects, aims to expand the formal repertoire available to architects, leading to better design solutions [12]. In this study, Schumacher discusses the possibility of applying scientific methods to architectural design through simulations of social functionality. By simulating a design's social functionality, parametric design can contribute to a more performance-oriented approach in architecture, considering factors such as user interaction, urban setting integration, and spatial-social relationships [12]. The simulation process allows for the assessment of a design's social performance relative to alternative options, emphasizing the importance of understanding how occupants interact with parametric-based spaces and their impact on the built environment [12].

3 The Relationship Between Human and Space

In the digital age, the implications of digital technologies on human perception and the blurring of boundaries between the physical and digital worlds, have led to a reevaluation of architectural materiality and subjectivity [13]. This shift involves a renewed focus on environmental factors, as well as a redefinition of the human experience in the built environment [13]. In this context, Antoine Picon emphasizes the disconnect between digital avant-gardes and environmental concerns, highlighting the need to integrate the two domains for a more holistic approach to architecture [13]. Consider how parametric approach can be used to respond to socio-cultural contexts and diverse human experiences and how designers can employ parametric strategies to create spaces that reflect the values, traditions, and identities of different communities, thereby, it is essential to foster a sense of belonging and cultural continuity.

3.1 The Perspectives from Taoism and Heideggerian

Derived by Martin Heidegger's essay *Building, Dwelling, Thinking*, published in 1951, which centers on the question "what is the state of dwelling in our precarious age?", this paper questions "what is the state and impact of 'architecture' in our digital age?" Heidegger defines dwelling as more than just a physical structure, a place to live. Dwelling is an essential aspect of human existence and involves a deep relationship between humans and the world around them. In Heideggerian philosophy of dwelling, the "fourfold" refers to the interdependent relationship between the four fundamental elements of being-in-the-world: earth, sky, divinities, and mortals. Heidegger argues that these elements are not separate entities but are inseparable and interconnected, forming a holistic understanding of being and dwelling [14]. However, Heidegger did not originate with the idea of contemplating the relationship between humans and the world. One might encounter a comparable notion in the *Tao Te Ching*, a seminal work of Daoism, established by the Chinese philosopher Lao-tzu but written down by later generations. In Taoism, four elements are human, earth, heaven, and Tao, following a fundamental principle: "Human follows the earth, the earth follows heaven, heaven follows Tao, and Tao follows nature [15]". This principle emphasizes the importance of living in harmony with nature and following the natural order of things. By following the rules of nature, one can achieve balance and peace in life. Moreover, it suggests that everything in the universe is interconnected and should be respected and treated with care. In essence, this phrase encourages people to live in harmony with nature and to respect the natural world around them. The importance of the relationship between human and external space in terms of nature, has been addressed both in Taoism and Heidegger's fourfold.

3.2   The "Bridge": a Symbolic Connection Between Human and Space

In the essay *Building, Dwelling, Thinking*, "bridge" is well explicated by Heidegger as more than a location but a symbol that connects both material and immaterial things. Thus "bridge" answers to the question "what is the relation between human and space?" In Heidegger's words, the bridge is "about the relationship between location and space, but also about the relation of humans and space… The bridge is a thing of this kind. The location lets the simple onefold of earth and sky, of the divinities and the mortals, into a site by erecting the site in spaces" [14]. He argues that a bridge serves as a symbolic connection between human and space. This notion resonates deeply within Chinese culture, where the concept of the lounge bridge embodies such philosophical depth.

Originating from the Song Dynasty, the lounge bridge or arched wooden lounge bridge 木拱廊桥, with its special form, served as more than a mere crossing point. In Zhejiang and Fujian provinces, the lounge bridge is a specific type of arched wooden bridge covered with a roof, and in some cases is also covered with wooden shadow claddings from top to bottom. Initially, the lounge bridge was designed to host and rest people who had to travel in mountains for long days in the past, while also serving as a temporary shelter from harsh weather conditions. Often, the timber structural elements of the lounge bridge, like beams and pillars, are frequently adorned with paintings and poems that convey the sentiments and dispositions of the travelers. Nowadays, lounge bridges, suspended between lush greenery on two banks, are still fully functional and present in everyday life. For instance, an ancient lounge bridge located in Zhejiang, China, known as the Lan Xi Lounge Bridge 兰溪桥(Fig. 1) exemplifies the concept of Heidegger's "bridge" by displaying numerous significant characteristics of a lounge bridge. Specifically, various window opening on the shadow claddings (hand fan-shaped or round) provide aesthetically pleasing frames for the views of the surrounding landscape. The lounge bridge presents both the materiality and immateriality of a space, as well as the relation between humans and the environment, in terms of tectonics, aesthetics, the use of local materials.

Fig. 1. Lan Xi Lounge Bridge, Zhejiang, China, 2023 (Photograph by Hongye Wu)

4   Conclusion and Discussion: Immateriality and Materiality of Architecture in Digital Age

The purpose of bringing Taoism and Heideggerian philosophy together is not to interpret the similarity on their thoughts but to bring out the importance of the relationship between nature and people, between space and people, from different cultures but with the same concern in a sense. The exploration on the question of the interaction between human and space aims to bring up that architecture has a profound impact on emotions, memories, and overall well-being, and should be designed with a deep understanding of human experience. "The architect must act as a composer that orchestrates space into a synchronization for function and beauty through the senses—and how the human body engages space is of prime importance. As the human body moves, sees, smells, touches, hears and even tastes within a space—the architecture comes to life [16]". Consequently, the paper revisits several relevant fundamental ideas from Taoism and Heideggerian philosophy, such as 'dwelling', "fourfold", "bridge" from Martin Heidegger on the one hand, four basic elements from Taoism on the other hand, aiming to reinforce the foundation and reveal the fog for the "architecture" for the future. Heidegger emphasizes that "the spaces, which we go through every day, are made for by locations; whose nature is grounded in things of the type of buildings. If we pay attention to these relationship between location and spaces, between space and space, we gain a clue to help us to think about the relationship between human and space [14]". This is a hint in a way on design thinking and on how to concern about the environment, as well as the relationship between human and its context. Digital architecture has played a significant role in redefining human experience and understanding of the physical world by transforming sensory perceptions, challenging traditional distinctions, and raising questions about the human condition in the digital age [8].

The advanced digital tools enable designers to visualize the design ideas in short time so that designers can better modify, rethink, realize their ideas, by reviewing the result of the variations. In terms of parametric design, generally, it implies that the parameters as key elements and variations in the process of design, architectural design in this case. As Schumacher argued that parametriscism's superiority is "social functionality/performativity" rather than technical functionality [7], Mark Burry suggests that the parametric inputs of architectural design should include multiple parameters, such as "environmental, political, social, cultural, practical, economic, theoretical, philosophical and behavioural parameters", and better to work parametrically across the full gamut of inputs so that there are abundant opportunities to enrich individual practice [3]. This contribution points out that, on the one hand, parametric design thinking has been practicing before the digital age, and on the other hand, parametric design approach in architecture can be both quantitative and qualitative. The advanced value of parametric design thinking, in this sense, is worthy of further exploring and sharing. Here, the purpose is to call for an interdisciplinary design approach, as well as to think differently, from different but interrelated fields. As Rivka Oxman emphasizes that, "parametric design thinking should exist as a shared design paradigm among all specific domains, producing holistic conceptual thinking processes from conception to fabrication. This is a new horizon of theory and pedagogy for the future of design thinking" [1]. In this sense, therefore, the model of parametric design thinking could be seen as a promising approach to future architectural design.

References

1. Rivka, O.: Thinking difference: theories and models of parametric design thinking. Des. Stud. **52**, 4–39 (2017)
2. Carpo, M.: Parametric notations: the birth of the non-standard. Archit. Des. **86**(2), 24–29 (2016)
3. Burry, M.: Antoni Gaudí and Frei Otto: essential precursors to the Parametricism manifesto. Archit. Des. **86**(2), 30–35 (2016)
4. Schumacher, P.: The Autopoiesis of Architecture, Volume I: A New Framework for Architecture. John Wiley & Sons (2011)
5. Davis, D.: A History of Parametric. www.danieldavis.com/a-history-of-parametric/ (2013). Accessed 14 Mar 2023
6. Schumacher, P.: Parametricism Style - Parametricist Manifesto. https://designmanifestos.org/patrik-schumacher-parametricism-as-style-parametricist-manifesto/ (2008) Accessed 13 Mar 2023
7. Schumacher, P.: Parametricism: The next decade. A+ u Archit. Urbanism **595**, 40–47 (2020)
8. De Wilde, P., de Souza, C.B.: Performative design strategies: the synthesis process of a woven complexity. In: Kanaani, M. (ed.) The Routledge Companion to Paradigms of Performativity in Design and Architecture, pp. 403–415. United Kingdom: Routledge, New York (2019)
9. Schumacher, P.: Parametricism 2.0: gearing up to impact the global built environment. Archit. Des. **86**(2), 8–17 (2016)
10. Schumacher, P.: Social performativity: architecture's contribution to societal progress. In: Kanaani, M. (ed.) The Routledge Companion to Paradigms of Performativity in Design and Architecture, pp. 13–31. United Kingdom: Routledge, New York (2019)
11. Gaha, I.S.: Parametric architectural design for a new city identity: materials, environments and new applications. J. Contemp. Urb. Aff. **7**(1), 122–138 (2023)

12. Schumacher, P.: From Intuition to Simulation. Positions: Unfolding Architectural Endeauvors. Edition Angewandte. Birkhaeuser, Basel (2020)
13. Picon, A.: Beyond digital avant-gardes: the materiality of architecture and its impact. Archit. Des. **90**(5), 118–125 (2020)
14. Heidegger, M.: Bauen Wohnen Denken: Vorträge und Aufsätze. Klett-Cotta (2022)
15. Tzu, L.: Tao Te Ching : A New Translation. Shambhala Publications Inc, William Scott Wilson (translator) (2013)
16. Spence, C.: Senses of place: architectural design for the multisensory mind. Cognitive research: principles and implications, **5**(1) (2020)

Branches of Architecture: Ways of Practice

Branches of Architecture: Ways of Practice

Santiago Gomes[✉]

DAD Department of Architecture and Design, Politecnico di Torino, Turin, Italy
santiago.gomes@polito.it

In the session dedicated to Ways of Practice, the contributions explored the domains of professional activity in architecture today. Starting from the investigation of the relationship between the different knowledge systems (not exclusively belonging to the disciplinary field of architecture) that characterise the work of architects in the contemporary condition [1], at least two levels, two complementary themes or questions emerge. On the one hand, the question of the redefinition of the very concept of architectural practice and the updating of the objectives of a professional knowledge that, while changing its physiognomy and expanding its boundaries [2], seems to maintain a certain relationship with the traditional education and teaching system [3]. On the other hand, a reflection on the role of academia, teaching, and design-based research in a panorama in the face of which the question of refuge in disciplinary autonomy or the dissolution of the figure of the architect within complex multidisciplinary processes could appear as the only possible paths.

The role of communities in the processes of evaluation, construction, and validation of design practice, as well as the interaction with stakeholders and thus their contributions to the project, are all issues that influence the two levels just mentioned. In this sense, the contributions have, through the presentation of case studies, investigated transversally from a methodological point of view the methods deployed by professionals and educators in order to succeed in integrating and holding together all these contributions, working on research topics, innovative functional programmes, the use and implications of new digital technologies and the two-way contamination between professional and academic research activities.

If these changes in everyday professional practice seem to be consolidated or seem to have been forcibly incorporated into the work of architects within architecture schools, the debate is animated and far from over. Several interventions, in fact, focus on the opportunity (or the possible effectiveness) of rethinking the curricula and the university educational offer in order to consolidate and develop the adaptive behaviour of the architect as a coordinator, solicitor and stimulator of design activities that go far beyond the boundaries that traditionally defined the architect's action within the processes of habitat transformation and global sustainability challenges.

References

1. Gutman, R.: Architectural Practice: A Critical View. Princeton Architectural Press, New York (1997)

© The Author(s) 2025
M. Barosio et al. (Eds.): EAAE AC 2023, SSDI 47, pp. 317–318, 2025.
https://doi.org/10.1007/978-3-031-71959-2_34

2. Harriss, H., Hyde, R., Marcaccio, R.: Architects after Architecture: Alternative Pathways for Practice. Routledge, New York (2020)
3. Voet, C., Schreurs, E., Thomas, H.: (eds.) The Hybrid Practitioner. Building, Teaching, Researching Architecture. Leuven University Press, Leuven (2022)

Level II Training and Development of Scientific and Didactic Content. The Case of Executive Master: Mountain-Able. Planning and Design for the Sustainable Development of Mountain

Emilia Corradi[✉]

Polytechnic of Milan, 20133 Milan, Italy
emilia.corradi@polimi.it

Abstract. In a fast-changing society, training architects to be one step ahead of the challenges that policies and regulations impose on architectural design has become increasingly necessary. Schools of architecture began to address the training of architects of the future to face new challenges. However, the scenario appears more complex for all those architects already working in the professional field and requires training in specific topics. This situation requires the development of progressively more specialized courses, especially for postgraduates. In addition to knowledge, adequate skills must be acquired to face the challenges posed by the EU Green Deal and the new European Bauhaus. The key is to create a higher level of training system that frames specific issues and themes, with a learning curve that always keeps the role of architecture at its centre as the process and outcome of the transformation of places and as a tool for serving residents and the community. The essay intends to illustrate the specific experience of the Mountain-Able: Planning and Design for the Sustainable Development of Mountain Executive Master project, activated by the Polytechnic of Milan's Department of Architecture and Urban Studies. The themes and issues addressed by the suggested specialist training course will necessarily combine themes of scale, multidisciplinarity, fragility, and community, which require the development of a very specific training platform to ensure students engage in hands-on specialist learning open to embracing contributions from other disciplines.

Keywords: Architectural training · Education · Multidisciplinarity · Community · Fragility

1 Premise

Currently, the university curriculum for an architectural degree often follows the path leaning towards methodological training in which the working method acquired is even more important than the hard and soft skills developed. The training of an architect combines technical and intellectual knowledge and often requires specialization in specific themes and aspects that will be needed after graduation to navigate an open and constantly evolving market, connected not only to the world of construction [1].

© The Author(s) 2025
M. Barosio et al. (Eds.): EAAE AC 2023, SSDI 47, pp. 319–328, 2025.
https://doi.org/10.1007/978-3-031-71959-2_35

Despite the belief that modern architectural training focuses on increasingly specialized segments [2], the need often emerges to combine multiple segments of knowledge and expertise in specific subjects. This need comes to the fore - confirmed by ARCHITECTURE'S AFTERLIFE project research consortium survey - [3], above all when the architect chooses not to operate directly in the architectural design or urban space fields but is tasked with addressing actions, policies, or business decisions in communities in which the architectural, urban space or territory and landscape planning is only part of the design process. This often happens when the architect is sent to work in unconventional social, territorial, and urban contexts, facing problems whose solution requires a combination of subjects and disciplines on very different scales. This aspect is, in turn, linked to the need for architects to adapt their knowledge and skills to demands that society, environment, and policies activated at various national and European levels pose with respect to the impact on architectural design on different scales. Choices are thus put in place after completing the university degree, and another level of education – postgraduate – appears on the horizon. This should dictate a highly specialized path, not least because of the need to activate and systematize the skills previously acquired during an architecture degree course that are attuned to approaches that are often multidisciplinary and go beyond the field of architecture.

Hence, the need arises for high-level postgraduate training in terms of quality of response and operational capacity, with a constantly increasing number of cases impacting the field of architecture. The objective is to find a way for trained architects who wish to broaden their professional opportunities or specialize not only in order to access adequate training opportunities provided by schools of architecture but also further training delivered by professional associations of architects. For example, the mission of the Council of Architects of Europe includes working towards an enabling practice for architects to guarantee the existence of a network and a community able to orient themselves and interact with the demands that EU policies and regulations impose on architecture [4]. As far as the innovation and culture promoted by the New European Bauhaus are concerned, with architectural design restored to its central role in the challenges of sustainability, beauty, and inclusiveness [5], the question arises of which skills – in addition to general abilities developed during first and second level university education – can be transmitted to those who intend to specialize in very different topics and scales of design embracing a broad transversal, cultural, technical, and economic dimension. The need to specialize or acquire adequate skills, in addition to knowledge, for handling the challenges posed by the EU Green Deal and the New European Bauhaus is thus a fundamental requirement to guarantee effective and continuing training over time for architects, to allow them to deal with the challenges of the future. When considering the traditional Bachelor's Degree and Master's Degree path, it is therefore increasingly necessary to think about activating subsequent higher-level training, like Executive Master courses formulated for architects already operating as professionals. Ongoing training after graduation poses another stumbling block in the architect's learning. It would be useful to activate a synergy for postgraduate training, too, between the European Association for Architectural Education and the Council of Architects of Europe, as had been proposed in 2015 through the Erasmus + project *Confronting Wicked Problems: Adapting Architectural Education to the New Situation in Europe* [6].

1.1 The Context

Observing the current scenario of Executive Masters offered by schools and universities in the field of architecture, we see a broad choice for postgraduate architect training addressing mainly architectural project management or the use of state-of-the-art technologies. Less frequently, there are Executive Masters that deal with architectural design innovation, especially in territorial or urban contexts, which require extensive multidiscipline and multiscale design approaches. Mountain or rural areas are emblematic of these settings and suffered from a critical problem of underuse, abandonment, and depopulation throughout Europe as there was a growing need to guarantee their resilience and preservation to safeguard residents and the cultural, historical, and architectural heritage they express. The Commission's communication to the European Parliament, the Council, the European Economic and Social Committee, and the Committee of the Regions states that rural areas account for almost 30% of the population and more than 83% of EU territory [7], of which about 30% is made up of rural mountain areas [8]. The same document illustrates a series of actions aiming to transform the EU's rural areas into stronger, more resilient, and connected districts by 2024 [9], enabling them to communicate with urban areas not in contrast but as an alternative model to metropolitan life. These actions impact the transformation of rural mountain space through new infrastructural works; regeneration of existing built heritage; achievement of energy and environmental sustainability; modernization of agricultural techniques for the economic and social evolution of the communities involved. With respect to these themes, there is a need to specialize architects – who usually operate in urban settings that are more robust in terms of resources and opportunities – to address the scales and themes that characterize rural districts, which are very different from those found in urban areas. Rural mountain districts have to combine architectural standards covering very different scales and disciplines, ranging from agricultural issues to energy, renewable sources, infrastructure, and community protection from natural risks such as floods, landslides, or earthquakes, not to mention those looming in the not-too-distant future deriving from lack of water resources. This long premise was required in order to illustrate the experience of the Mountain-Able: Planning and Design for Sustainable Development of Mountain project.

2 The Mountain-Able: Planning and Design for Sustainable Development of Mountain Executive Master

We will focus on one particular experience in this contribution, linked to the development of the "Mountain-Able: Planning and Design for the Sustainable Development of Mountain" Executive Master, put in place by the Polytechnic of Milan's Department of Architecture and Urban Studies. The project came to be in a specific scientific and training setting, relating to the Excellence Project developed by the aforementioned Department of Architecture and Urban Studies (DAStU). The Department is one of one hundred and eighty in Italy to have received the Ministry of Research award as a department of excellence in the approach to issues of territorial fragility for the five-year period 2018–22 and for the five-year period 2023–27. The DAStU conducts research, design experimentation, and didactic instruction in architectural and urban design, territorial planning and land governance, urban policies, conservation and intervention on the built and natural heritage, historical interpretation and criticism of architecture and the city [10]. As part of the research venture, the need emerged for training architects to be specialists in issues related to emergencies, which is ever more typical of fragile areas like rural and mountain districts, often given scant consideration in research and teaching.

The objective of the Executive Master, therefore, is to specialize architects in the themes of design and planning for mountain and rural areas, focusing on the various sectors of fragility, from marginalization to abandonment of built and agricultural heritage, from prevention of natural risks to the decline of cultural memory. The Master was devised for architects and public administrators working for regional, provincial and municipal government, consortiums, institutions, subsidiary companies, etc., as well as all those professional profiles engaged in various ways in complex contexts and in charge of architectural design in rural and mountain territories. In these settings, the architect often serves as a social facilitator and as the organizer of procedures and actions shared with the community, which then flow into the regeneration project. These are, therefore, professional figures who activate processes seeking qualitative development of space, environment, and territory in complex factor combinations [11].

In Italy, it is not unusual to find professional figures being asked to assess the quality of a project despite having no specific technical or multidisciplinary skills or awareness of the interactions between the project's various components. Their task is often to verify the correctness of the procedural process; less frequently, they may express opinions on the quality of architecture in terms defined by the New European Bauhaus. Conversely, some architects are unfamiliar with the scales and needs of rural and mountain districts and struggle to bring forth adequate projects in terms of quality and sustainability. Architectural, urban space, and landscape design in rural and mountain communities include multiple cultural, technical, legislative, and economic aspects expressed on many scales. In these areas, the architect's role is often to build visions and perspectives for the regeneration and enhancement of the places and the communities that represent them. This vision must take into account the relationship established with the assortment of elements making up rural and mountain territories, which are road, hydroelectric, and sports infrastructure. But also, what is found in the agricultural landscape – small buildings for storing equipment or sheltering livestock, crop terraces, tiny, inhabited centres, and historical and architectural monuments that bear witness to a past rich in art and culture. A further factor is that of maintenance and management of the territory, increasingly impacted by the effects of climate change with the consequent repercussions on resident communities who are then forced to live with the impact of the loss of public services, abandonment of built heritage, depletion of environmental resources, and loss of memory.

These factors revolve around the decreasing number of communities, which must become part of a regeneration process as a shared system. So, what comes to the fore is the need to train both architects who operate not strictly in the construction field but who want to address projects for the transformation of space as a complex process in which policy, space, community, financing, anthropological, social and cultural aspects all play a role. But also, architects who intend to deal with the design of architecture and urban or landscape space in contexts characterized by fragilities like loss of services, abandonment, natural risks, depopulation, and demise of identity located, nonetheless, in precious natural and environmental settings. The former can operate through specific training to become the bedrock for these territories' harmonious and sustainable development. At the same time, the latter, with other figures, will undertake projects on different scales, addressing different specialist themes.

2.1 Structure of the Executive Master

It took two years of teamwork by the Director, Professor Emilia Corradi and Deputy Directors, respectively Professor Annunziata Maria Oteri and Professors Paolo Bozzuto, to develop the Executive Master contents. The structure of the Executive Master had to act simultaneously on two mutually integrated aspects, which were then merged into the didactic platform. The first aspect of managing the Master involved administrative procedures and the accreditation process for the recognition of the qualification acquired; the second aspect concerned learning and scientific content. With regard to administrative procedures, the Master had to take into account both the regulations outlined by Italy's Ministry of University and Research, which govern and normalize higher-level training, and the Polytechnic of Milan's indications for the accreditation of Level I and Level II master's degrees. This was a requirement for obtaining legal recognition of the Executive Master, based on compliance with Ministerial Decree 270/2004 conditions, specifying a duration of at least one year for these courses, equal to 60 ECTS, or at least 1,500 student hours [12]. A further administrative aspect to be taken into account was the budget calculation, verifying that it was financially workable to complete the planned activities, with a projection of a hypothetical number of enrollees. It is crucial to decide how many faculty members to involve, based on their skills and qualifications, because it is an aspect that a candidate who intends to enrol in and attend an Executive Master course will consider very carefully. These elements, in turn, made it possible to identify a format for the Mountain-Able Master, contributing to the definition of its didactic and formative contents.

Looking at the didactic and scientific structure, the first step to developing the Master's was to outline the central theme of sustainable development for mountain districts with regard to the research and teaching standards established by Polytechnic of Milan's DAStU and AUIC (School of Urban Architecture and Construction Engineering). This first step was followed by a benchmark analysis of the executive masters on offer in Italian and European architecture schools. The study contributed to estimating the potential of the proposed Mountain-Able Executive Master's didactic and scientific configuration. Results confirmed the need to establish a master's degree that would deliver integrated and multidisciplinary planning for mountain locations with reference to both national and European policies and programming. This process significantly impacted pinpointing the international dimension of themes connected to the potential application for the qualification obtained by students. Subsequent research was conducted to identify the target audience who would be interested in the Executive Master. During this phase, a significant role was played by a series of meetings and interviews undertaken with public offices like the Ministry of Culture or Lombardy regional authority, tasked with mountain district planning and policies, and also with representatives of national, regional, and local associations or institutions, as well as with businesses and companies operating in mountain districts.

The phase led to setting up the scientific committee and opening the Executive Master to sponsors and partners who support it in various capacities. The final step was to develop the didactic platform, whose primary objective was to provide a training and specialization path that focused at all times on the role of architecture as a process and outcome of the transformation of places as the way to serve residents and the community. The planned didactic contents develop mainly around an operational framework for achieving the general training objectives previously indicated and summarized in the diagram below (Fig. 1).

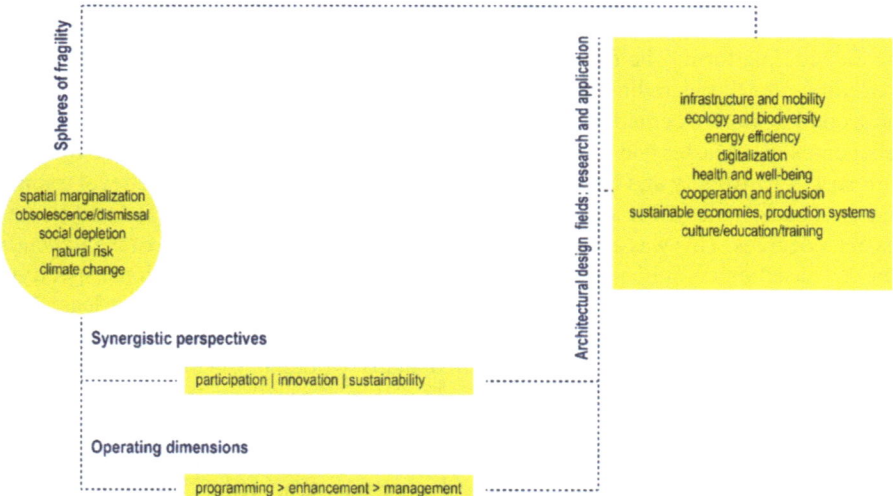

Fig. 1. Executive Master Mountain-Able. Planning and design for sustainable mountain development: topics and objectives, Milan 2023. © Emilia Corradi.

2.2 Didactic Structure and Contents with Respect to Learning Objectives

From a learning perspective, the didactic structure does not intend to train a professional who embodies every single skill. The objective is to train a problem-solving figure to develop a specific awareness of the fact that the creative process is a fundamental part of learning about the themes and problems of mountain districts. The creative process is reinforced by the technical and cultural learning acquired during university studies and stimulated by the Master. From the standpoint of the didactic content of learning objectives, the teaching structure comprises two macro-scales: *projects and policies on the territorial scale and projects and policies at the local level.* The former address issues like accessibility, organization of social and health services, as well as concepts inherent to the environmental sustainability of state-of-the-art economic activities and socio-economic interaction and partnerships with metropolitan districts. The latter address issues, projects and policies at the local level, which can be summed up as strategies for transforming and enhancing communities' physical and cultural heritage through recovery and development actions. This learning structure integrates typical architectural planning disciplines with telemedicine, economics, law, anthropology, geography, and art history to provide adequate tools for contributing to the sustainable transformation of complex mountain districts. The platform considers the project's centrality to be a creative and technical process serving as a pedagogical foundation and fusion of the spatial, landscape, technical, cultural, environmental, economic, legal, and procedural dimensions that affect the development and regeneration of mountain and rural territories and communities.

Through the pedagogical teaching model based on the driven design path, the training approach integrates theory with practice applied to learning by doing. In this respect, the Master opted for a credit structure that supports this didactic approach, dedicating many of the 1,500 student hours of engagement – required by law – to seminar activities, including an international workshop and internships in public administration, agencies, companies, and institutions that are partners in the Master. In a certain sense, this decision influences the didactic and pedagogical structure, especially with regard to the need to integrate theoretical and hands-on training aspects and the understanding of operational procedures.

The Mountain-Able Master is set out as theoretical modules, thematic design laboratories, workshops, seminars, internships, and degree theses. The theoretical modules award 16 ECTS and include a package of 20 modules whose contents address the different themes in a multidisciplinary manner (Fig. 2). Professors provide the theoretical lessons and exercises, but also expert professionals brought on board from the mountain world and public administrations, or companies operating in various capacities in this complex reality. There are economists, experts in co-design and participatory processes, experts in energy communities, territorial planning, jurists, archaeologists, art and architecture historians, experts in climate change, risk, communication, and marketing, and experts in territorial and community policies and management. The theoretical modules help define the field of problems within which the professional must move, both as a designer and as a facilitator.

The three thematic design laboratories provide a total of 15 ECTS. The thematic design laboratories are structured to include experts who illustrate best practices and, with the participants in the Master, identify their methods of application in several areas based on the themes and scales chosen, especially landscape scale, urban scale, and architectural scale. Thematic design laboratories will be able to apply the knowledge acquired in the theoretical modules thanks to an integrated project that considers different scales and spatial, social, participatory, and cultural components, considering extant material and immaterial heritage.

The workshop awards 12 ECTS and is implemented in a mountain area where Master faculty and participants reside for a week. Stakeholders, administrators, experts, and residents are also present, as are international counterparts who can illustrate experiences in similar contexts. The latter aspect is considered extremely important as comparison with other European experiences can be helpful in understanding how to activate partnerships

Theory and tools 16 ECTS

Methodology for investigating
the processes of historical evolution of territories

Integrated policies
of mountain areas Infrastructure, transport, accessibility and mobility. Sustainable mobility and transport planning in mountain areas

Prevention and mitigation strategies
for damage from hydrogeological and environmental risks

Anti-seismic
Projects and strategies for improvement of widespread heritage, best practices and case studies

Networks and healthcare and communities
Networks, technologies and tools for digital communication, telemedicine distance learning

Communication and digitalisation
Networks,technologies and tools for digital communication, telemedicine and distance learning

Energy efficiency of mountain building heritage
Sustainable production and distribution of energy in mountain areas

Co-design analysis and techniques
Community involvement processes

Co-design analysis and techniques
Design and programming of co-design strategies

Programming policies
for regeneration actions

Design of cultural networks in mountain areas
Historical characteristics and cultural heritage

Design of cultural networks in mountain areas
Physical heritage and its transformation

Economic, procedural and legal feasibility
of the reactivation processes of mountain areas

Projects and tools
for the reuse of widespread heritage in mountain areas, best practices and case studies

Enhancement of existing physical networks
Types of networks, analysis and mapping systems

Enhancement of existing physical networks
Strategies and projects

Analysis and mapping criteria
Sustainable design of the landscape and urban space of mountain areas and centres, best practices and case studies.

Interpretative readings
Sustainable design of the landscape and urban space of mountain areas and centres, best practices and case studies.

Sustainability
Sustainable design of the landscape and urban space of mountain areas and centres, best practices and case studies

Risk prevention
Planning and programming of sport and sustainable tourism in mountain areas, best practices and case studies

FINAL THESIS 5 ECTS

INTERNSHIP 12 ECTS

WORKSHOP 12 ECTS

DESIGN LAB 15 ECTS

Fig. 2. Executive Master Mountain-Able. Planning and design for sustainable mountain development: didactic structure, Milan 2023. © Emilia Corradi.

and structure exchange networks in the context of European projects for participation in calls for funding or international cooperation.

The internship awards 12 ECTS and consists of a period of practical experience hosted by institutions, public administrations, or agencies connected to the theme of the Master. It aims to offer an understanding of procedural and operational practices in a workplace setting.

The final thesis for 5 ECTS is when the Master comes together, each student developing their own paper. The thesis involves drafting a critical essay illustrating each candidate's specific topics of interest, describing a design or procedural path through the knowledge and skills acquired.

3 Conclusion

The path of inductive research developed as the underpinning of the Mountain-Able: Planning and Design for the Sustainable Development of Mountain Executive Master allowed us to spotlight the themes and problems addressed, beginning with a wide range of data. The specific training path suggested for multidisciplinary and multiscale themes inherent to a series of fragilities required the development of a highly specialized learning platform to ensure participants received adequate practical and technical training and were also open to contributions from other disciplines.

The ongoing relationship with stakeholders fully highlighted the possible synergies that can be activated between schools of architecture and players in the territorial transformation process. The result is the chance to implement a dynamic form of training capable of responding to social changes and fully integrating with community policies and strategies. The topic of continuing higher education represents a significant opportunity to improve the educational prospects offered by architecture schools and think about specific shared paths that allow architects to implement a proactive approach appropriate for future challenges. At the same time, continuous learning models must be defined for professional practice also, allowing the figure of the architect to adapt to the changes required by the professional market at increasingly fast speed. Professionalizing Master's degrees can be a positive way of offering extremely effective educational models quickly. These can represent an interesting way to combine education, research, and professional skills, above all because, through a learning-by-doing approach, they can specialize as professionals so they can meet job market demands. At the same time, they also represent a multidisciplinary education model that can overcome sometimes inflexible protocols implemented by state education systems.

The other aspect of executive masters is that they can activate effective synergies not only with the Architects' Council of Europe but also with administrations, companies, and operators in the various sectors, narrowing the gap between academic training, research, and professional practice in the field of architecture [13]. In this respect, it would be appropriate to launch a European-wide survey of executive masters within the EAAE, assessing learning in and updating the architect's professional practice. The survey could facilitate the circulation of models and good practices to be shared in architecture schools and introduce ways to evaluate the effectiveness of a multidisciplinary but also a hybrid path, navigating theory and professional practice and bringing professional résumés up to speed.

References

1. Harris, H., Barosio, M., Sentieri, C.: Architecture's afterlife: the multi-sector impact of an architectural qualification. In: Adil, Z. (ed.) A Focus on Pedagogy: Teaching, Learning and Research in the Modern Academy, pp. 280–289. AMPS Proceedings Series 28 (2021)

2. Frank, O.H.: The changing roles of the architect. https://www.eaae.be/eaae-academies/edu cation-academy/themes/changing-roles-architect/. Accessed 18 Jan 2024
3. Architecture's Afterlife. http://architectures-afterlife.com. Accessed 18 Jan 2024
4. ACE Architects' Council of Europe. https://www.ace-cae.eu/about-us/missions-values/. Accessed 23 Oct 2023
5. EU, New European Bauhaus. https://new-european-bauhaus.europa.eu/index_en. Accessed 23 Oct 2023
6. Erasmus+, Confronting wicked problems: adapting architectural education to the new situation in europe, Annexe S3. https://www.eaae.be/wp-content/uploads/2017/04/Erasmus-Pro ject_CWP_11_Annexe-S.3_lr.pdf, 4, Accessed 19 Jan 2024
7. EU.   https://eur-lex.europa.eu/legal-content/EN/TXT/?uri=CELEX%3A52021SC0166/. Accessed 23 Oct 2023
8. EU.   https://eur-lex.europa.eu/legal-content/EN/TXT/HTML/?uri=CELEX:52021DC0345. Accessed 23 Oct 2023
9. EU.   https://eurlex.europa.eu/legalcontent/IT/TXT/PDF/?uri=CELEX:51995IR0142. Accessed 23 Oct 2023
10. Trigg, S.: The architect-organizer. In: Harriss, H. et al. (ed.) Architects after Architecture. Alternative Pathways for Practice, pp. 122–126. Routledge, New York (2021)
11. DAStU Progetto Eccellenza. https://www.eccellenza.dastu.polimi.it/. Accessed 23 Oct 2023
12. Politecnico di Milano Regolamento dei Corsi di Master Universitari di I e II livello, Corsi di Perfezionamento e Formazione Continua. https://www.normativa.polimi.it/fileadmin/user_u pload/regolamenti/studenti/Regolamento_Corsi_Master_I_e_II_livello.pdf. Accessed 23 Oct 2023
13. Ruhi-Sipahioglu, I., et al.: e-FIADE: Exploring the Field of Interaction in Architectural Design Education. PROJECT REPORT O2 -MAPPING AND ANALYSIS OF INTERN-SHIPS (2018). https://www.researchgate.net/publication/336778418_eFIADE_Exploring_ the_Field_of_Interaction_in_Architectural_Design_Education_PROJECT_REPORT_O2_ MAPPING_AND_ANALYSIS_OF_INTERNSHIPS. Accessed 20 Jan 2024

The Glass House Revisited

Stamatina Kousidi[(✉)]

Department of Architecture and Urban Studies, Politecnico di Milano, Milan, Italy
stamatina.kousidi@polimi.it

Abstract. The paper draws upon a pedagogical project that addressed the design of spaces for cohabitation, of new alliances between buildings and the natural environment, in the context of the contemporary city. First, it explores how the fantasy of the large-span, glazed, vegetated environment has shaped visionary projects, marking a shift of attention from the physical to the physiological and from the tangible to the intangible qualities of space which resonates with contemporary concerns on design for sustainability. Following the evolution of the glass house from a place of nature preservation to a vehicle of experimentation into new ways of inhabiting the city, it then examines the contemporary relevance of such a building type and how it has fostered new architectural narratives on the co-existence of people, buildings and plants. Second, the paper presents the methodology practiced in a postgraduate design studio at the AUIC School/Politecnico di Milano which aimed to raise awareness among students about the relational dimension of architecture and the reciprocal exchange between design and research. It discusses how, by revisiting the glass house figure, the studio output set out to generate new conceptual, aesthetic and design definitions of the architecture of the in-between, of spaces of transition between the natural and the manmade, inside and outside.

Keywords: Design Studio · Glasshouse · Collective Housing · Building Envelope · Design Methodology · Cohabitation

1 Genealogy: The Evolving Concept of Human-Nature Cohabitation

The glasshouse model grew pertinent to early twentieth-century experimentations which cast a special attention on the non-physical aspects of space, connected to thermal comfort, hygiene and concepts of health. These experimentations drew upon prior architectural visions that saw the inclusion of greenery in the all-glass structure whose design, construction and maintenance oscillated between horticulture, architecture and engineering, highlighting issues of environmental management [1–3]. The indoor patio or winter garden, which formed an integral part of the modern interior, marked the evolution of the glasshouse from a place of constructing aesthetic experiences to an incubator of novel approaches to the design of the built environment [4].

The phenomenon that saw the hybridization of residential and green spaces continued to manifest itself from the late 1960s onwards, as concerns about the environmental

M. Barosio et al. (Eds.): EAAE AC 2023, SSDI 47, pp. 329–338, 2025.
https://doi.org/10.1007/978-3-031-71959-2_36

impact of architecture proliferated. Office, university and commercial buildings, such as Cedric Price's project for a glasshouse in Parc de la Villette in Paris, blended the boundaries between the natural and the manmade (Fig. 1). The latter example incorporated glass-roofed vegetated atria, partly integrated with human activities and equipped with adjustable blinds for indoor heating and ventilation control. The sketch depicts a section of the building permeated by heat, air and energy flows (Fig. 2). It is telling of the then architectural experimentation into new confluences between internal green spaces and program, aspects of air quality and control. Moreover, another stream of experimentation in those days speculated on the definition of 'artificial ecologies', drawing on the glasshouse as the replica of another, ideal and constant climate, in search for a *symbiotic* relationship between human and non-human organisms.

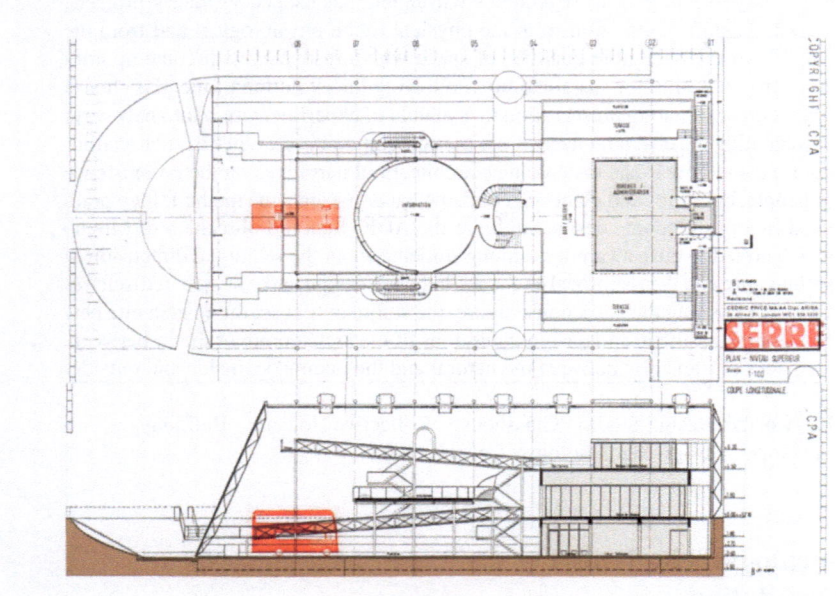

Fig. 1. Cedric Price, Serre (2), upper level plan and longitudinal section, 1988–1990, graphic appliqué film over electrostatic print on paper, mounted on paper with red ink stamp, 29,8 × 42,1 cm. DR2004:0533:004. Canadian Centre for Architecture (drawings © Cedric Price fonds/Canadian Centre for Architecture)

"I thought of referring to communication networks as something fluid, like water streams, to produce an artificial nature rather than architecture," Toyo Ito described with reference to the design process of the Sendai Médiathèque (1998–2001) [5]. If modernist architecture allowed for an articulated space between building and landscape, today the boundaries between the artificial and natural, the urban and the sylvan, are folding in, influencing the architectural project. Contemporary housing design projects, which range from Lacaton and Vassal's Exhibition Hall in Paris (2006) to Baukunst

+ Bruther's project for the ZHAW campus in Winterthur (2018) and from Bruther's Super-L – 150 Housing Units in Eysines (2013) to Atelier Kempe Thill's winter garden housing project in Amberes (2015), continue to build upon the glazed structure for the cultivation, preservation and display of tender flowers, plants or biomes. They approach the integration of the building envelope with greenery, in a way that the latter engages with human activity.

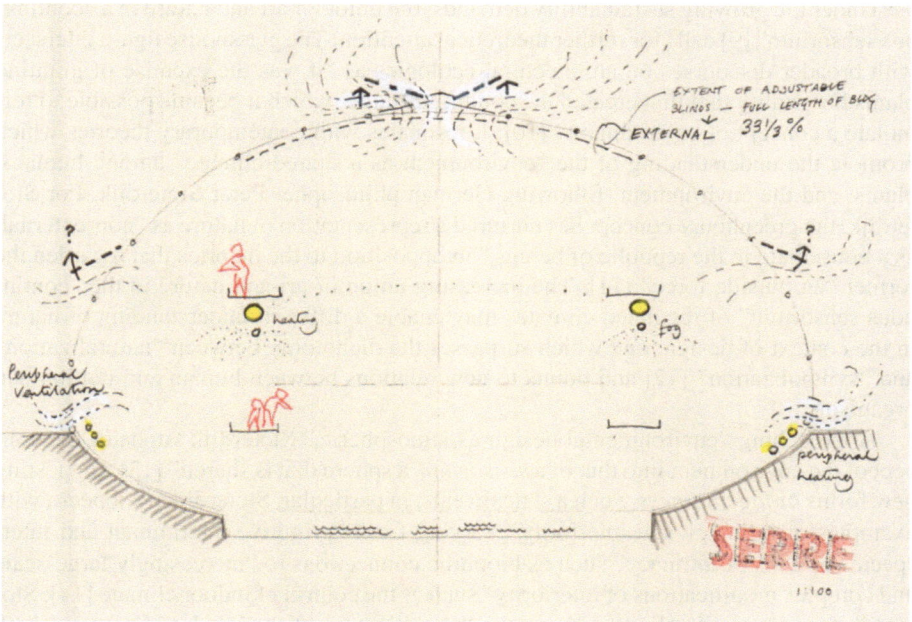

Fig. 2. Cedric Price, Serre (2), sketch showing adjustable blinds, heating and ventilation, 1988–1990, ink, graphite, white paint and coloured pencil over electrostatic print on heavy yellow paper, 21,1 × 29,7 cm. DR2004:0558:003. Canadian Centre for Architecture (diagram © Cedric Price fonds/Canadian Centre for Architecture)

From new construction to transformation projects, and from hybrid-use to urban farming buildings, they point to a phenomenon of *multinaturalisation* of the built environment, in which "green does not stop at a building's surface: It also penetrates the interior space, to give the impression of living everywhere with nature" [6]. Testifying to the fact that "green additions have taken on various forms that continue to extend perceptions of the term" [7], Penelope Dean makes reference to projects which have set out to bring, through punctual interventions, a positive impact on living space at the scale of the city, such as Atelier Bow-Wow's Void Metabolism, Emilio Ambasz's Green Town and Toyo Ito and Associates' Parque de la Gavia.

In certain cases, as in the work of architects Anne Lacaton and Jean-Philippe Vassal, the glasshouse figure is valued specifically for issues of material efficiency, aspects of comfort and climatic behavior. For the architects, "the potential of technology lies […] in its ability to be reprogrammed and combined with other things" [8], not on its mere reappropriation. In this context, the need to revise traditional means of representation and to connect building with processes that are external to it – the movement of its users, the flow of air, the angle of sunlight, energy flows – comes to the fore, reminding us that environmental sustainability is equally a representational and a design issue. The climatic drawings and section diagrams deployed by numerous projects of architects Lacaton and Vassal allude to the increasingly changing definition of the architectural agency which is not limited to physical and material aspects of space alone.

Under the growing sustainability demands, the union of art and nature in a "continuous sensorium" [9] calls for further theoretical attention. The glasshouse figure intersects with broader discourses on architectural ecologies as "it was the exercise of granting plants hospitality that first created the conditions under which it became possible to formulate a concept of environment" [10]. It resonates with contemporary theories which promote the understanding of the "environment as a shared climate" among humans, plants, and the environment, following German philosopher Peter Sloterdijk. For Sloterdijk, the greenhouse concept has nurtured a representation of nature as "non-external, as a housemate in the republic of beings," in opposition to the theories that regarded the former "an outside force" [11]. The increasing union of art and nature in the "continuous sensorium" of regulated climates may enable a different understanding of nature in the context of design, one which surpasses the dichotomy between "naturalization" and "symbolization" [12] and points to new relations between human and non-human organisms.

Approaching "environmental design as atmospheric," Sloterdijk "updates the concept of the environment into that of a *sensorium*, a sphere that is shared" [13], suggesting new forms of *togetherness*. Such a state reveals, in particular, Sloterdijk's "concern with examples of intimacy and interiority," varying from "primitive interhuman and interspecies notions of intimacy" such as biophilic connections to "increasingly large-scale and complex modifications of interiority" such as the control of indoor climate [14]. Sloterdijk focuses a critical attention on the "climatization of the inhabited space" which entails "envisaging the anthropogenic climate in all its thematic intrusiveness" following different degrees of environmental appropriation [15] that ring all the more familiar today as contemporary societies are confronted with the fragility of nature.

It renders explicit the quality, design, and agency of air, after the proliferation of "zones of a carefully manipulated climate, flooded with natural sunlight, overgrown with plants, and populated with humans" which, as cultural theorist Eva Horn remarks, "[represent] artificial atmospheres that experiment on the artificiality of nature itself" [16]. As the boundaries between nature and artifice are increasingly folding in, attention shifts away from static forms and solid volumes and is guided towards the design of envelopes, spaces of flows, and atmospheres with the aim to bridge quantitative and qualitative design criteria.

2 Pedagogy: Addressing Collective Housing Design as an Interface Between the Natural and the Manmade

Framed by these premises, the postgraduate design studio titled "The Architecture of the In-Between," at the School of Architecture, Urban Planning, Construction Engineering (AUIC), Politecnico di Milano, addressed the intermediate space as a new interface between built and natural environments. The studio placed a particular attention on the integration of architecture with nature – of open green spaces in collective housing design – reflecting on its ability to enhance environmental performance, notions of engagement and care. The studio explored how such integration may nurture new critical narratives and foster a critical reflection on the design of spaces for cohabitation in the contemporary city. It recognized in the figure of the greenhouse an opportunity for design experimentation to address current environmental issues and the shift from the physical to the physiological qualities of space, speculating on its future stance.

The first phase of the studio asked students to undertake a case study analysis concerning projects that have problematized the dichotomy between inside and outside. The notion of inhabiting spheres of different environmental qualities, in a state of co-presence, coevolution, co-breathing between human and non-human organisms, suggests the construction of "an environment of relationality and interrelational movements" [17]. The tracing of a genealogy of collective housing projects which have addressed the open space, landscape and natural environment in an explicit manner, revealed an evolving and multifaceted building typology, while challenging students to contemplate such integration in connection with contemporary social and ecological issues. It drew a novel attention to the character of liminal spaces, such as loggias, terraces, winter gardens, glazed atria and galleries, inherent to modernist and contemporary projects which have

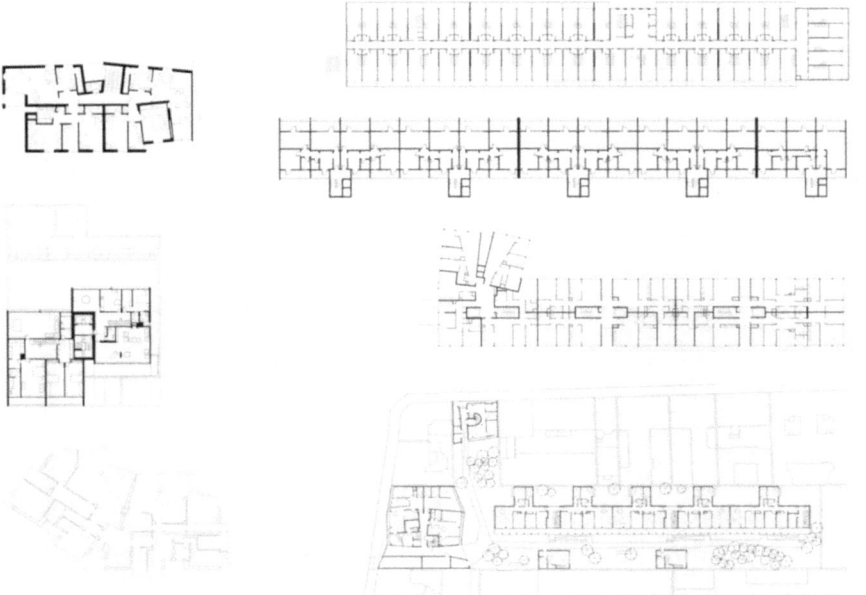

Fig. 3. Case study atlas: floor plan design (selection). "Thresholds. Architecture of the In-Between," Architectural Design Thematic Module, Architecture of Interiors Design Studio (section B), AUIC School, Politecnico di Milano, a.a. 2022–23.

attempted to define more efficient, sophisticated, non-mechanical means for environmental control. The resulting case study atlas (Fig. 3) underpinned a critical comparative analysis of design precedents, exploring their potential to generate new conceptual definitions of intermediate spaces.

It reflected the studio's engagement with design discourses and practices that have interrogated architectural form *after* performance, coining novel approaches to the relation between aesthetic perception and social concerns. The critical comparative analysis of design precedents led to the tracing of controversial aspects regarding the case studies at stake, embracing the hypothesis that "learning about architecture by mapping controversies can cultivate a specific attention to the performativity of design and can ultimately result in better design" [18].

The impressions from the studied project were then integrated into a conceptual collage-manifesto: one of the adopted analytical tools, alongside processes of redrawing, schematic representation, photographic documentation and writing. The collages highlighted the manifold declinations of spaces of sequence, transition, and continuity between public and private, natural and artificial domains (Figs. 4 and 5). They addressed in-between spaces as zones in their own right, which organize transitions between the respective domains, pointing to the conceptualization of architecture as an articulated system of mediation with open boundaries. Crossing between historical precedents analysis and design speculation, the design studio was practiced in its ability to offer "a productive environment to conduct research, by engaging in wide-ranging networks, adapting seemingly determined technologies, and testing didactic structures and methodological approaches" [19].

The deployed tools included but were not limited to archival research, on-site surveys, observation, graphic representations and concept mapping, conceiving of the studio environment as "a workshop, a platform for debate, a synthesizer of ideas and concepts, as it takes advantage of the expertise of a wide range of individuals and fields of interest" [20]. The design proposals then set out to explore potential future scenarios of high-density dwelling which integrate the exterior open space as an integral part of the

Fig. 4. Efimova A., Mijatov D., Vennitti J.M., Vujinovic K., Case study: manifesto – Francis Soler, 'Suite Sans Fin', rue Durkheim, Paris (1994–97). "Thresholds. Architecture of the In-Between," Architectural Design Thematic Module, Architecture of Interiors Design Studio (section B), AUIC School, Politecnico di Milano, a.a. 2022–23.

domestic environment (Figs. 6 and 7). Focusing on the definition of skins, filters, membranes and surfaces, the adopted studio method raised awareness among students about the relational dimension of architecture and the need to re-think patterns of cohabitation in the contemporary city.

To address the issue of design for sustainability, what emerges as important is our relationship to natural objects rather than their understanding as performative apparatuses in support of our increasingly regulated environments. It involves developing design proposals which do not merely attempt "to mitigate a building's impact on natural systems, but [which seek], at least rhetorically, to become a part of those systems" [21], suggesting new hybrids between architecture and nature. The studio outcome ultimately set out to explore the implications into form, space and materiality that such notion entails. The various recent intersections between biology and building nurture the belief that "the concepts of nature and architecture are not separable but linked. The binary opposition between the natural and the artificial is increasingly called into question, conceiving of plants, flowers, and biomes as central elements of new design scenaria: "nature is just as designed as design is natural; life is planned in the same way that the plan is something alive" [22]. In such a context, architecture is not solely to be understood as the theory and practice of a singular building or the spatial design of our environment, but extends to encompass design, planning and visualisation of politics, economy, environment, future and human life in general" [23].

Fig. 5. Bayraktar B.N., Garre A.C., Saison L.I., Yesilyurt A.H., Case study: manifesto – Angelo Mangiarotti, 'La Balossa', Monza, (1972). "Thresholds. Architecture of the In-Between," Architectural Design Thematic Module, Architecture of Interiors Design Studio (section B), AUIC School, Politecnico di Milano, a.a. 2022–23.

Seen from this perspective, the figure of the glasshouse underlines the urgency of safeguarding natural organisms and environments, enabling us to reimagine architecture as part and as an expression of nature, as something that emerges from within the latter, instead of opposing it. The condition of *living together* introduces a new understanding of the nature-culture oppositional relationship, fostering novel definitions of form, performance and aesthetic perception. Under the growing sustainability demands, the notion of aesthetics addresses "the relationship between sensory perception (the subjective) and quantifiable measures (the objective), and furthermore, they address the role of architectonics in informing the relationship between the expression of material culture and the environment" [24]: it emerges as "a discipline of reflecting on art as mediation between culture and nature" [25]. Initially a place for contemplation and retreat from the industrialized city, the glasshouse figure emerges today as a multifaceted design notion, as an operative and symbolic subject matter alike. In a context that sees design for sustainability lacking paradigmatic icons, revisiting the glasshouse as a symbol of a new aesthetic perception connected to collective housing may hold the key in establishing new connections between green objects and buildings. It re-affirms the need to define new means of aesthetic expression mediated through the project. It points, on the one hand, to a design stance which associates nature with the animation of culture and its symbols and which addresses, on the other, issues of environmental performance and the fragile ecosystems on which our living spaces depend.

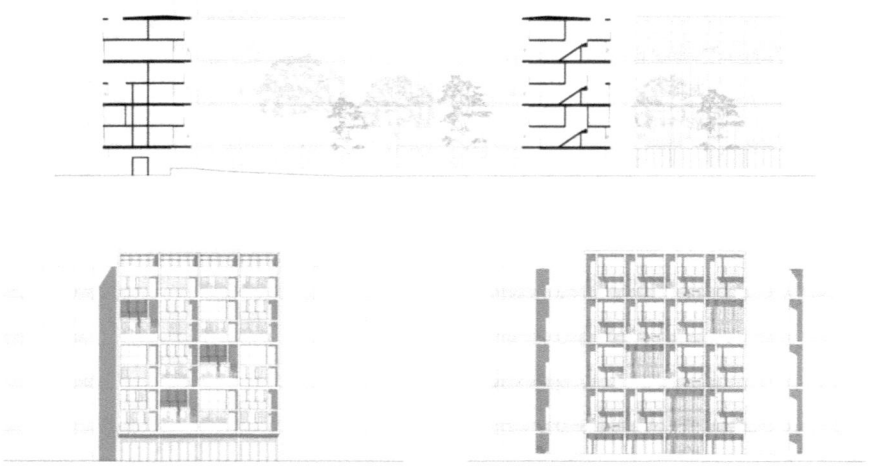

Fig. 6. Anversa A., Köhn M.K., Lo Vecchio J., Vignotti M., Project: section and elevations. "Thresholds. Architecture of the In-Between," Architectural Design Thematic Module, Architecture of Interiors Design Studio (section B), AUIC School, Politecnico di Milano, a.a. 2022–23.

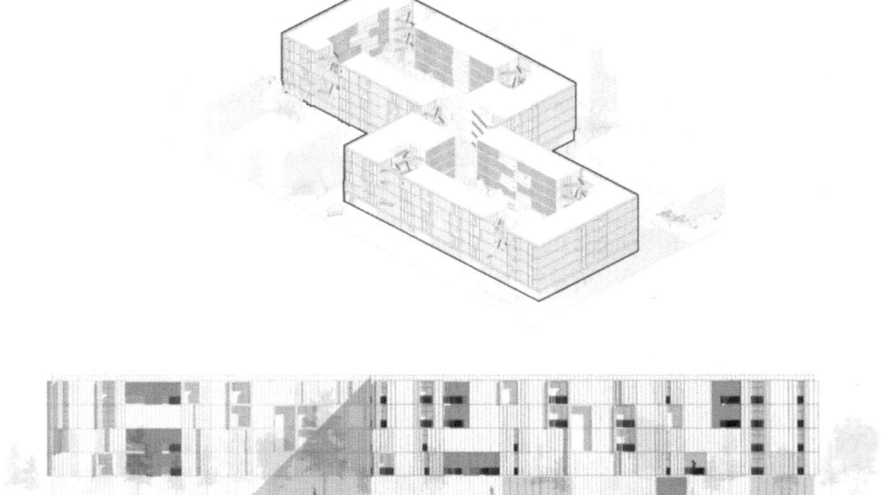

Fig. 7. Albin J., Morozova P., Pernot M.J.E., Rodríguez C.W.L., Veeramuthaiah N., Project: axonometric and elevation. "Thresholds. Architecture of the In-Between," Architectural Design Thematic Module, Architecture of Interiors Design Studio (section B), AUIC School, Politecnico di Milano, a.a. 2022–23.

References

1. Hix, J.: The Glasshouse. Phaidon Press, London, New York (1996)
2. Sparke, P.: Nature Inside: Plants and Flowers in the Modern Interior. Yale University Press, New Haven, London (2021)
3. Stein, A., Virts, N.: The Conservatory. Gardens Under Glass. Princeton Architectural Press, New York (2020)
4. Kousidi, S.: On greenhouses and the making of atmospheres. Ardeth **12**, 101–120 (2023)
5. Sakamoto, T., Ferre, A.: Toyo Ito Sendai Mediatheque. Actar, Barcelona, p. 175 (2003)
6. Zardini, M., Borasi, G.: Demedicalize Architecture. In: Id. (eds.) Imperfect Health: The Medicalization of Architecture, pp. 15–37 (here p. 19). Lars Müller Publishers, Zürich (2012)
7. Dean, P.: Under cover of green. In: Cuff D., Sherman R. (eds.) Fast–Forward Urbanism. Rethinking Architecture's Engagement with the City, pp. 62–74 (here p. 67). Princeton Architectural Press, New York (2011)
8. Ruby, I., Ruby, A.: Arquitectura naif. Notas sobre el trabajo de Lacaton & Vassal | Naïve Architecture: Notes on the Work of Lacaton and Vassal. In: Id. (eds.) Lacaton & Vassal – 2G Books, pp. 11–23 (here p. 18). G. Gili, Barcelona (2006)
9. Latour, B.: Air. In: Jones C.A. (ed.) Sensorium: Embodied Experience, Technology and Contemporary Art, pp. 104–107 (here p. 107). MIT Press, Cambridge, Massachusetts (2006)
10. Sloterdijk, P.: Atmospheric politics. In: Latour B., Weibel P. (eds.) Making Things Public. Atmospheres of Democracy, pp. 944–951 (here p. 945). MIT Press, Cambridge, Massachusetts (2005)
11. Sloterdijk, P.: Foams. Spheres III: Plural Spherology. Semiotext(e), South Pasadena, pp. 458–459 (2016)

12. Latour, B.: Air. In: Jones, C.A. (ed.) Sensorium: Embodied Experience, Technology and Contemporary Art, pp. 104–107 (here p. 107). MIT Press, Cambridge, Massachusetts (2006)
13. Blackman, L., Harbord, J.: Technologies of mediation and the affective. In: Hauptmann D., Neidich W. (eds.) Cognitive Architecture: From Bio-politics to Noo-politics; Architecture and Mind in the Age of Communication and Information, pp. 302–323 (here p. 313). 010 Publishers, Rotterdam (2010)
14. Lee, T., Wakefield-Rann, R.: Design philosophy and poetic thinking: peter sloterdijk's metaphorical explorations of the interior. Hum. Ecol. Rev. **24**(2), 153–170 (here p. 159) (2018)
15. Sloterdijk, P.: Foams. Spheres III: Plural Spherology. Semiotext(e), South Pasadena, p. 461 (2016)
16. Horn, E.: Air conditioning: taming the climate as a dream of civilization. In: Graham J (ed.) Climates: Architecture and the Planetary Imagination, pp. 233–242 (here p. 240). Columbia Books on Architecture; Lars Müller Publishers, New York; Zurich (2016)
17. Bruno, G.: Atmospheres of Projection: Environmentality in Art and Screen Media, p. 286. The University of Chicago Press, Chicago (2022)
18. Yaneva, A.: The new studio. In: Silberberger J. (ed.) Against and for Method. Revisiting Architectural Design as Research, pp. 171–186 (here p. 183). Gta Verlag/ETH, Zurich (2021)
19. Geiser, R.: Explorations in Architecture: Teaching, Design, Research. Birkhäuser, Basel, p. 11 (2008)
20. Ivi, p. 10
21. Barber, D., Putalik, A.: Forest, tower, city: rethinking the green machine aesthetic. Harvard Des. Mag. **45**, 234–243 (here p. 236) (2018)
22. Ursprung, P.: Nature and architecture. In: Mateo J.L. (ed.) Natural Metaphor. An Anthology of Essays on Architecture and Nature, pp. 10–21 (here p. 13). ETH; Actar Publishers, Zurich; Barcelona (2007)
23. Ibid
24. Lee, S.: (ed.) Aesthetics of Sustainable Architecture. 010 Publishers, Rotterdam, p.10 (2011)
25. Ibid

Participation of Stakeholders in Open Architectural and Urban Planning Competitions. Procedure Model and Application in Croatian Context

Rene Lisac[✉] and Kristina Careva

Faculty of Architecture, University of Zagreb, Kačićeva 26, 10000 Zagreb, Croatia
{rlisac,kcareva}@arhitekt.hr

Abstract. The participation of society in public spaces issues, though advocated for decades, experiences an increase in recent years. Citizen participation in Croatian spatial planning appears to be mostly declarative and formally takes place in the final stages of planning. On the contrary, the full benefits of integrative planning would be if participation of all sectors (civil, academic, economy and management structures) is continuous, especially important in the initial planning phase - forming the basic intentions.

The selection of the best spatial solution by architectural and urban planning competitions has a long and fruitful tradition in European societies. It represents a good yet hermetic method in searching for the most appropriate, most innovative and overall best solution whose authors remain anonymous till the very end of the selection.

The question is how to combine these two widely accepted and proven procedures into one, to acquire the best creative solution and raise awareness and involvement of the community?

The article will present recent case studies of participation in open competitions in Poland, Germany, Norway, and Croatia. The contribution of this paper is the effort to present systematically the ways of participation and how they are linked to individual groups of stakeholders within the competition procedure. Through the comparative analysis, highlighting benefits and challenges, the combined procedure model is proposed and applied in the Croatian context.

Keywords: Participation · Architectural competition · Urban Planning competition · Quadruple Helix

1 Introduction

The procedure of architectural and urban planning competitions represents a long and fruitful tradition in European societies. The selection of the best spatial solution through these competitions is a good but hermetic method to search for the most suitable, innovative, and overall best solution, whose authors remain anonymous until the very end of the selection.

© The Author(s) 2025
M. Barosio et al. (Eds.): EAAE AC 2023, SSDI 47, pp. 339–348, 2025.
https://doi.org/10.1007/978-3-031-71959-2_37

Regarding community participation in public space issues, it has been advocated for decades and has been experiencing an increase in recent years. Citizen participation in spatial planning seems to be mostly declarative and formally takes place in the final stages of planning. On the contrary, the full benefit of inclusive planning would be given if the participation of all sectors (civil society, academic, business, and administrative structures) is continuous, which is especially important in the initial stages of planning - when the basic intentions are formulated.

As for the mode of participation, we used an adapted version of Arnstein's participatory ladder [1] and considered those in the lower and middle part of the scale. Specifically, they are: informing; survey; interview; interactive discussion; creative discussion; and participation in realisation. As the level of participation increases, so does the involvement of the participants, both in terms of the amount of time that needs to be devoted to work, and in terms of intellectual effort as well as motivation. It is therefore not surprising that the number of participants decreases as the level of participation increases (Fig. 1).

Fig. 1. A scheme of inverse proportionality of the number of participants and their involvement, Zagreb 2023 (Scheme by authors)

Quadruple helix is a widely accepted model for conducting inclusive innovative processes with not only civil society [2]. Using a moderator as a neutral and objective subject, it brings together representatives of four groups: decision makers from management structures, experts on specific issues from the academic community, investors from the business sector, and citizens and associations as direct beneficiaries [3]. Observing the process, such as this research, through the Quad Helix model, can ensure appropriate and continuous involvement of all relevant stakeholders.

The question is how to intertwine the procedure of architectural and urban planning competitions and community participation methods to achieve quality process, without each of them losing its valuable qualities. In particular, the task is to include participation in the tender process without impairing the proven quality of the independent selection of anonymous solution proposals. The benefits we see that could come from this are community involvement that would result in public maturity and participatory data collection. Of course, there are also threats, mainly how to ensure the anonymity of the contestants and maintain professional integrity.

The goal of the research presented in this article is the formation of a procedural model that shows when and in which form it is possible to integrate some of the participatory methods within the architectural and urban planning competition. The method derives from the analysis and comparison of examples of such practice, considering the positive effects and negative implications.

2   The Examples of Citizens' Participation in Open Architectural and Urban Planning Competitions

Although examples involving community participation in the conduct of architectural and urban planning design competitions are occasional, some research that tackle this topic are available within the European context. Europe shares the same issues related to participation in competitions, as current research shows.

The article by Bern and Røe from Norway [4] points out that, although competitions can arouse significant community interest, the institution of architectural and urban planning competitions cannot easily incorporate participation in its implementation. Through interviews with architects both anonymity and professional integrity issues were raised. The authors argue that the best practice is to involve the interested public as soon as possible, in the preparation of the competition task and in defining the problems and requirements. That is when the opinion of the wider public has an impact on competitors and their solutions without compromising anonymity, which is one of the valuable features of the competition process.

Some of the examples implemented a parallel assignments model, where different programmes were given as a task to different architectural offices that were invited to participate. Although this model can provide interesting insights from proposed solutions and involve participation on the greater level, it is not applicable where public procurement and competition obligations are mandatory. That is why we will not include those examples in further analysis.

Kowalczyk in his article [5] points out that it is impossible to reconcile presenting works to a wider audience before the final decision of the jury and maintaining anonymity. He demonstrates the anonymity issue through several illustrative examples that served as solid case studies for this research. Cases are chosen from some EU countries with diverse attempts to involve participation in different competition stages.

2.1   Design of the Main Square in Koprivnica, Croatia

In the Croatian context, the example of the competition for city square in Koprivnica from 2019 stands out, in which the opinion of the public was analysed and included in the competition program through the project terms of reference and the attached documentation [6]. Citizens' opinion was questioned on three levels: local stakeholder groups, internet questionnaires and field surveys. Such comprehensive involvement in the initial phase is proven to be not only possible but strongly recommended (Fig. 2).

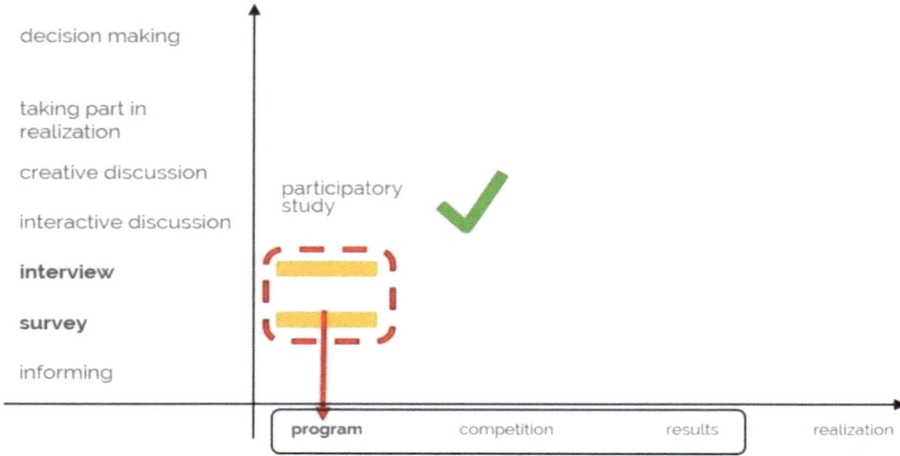

Fig. 2. Applied participation in the competition procedure for the design of the main square in Koprivnica, Croatia, 2023 (Scheme by authors)

2.2   Reconstruction of the Aeja in Komorow, Poland

For the Aleja Marii Dabrowskiej reconstruction in Komorow, Poland, a different formula was adopted. Delegates from interested groups were appointed and involved in the competition process as external experts, whose opinion on the individual works the jury must consider when making its decision. In this case wide participation was not ensured, but anonymity in the judging process was preserved [5] (Fig. 3).

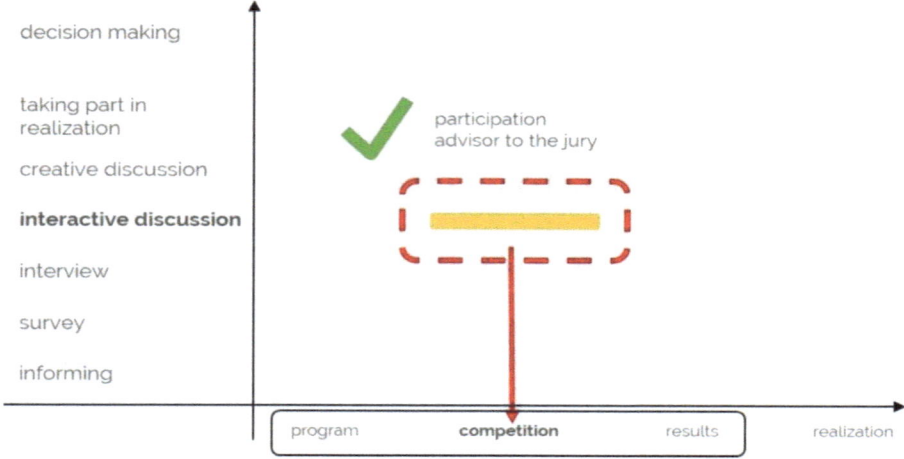

Fig. 3. Applied participation in the competition procedure for the reconstruction of the Aeja in Komorow, Poland, 2023 (Scheme by authors)

2.3   Construction of the Cultural Centre in Wolfsburg, Germany

Competition for the Cultural centre in Wolfsburg, Germany is an interesting example of a two-stage competition with an exhibition of the entries between stages with the introduction of the participation. After the 1st stage an open exhibition was organized in which local residents could participate. Citizens' comments were adopted by the jury and transformed into recommendations for the more detailed 2nd phase of the competition. Keeping anonymity while exhibiting the entries publicly presented significant organizational difficulty and risk. Still, the process was successfully carried out, with wide participation in the middle of the competition process [5] (Fig. 4).

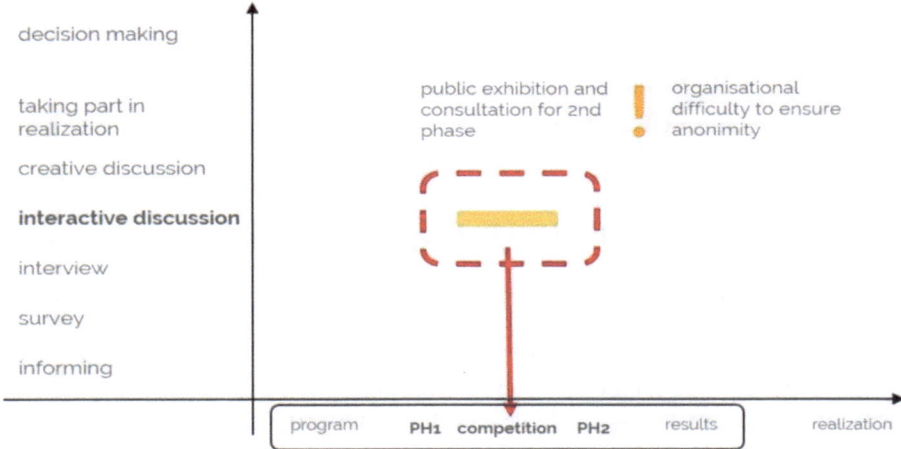

Fig. 4. Applied participation in the competition procedure for the construction of the cultural centre in Wolfsburg, Germany, 2023 (Scheme by authors)

2.4   Spatial Concept of the Central Square in Warsaw, Poland

Another interesting and somehow controversial example is competition for the spatial concept of Warsaw Central Square. Operated on the principle of maintaining anonymity and awarding five equal prizes from which the public was to choose the best solution through open consultation. This solution has been criticized by many architects that actually anonymity was not guaranteed in this way [5] (Fig. 5).

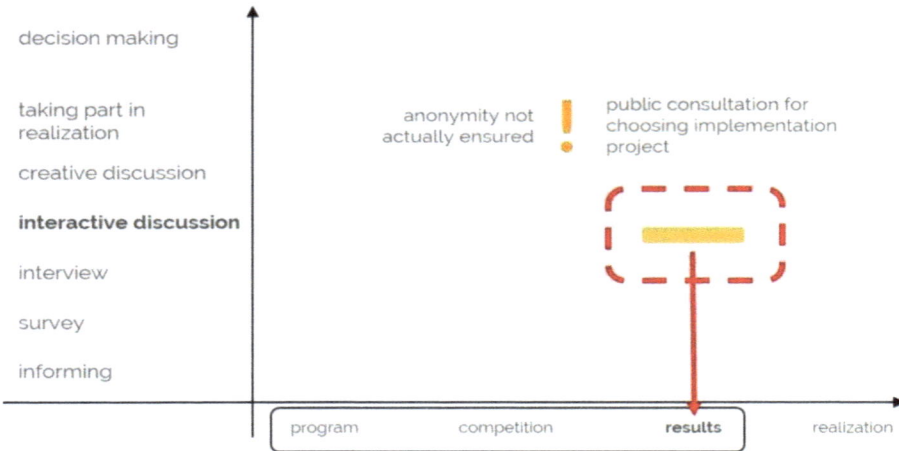

Fig. 5. Applied participation in the competition procedure for the spatial concept of the central square in Warsaw, Poland, 2023 (Scheme by authors)

3   Design of the Main Square in Petrinja, Croatia

Ongoing task, which is a part of this research, is the development of a project program for an architectural and urban planning competition for the project of the main square in the Croatian city of Petrinja. The City suffered significant damage in the earthquakes of 2020, especially the central zone, and establishing a new main square as a hearth of public life in the city was deemed fundamental for restoring faith in the normalization of life in Petrinja. Therefore, participation was an essential element of the process, as well as gaining the best solution through a competition (Fig. 6).

The creation of the competition program was preceded by a participatory study that included: informing the community, citizens' survey, interviewing experts who are familiar with Petrinja and/or the area of the square, and interactive discussions with the city's management structures [7].

Informing the community represents the first rung of the participatory ladder, the beginning of raising awareness of the possibility and need of involving interested citizens in the planning of the built environment. In the case of Petrinja, the public was already very interested in redesigning the main town square through several activities of non-governmental organisations as well as multiple inquiries from citizens to the town administration about the normalization of life in the town after the devastating earthquake.

During the spring of 2022, at the invitation of the City of Petrinja, the team from the Faculty of Architecture, University of Zagreb launched a complete process for programming the redesign of the city's central public space at the location of Croatian Veterans Square in Petrinja. The basis for the creation of a competition program for the Square and its surrounding area is a participatory study that included representatives of three relevant sectors: administrative, civil, and professional; through three rungs of the participatory ladder: survey, interviews, and interactive discussions.

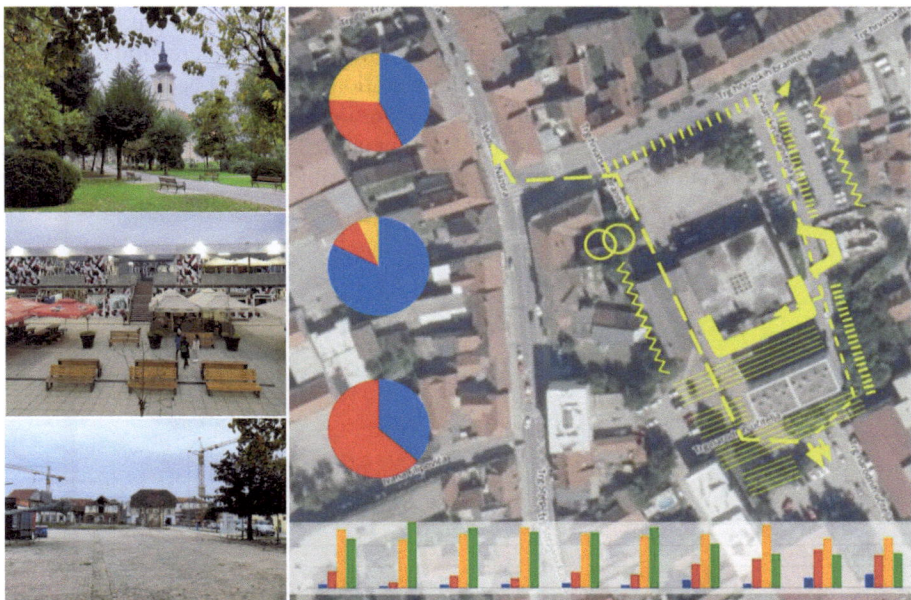

Fig. 6. Public spaces in Petrinja after earthquake and participation study results, 2023 (Scheme by authors, photographs by Gabriel Nikolić, Franka Omazić, Petra Omazić)

An online questionnaire was distributed among the residents of Petrinja, which at the same time informed them about the intention of redesigning the main square and asked important questions for viewing the space from the perspective of citizens. By surveying through a questionnaire, participation moved to the second rung of the participatory ladder. In the period from March 25 to 30, 2022, in just six days, 300 people answered the questionnaire, which represents a significant random sample. The questionnaire collected general information about the respondents, their opinion about life in Petrinja in general, as well as their opinion about the centre of the City of Petrinja and the main town square.

The third rung of the participatory ladder was reached in spring 2023 through inter-views with important experts who actively participated or are still involved in planning and designing public urban spaces in the City of Petrinja. The expert insights of urban planners, landscape architects and architects are thus included in the research.

Interactive discussions, as the fourth rung of the participatory ladder, were held with members of the city administration of the City of Petrinja on several occasions in the period from April 2022 to May 2023 in the premises of the Petrinje city administration. Through in-depth discussions, an attempt was made to cover all topics of importance for the area of the main city square.

All data were processed, structured and interpreted in a participatory study that was submitted to the City of Petrinja in May 2023, and in which all significant elements of the intention for the urban development of the square were presented, coordinated with representatives of three relevant sectors.

The essential insights gained from the study are incorporated into the project program and represent a significant level of citizen participation in the creation of the task set before the contestants. Comprehensive participation model will be applied and will include public presentations and discussions both for the programme as well as competition results. Also, a representative for involvement will be included as a member of the jury, to ensure continuous participation as much as possible (Fig. 7).

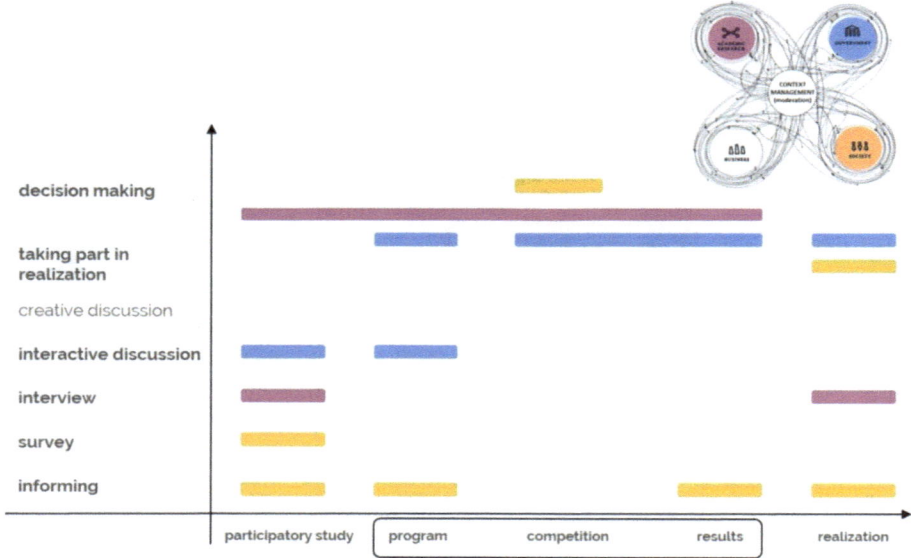

Fig. 7. Expected course of applied participation in the competition procedure for the main square in Petrinja, Croatia, 2023 (Scheme by authors)

4 Conclusion

Introducing comprehensive participation into procedures of architectural and urban planning competitions is not a straightforward process. As seen from the examples, conflicts in terms of continuously involving citizens and at the same time keeping anonymity emerge in the central phases of the competition, until the jury reaches its decision and announces results. Also, in the part when architects design their projects, keeping integrity of the profession (allowing architects to do their work) was advocated. The goal of this research was to find all possibilities of participation, without compromising anonymity and professional integrity mentioned above, considering that participation has several options and levels on the participatory ladder and bearing in mind that different design tasks have different needs for participation.

Roughly, we can indicate three types of participation within the procedure of architectural and urban planning competitions: before (in the preparation of the competition program); centrally positioned (in making decisions on the ranking of the submitted works); and after (in communicating the decision).

Involving by all methods in the first phases before articulating an architectural program is not only possible but strongly recommended. Input from these participatory studies becomes the backbone of the competition brief / programme, and ensures that strong involvement from the very beginning is achieved.

In the central phase of the competition, while judging entries, wide participation compromises anonymity, but introducing a valid public representative as a member of jury or advisor to the jury, remains a reasonable option. Still, examples show that wide participation for certain cases in this phase can be practiced and successful, but with great care for the process and critical anonymity issue.

After the results are announced, public presentations or exhibitions are necessary to communicate the winning solution and bring it closer to the interested and previously involved stakeholders.

In some planning cases where architectural and urban planning competition and anonymity are not mandatory, stronger involvement in the direction of co-creating can be implemented. Models like parallel assignments given to different teams, co-creative workshops and similar are available, still this is not in the framework of this article, and these alternative models present potential for future research.

In conclusion, every design task and competition is different and requires its individual participation approach. This depends on several factors, for example how interested the stakeholders are or how high is their capacity for involvement. In terms of different tasks, some are easy to be involved in since they are familiar to the wider public, some are maybe too complex and are better to be solved in professional discourse. Sometimes, participation is requested to strengthen the community, improve relations between stakeholders or increase social cohesion.

References

1. Arnstein, S.R.: A ladder of citizen participation. J. Am. Plann. Assoc. **35**(4), 216–224 (1969)
2. Arnkil, R., Järvensivu, A., Koski, P., Piirainen, T.: Exploring quadruple helix, Tampereen yliopistopaino Oy Juvenes Print: Tampere CLIQ, p.70 (2010). http://urn.fi/urn:isbn:978-951-44-8209-0. Accessed 26 Sep 2023
3. Cavallini, S., Soldi, R., Friedl, J., Volpe, M.: Using the quadruple helix approach to accelerate the transfer of research and innovation results to regional growth Eur. Union Publ. (2016). https://doi.org/10.2863/408040
4. Bern, A., Røe, P.G.: Architectural competitions and public participation. In: Cities 127 (2022). https://www.sciencedirect.com/science/article/pii/S026427512200169X?via%3Dihub. Accessed 07 Sep 2023
5. Kowalczyk, M.: Architectural design contest with social participation as a part of building culture in Europe. J. Educ. Cult. Soc. **2**, 195–200 (2018)
6. Ekovjesnik/URBACT Hrvatska, Koprivnica očitala bukvicu Zagrebu (2019). https://www.eko vjesnik.hr/clanak/1954/koprivnica-ocitala-bukvicu-zagrebu. Accessed 07 Sep 2023
7. Careva, K., Lisac, R.: Participativna studija – uređenje glavnog gradskog Trga hrvatskih branitelja u Petrinji, Faculty of architecture, University of Zagreb (2023)

Architectural practice in the Digital Age: Balancing Adoption and Adaptation

Damir Mance[✉]

Faculty of Architecture, University of Zagreb, 10000 Zagreb, Croatia
dmance@arhitekt.hr

Abstract. The digital age presents both challenges and opportunities for the architectural profession. Architects must navigate the integration of new technologies, such as artificial intelligence and digital design tools while adapting to the evolving scope of their practice. This paper investigates the balance between adoption and adaptation in architectural practice, exploring strategies for architects to effectively engage with the digital landscape while maintaining the integrity of their profession. The outcomes of fully embracing new technologies and methodologies are contrasted with the approach of modifying existing practices to accommodate digital advancements, considering the implications for the architectural profession.

In addressing these issues, architectural education and research play crucial roles in preparing architects for the digital age by examining the relationship between design-based research, academic inquiry, and professional practice. This study investigates the advantages and challenges of digital technologies integration in design, decision-making, and resource allocation, as well as its influence on traditional professional boundaries and skillsets.

Drawing from an analysis of the current scope of architectural services, this study aims to provide architects (practicing and teaching) with a comprehensive understanding of the opportunities and challenges that the digital age presents, enabling them to make well-informed decisions about the future of their practice.

Keywords: Digital Technology · Architectural Education · Architecture · Adoption · Adaptation · Digital Transformation

1 Introduction

1.1 Digital World of Architecture

Architectural practice is undergoing a digital transformation, a transition that has been met with significant resistance as professionals grapple with integrating new technologies into established workflows. It appears that we are still in the early stages of implementation despite the impact of digital technologies on the nature of professional services. Insights from the Architects' Council of Europe (ACE) Sector Studies conducted between 2018 and 2022 indicate increased use of Building Information Modeling (BIM) between 2018 and 2020; however, its usage seems to have decreased since. We can identify the highest adoption rates in 3D modeling and rendering tools, hovering around

M. Barosio et al. (Eds.): EAAE AC 2023, SSDI 47, pp. 349–361, 2025.
https://doi.org/10.1007/978-3-031-71959-2_38

50%, and BIM at approximately 25%. Other tools, such as common-data environments, augmented reality (AR), virtual reality (VR), and 3D printing, are significantly trailing, with adoption rates of 15% or less [1].

The scenario in architectural education appears even more critical. The introduction of digital planning methods in architecture courses seems to be in its infancy. There is an urgent call to articulate expectations and strategize the integration of digital planning into architecture curricula [2]. That necessitates a profession-wide effort to define the minimum level of digital literacy students should learn through courses and studios at educational institutions.

The "Digital Planning in University Education" guide by the Federal Chamber of Architects in Germany (BAK) raises several relevant questions on what digital knowledge, skills, and competencies are essential for architects entering the third millennium, why, and how to integrate digital methods into the design process. [2] At the same time, digital methods integration requires the teachers to identify what advantages digital technologies offer compared to traditional methods, balancing the risk that faculty courses and studios could transform into software training sessions. That transformation could be incongruent with the principles of the Bologna system, which emphasizes the importance of experimental and innovative teaching content.

Our sector is notably atomized [3]. This fragmentation presents a challenge in adopting new digital technologies, and small and medium-sized enterprises (SMEs), which dominate our field, are at risk of being left behind. The American Institute of Architects Firm Survey Report highlights that 75.2% of architectural firms are considered small companies (1–9 employees) and account for only 18% of the total staff and 12.8% of billings in 2019. [4]. The situation in the European market is even more alarming, with the ACE reporting that 96.1% of firms are small-sized (1–10 employees), including a staggering 62.2% solo practices, 16.2% two-person offices, and 13.7% with 3–5 employees [5]. Digital solutions, often designed with larger enterprises in mind, are challenging to scale down to SMEs due to cost, lack of human resources, or a lack of digital competency necessary for adoption.

Aside from these challenges, the digital world is evolving at an unprecedented pace, with a 100-million-fold increase in computing power across various AI systems domains over the past decade [6]. Digital tools, particularly those integrated into the Building Information Modeling (BIM) process and the emerging suite of AI-enabled tools, could significantly enhance architectural processes and the overall quality of architecture. The AEC AI Hub [7], initiated by Stjepan Mikulić, sheds light on these tools in the architecture, engineering, and construction (AEC) industry, presenting both their potential and limitations. As we enter a new phase of the digital revolution marked by the Internet of Things, deep learning, and artificial intelligence, it is vital to recognize this as a continuation of the technological evolution in architecture, not a disruption. These advancements, especially AI text transformers and AI image generators, are reshaping the design processes, offering innovative approaches and efficient design exploration. While both bring the risk of generating mediocre content, they also provide opportunities for high-quality, innovative work, making it essential for architects to stay informed and proficient with these tools. The digital transformation in the building industry, exemplified by BIM, underscores the need for architectural services and education systems to adapt more effectively to these technological advancements. Benefits of BIM, such as advanced project data management and cost control, highlight the importance of data structure and management skills not emphasized in standard architectural education.

This rapid development widens the gap between small and large companies, potentially intensifying the SMEs' challenges. Even in large firms, issues such as data inconsistency across projects persist, a point highlighted by Martha Tsigkari and Sherif Tarabishi of Foster + Partners [8]. Paradoxically, this inconsistency could be providing a lifeline for smaller firms. However, this paper posits that now is the crucial time for the architectural community to engage proactively, offering insights and strategies to navigate this digital landscape.

1.2 Balancing Adoption and Adaptation

This research explores how the distinctions between adoption and adaptation can significantly influence outcomes and implications within the architectural profession. This inquiry is especially pertinent in environments dominated by SMEs. In the context of this paper, I will define **adoption** as selecting, implementing, and embracing new digital technologies, tools, and methodologies within architectural practice. Conversely, **adaptation** refers to adjusting existing practices, tools, and methods to accommodate the shifts brought about by the digital era. While we cannot guarantee adaptation only by adopting new technologies, it is often a crucial step in the broader change process.

But what influences these choices, and how are they manifested within the architectural practices? Moving beyond the classic adopter categorization curve based on innovativeness as outlined by Everett M. Rogers, Moore introduces a nuanced perspective, identifying psychological gaps between different adopter groups. The Chasm, as he terms it, is highlighted as a critical gap in this context. To the left of the Chasm, we find Innovators and Early Adopters - individuals motivated by the prospect of revolutionary change, architects typically characterized as risk-takers, prioritizing long-term impact over immediate practicality. On the opposite side of the Chasm are the Early Majority and Late Majority adopter groups [9].

Our focus, particularly for broader adoption, is on the Early Majority group. These architects are open to new technology, simultaneously seeking incremental improvements of existing practices, tools, and methods. They could play a pivotal role in guiding the architectural community through the digital transition, striking a balance between embracing innovation and preserving professional integrity. The challenge, and at the same time the goal, is to make the digital transformation accessible and advantageous for the majority.

1.3 Hypothesis

Within the architectural profession and education, there is a visible tension between the slow-paced evolution of systems and practices and the rapid progress characterizing the contemporary digital era. This dichotomy is particularly pronounced due to the atomized nature of the market, leaving architects in an uncertain position regarding the sustainable future of the profession. The building industry sector, by contrast, seems to be outpacing architects, intensifying the pressure on the profession.

Notably, resistance to change is a persistent issue rooted in a complex web of psychological factors, including fear, emotional responses, and individual biases. This research posits that these psychological elements play a critical role in shaping decision-making processes for architects and their ability to adapt to change. The study maps these psychological factors against the industry-specific challenges that are most influential in the decision-making processes of architects.

Multi-criteria analysis is conducted based on the data collected from three surveys to understand these complex challenges, providing architects and educators with insights that could help navigate these uncertain times and contribute to a more flexible and resilient profession.

2   Challenges of Digital Transformation

Resistance to change among people appears to originate from various factors, which we could categorize into two main groups: individual and situational factors [10], where individual factors represent the behavior caused by personal features, and situational factors represent behaviors caused by the environment. While these factors in the research by Darmawan and Azizah initially describe employee resistance within an organization, they can also apply to a market characterized by its atomized nature, where employers often double as employees. Several factors from each group significantly affect adoption in our environment and are relevant to this paper.

We can identify six main **individual** factors: lack of confidence due to insufficient training and resources, fear of failure, increased stress, feelings of uncertainty, low motivation, and poor self-efficacy. Out of **situational** factors, we can identify high ambiguity and inadequate information, inadequate communication and organizational silence, lack of participation in change processes, insufficient work integrity, an ever-increasing sense of job insecurity, and a weak organizational culture with professional associations and universities failing to be transparent leaders.

Additionally, often unrealistic timelines for adoption and existing organizational culture and norms play a crucial role at the organizational resistance level.

Finally, the endowment effect, a well-documented cognitive bias, plays a crucial role, revealing itself as a significant resistance factor. Experts, architects in our case, tend to overvalue their established knowledge and show reluctance toward adopting new methodologies. This reluctance becomes particularly pronounced in situations that demand significant technological advancements and a transformation in organizational standards, all necessitated by digitalization.

Within the specific context of our field, I have identified five main areas of industry-specific challenges.

Market Dynamics: Issues of productivity and market fragmentation strain the architectural profession. Efforts are continuously underway to deregulate the profession at a European level, as highlighted in the ACE The Economic Benefits of Regulation in Architectural Services report [3].

Historical Processes: The long-standing traditions influence the architectural profession, which has ingrained the way of doing business. Altering these established methods carries associated costs, contributing to resistance against change.

Collaboration: The atomized nature of organizational structures in the architectural field hinders effective teamwork. A notable issue is the *collaboration paradox*, which refers to the inability to achieve real-time collaboration in such a fragmented market, amplified by limited human resources. Here, we can also observe the form of the *productivity paradox* (also the *Solow paradox*) in a situation where IT collaborative tools designed to enhance efficiency consume more time than they save if they are not utilized adequately or are misused.

Education and Research: There is a noticeable delay in adopting new technologies within educational institutions and research bodies [2]. Moreover, there is inadequate dissemination of knowledge in schools, universities, and through continuing professional development programs within professional associations, which is especially accented on an interdisciplinary level [11].

Complexity and Variability: A diverse array of stakeholders and constantly changing regulations characterize the architectural field, which leads to high levels of volatility, uncertainty, complexity, and ambiguity (VUCA) [12].

Each of these areas has unique issues, ranging from market volatility to the challenges of real-time collaboration, all of which contribute to the overall resistance to change in the industry.

3 Survey

To validate my initial hypothesis, I conducted a survey targeting architects in Croatia to gain insight into the adoption of technology within our profession, customized for three distinct groups: practicing licensed architects, architecture students, and architecture faculty members, with minor adjustments in questions to suit specifics of each group. The distribution of the survey was as follows:

Practicing Licensed Architects: Disseminated through the Croatian Chamber of Architects, the survey saw participation from 120 out of 2,834 active members. Among the respondents, 84 categorized themselves as employers.

Architecture Students: Conducted at the University of Zagreb, Faculty of Architecture, where undergraduate and graduate programs participated, with 130 out of 862 students responding. That included 43 Master's students and 87 Bachelor's students.

Architecture Faculty Members: At the same University, 32 out of 106 teachers participated.

The survey aims to analyze the current level of integration of new technologies among practicing architects, students, and educators. Apart from general demographic data, a self-assessment of respondents' knowledge of computers, data protection, and cloud storage is collected. These questions, deemed irrelevant to this analysis, will form the basis for broader research in the future.

The questions relevant to this paper delve into three topics: **integration, education,** and **frequency of use** of different digital technologies in the work environment. The survey covers the most frequently used digital tools, categorized into eight major software groups. In the following diagrams, the alpha-numeric symbol within specific groups represents the software tool, as is presented in Table 1. Additionally, two hardware categories, Virtual/Augmented reality (VR/AR) and 3D printing, are assessed.

Table 1. List of software tools

Tool group		Software tools		
BIM	a1	Autodesk Construction Cloud	a4	Archicad
	a2	Graphisoft BIMcloud SaaS	a5	Revit
	a3	Allplan		
CAD	b1	AutoCAD		
	b2	Rhinoceros		
	b3	Sketchup		
Office	c1	Google Docs	c4	Microsoft Word
	c2	Google Sheets		
	c3	Microsoft Excel		
Photo Editing and Graphic Design	d1	Adobe Acrobat Professional	d4	AdobeInDesign
	d2	Adobe Acrobat	d5	Adobe Photoshop
	d3	Adobe Illustrator	d6	CorelDRAW

Tool group		Software tools		
Visualizations	e1	3DS Max	e4	Lumion
	e2	Blender	e5	Twinmotion
	e3	Enscape		
Programming	f1	Dynamo	f4	Other programming tools
	f2	Grasshopper	f5	Phyton
	f3	Basic/Visual Basic		
AI and Data analysis	g1	Microsoft PowerBI	g4	OpenAI – Dall-E
	g2	Midjourney	g5	Stable Diffusion
	g3	OpenAI-Chat GPT		
Urban Planning	h1	Autodesk Forma	h4	Map 3D
	h2	ESRI ArcGIS	h5	QGIS
	h3	ISPU		

3.1 Integration

Participants responded to five questions in this category.

In **Question 1**, participants self-assessed the digital tools integration, regarding their familiarity and usage of various software tools in their work through the 7-level scale: ? = never heard of; 0 = did not use; 1 = installed and opened; 2 = introduced to interface and tried elementary functions; 3 = used elementary functions in work; 4 = used advanced functions in work; 5 = expert with certificate (Fig. 1).

In multiple-choice **Question 2**, participants indicated why they had not integrated specific software tool groups into their work through multiple-choice answers, such as I use the tools intensively, I do not use/need these tools, lack of support, lack of time for learning new tools, I feel that technology limits my creativity, fear of negative impact on my obligations, high price.

In **Question 3**, participants indicated the integration of virtual/augmented reality technologies, and in **Question 4**, if they had incorporated 3D printing into their work, teaching, or studio assignments.

In **Question 5**, only practicing architects were asked if they had developed a digital transformation strategy for their company, and those participants who responded positively selected the activities planned by this strategy in multiple-choice **Question 6**. The participants chose from various digital transformation activities: implementation of BIM, website creation, social networks management, development of proprietary add-ons in the field of BIM, improvement of cybernetic security, integration of VR/AR technologies, development of applications for E-commerce, development of proprietary software/tools in the field of artificial intelligence, and integration of 3D printing technologies.

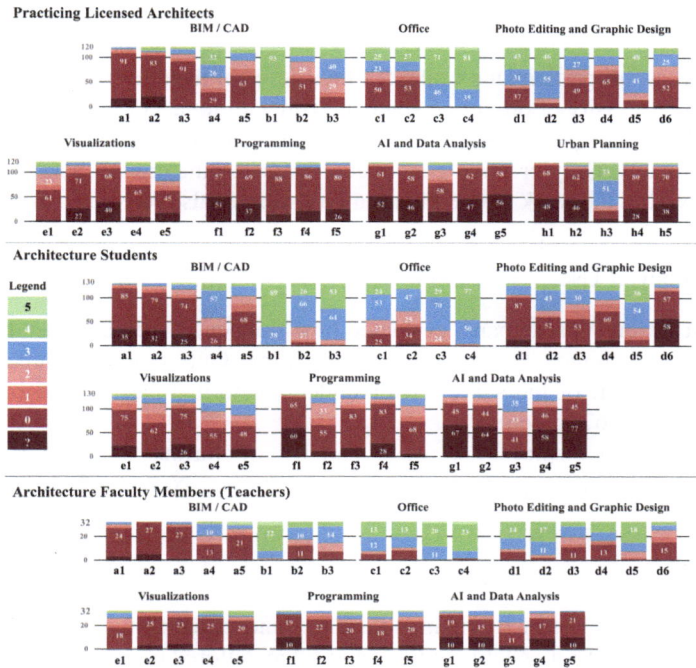

Fig. 1. Question 1 – self assessment of the digital tools integration (survey data diagrams), 2023, (by Damir Mance, previously unpublished)

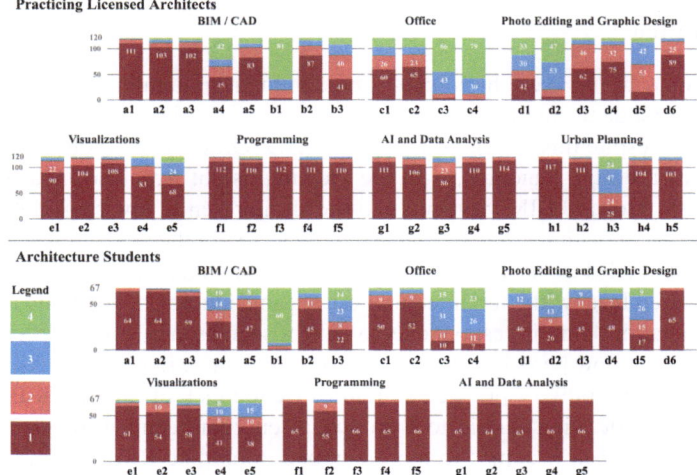

Fig. 2. Question 9 – Usage frequency of software tools (survey data diagrams), 2023, (by Damir Mance, previously unpublished)

3.2 Education

In this category, all participants evaluated their level of agreement with the statement regarding the necessity of providing students with additional education on digital tools and technologies in **Question 7**, rated on a scale from 1-Disagree Completely to 5-Agree Completely. Only practicing architects evaluated the digital literacy of students and interns within the work environment in **Question 8**, also rated on a 5-point scale.

3.3 Frequency of Use

Through **Question 9** participants evaluated how frequently they use specific software tools in design assignments within the office setting. Only practicing architects were included in this part of the survey, with students who had participated in internship programs (totaling 67 students).

Figure (Fig. 2) displays the usage patterns of various software tools, where participants rated their frequency of use on a scale from 1-Never to 4-Daily.

4 Analysis

I used the Microsoft Power BI engine to analyze and visualize the data collected from the surveys, allowing for specific options and filters tailored to each respondent group. This chapter provides insights based on the key findings from the eight major digital tool categories: BIM, CAD, Office, Programming, Visualizations, Photo and Graphic Design, AI and Data Analysis, and Urban Planning.

4.1 Integration

The visual data from Fig. 1 highlights a predominant unfamiliarity with many tools among the respondents, marked by the red bars indicating a lack of tool use in the work environment. Despite this, essential tools like CAD, Office, and Adobe Suite maintain their status as industry standards, with the growing use of Archicad indicating a slow but positive shift toward BIM. Students showed similar trends, favoring Rhinoceros and Sketchup, albeit at beginner levels, which also reflected in the skills of teachers.

The usage of BIM tools has slightly improved compared to past studies, but it is still not at the desired level within architectural offices and education. We could attribute the lack of certifications among respondents to factors such as employer indifference, cost, time commitment, low perceived value, and overconfidence in personal skills.

A surprising 40–60% of practicing architects cited lack of time for learning as the main reason for not using certain tool groups, a sentiment echoed by 30–75% of students, especially when it comes to more complex tools like BIM, programming, and AI. Teachers shared similar concerns, with the added fear of courses transforming into software training sessions.

VR/AR and 3D printing usage remains below 15%, except for students engaging with these technologies in their free time, at 24.6% and 36.2%, respectively. Respondents mostly use these tools for design presentations to clients.

Regarding digital transformation strategies, only 13.1% of participants confirmed their development, predominantly at early stages and revolving around BIM implementation, website creation, and social network management.

4.2 Education

Over 71% of all participants firmly believe that students should receive additional education on digital tools and technologies. Practicing architects rated students' digital literacy at an average level, suggesting that professional practices are more advanced in applying certain digital technologies than current academic curriculums.

Additionally, participants responded to specific questions related to learning material sources and the potential benefits of digital tools in architectural practice. The majority acknowledged the significant advantages digital transformation could bring to the field.

4.3 Frequency of Use

Data on tool usage frequency reaffirmed the dependence on CAD, Office tools, and the Adobe suite, with limited utilization of other technologies. Predominantly selected were Never and Rarely categories, indicating a substantial untapped potential in various digital tools. Students' responses mostly mirrored those of practicing architects, with notable differences in the less frequent use of Office tools and Adobe Acrobat and more frequent use of SketchUp and Photoshop, likely due to their educational and studio work contexts (Fig. 2).

5 Conclusion

Survey data highlights significant skill gaps, particularly in emerging technologies like BIM, AI, and data analysis, underscoring the need for focused training and education. Traditional tools such as CAD, Office applications, and the Adobe Suite continue to be the mainstay of architectural practice, signifying resistance to change while providing a foundational basis for integrating new technologies. The findings from the survey bring us to a pivotal question: What role can educators play in addressing the evident skill gaps in digital literacy within the architectural profession?

Today, only a small percentage of SMEs have embraced advanced technologies or developed a digital transformation strategy, representing a niche of industry innovators poised to lead the change. Being architects, adapting to these changes is crucial to remain relevant and competitive. In a market characterized by a predominance of micro-sized architectural offices, the sustainability of the architectural profession is at stake. Large architectural companies typically employ 2–4% of digitally advanced staff, focusing on tech implementation and research. That is not a feasible model for smaller firms, where dedicating architects solely to tech subjects is considered a waste of resources. While outsourcing is an option, it often results in time delays and losses in an increasingly productive environment. Due to the inherent limitations of SMEs, they cannot carry the responsibility for digital transformation. Therefore, it is imperative to promptly adjust architectural school curricula to meet the growing demand for digitally literate architects.

While the lack of education in schools and faculties is a significant issue, there is no need to educate all users to an expert level. Larger offices typically reserve advanced and expert tool use for a small percentage of their staff, indicating a possible pathway for educational institutions. Implementing minimum standards for various digital tools and technologies in the curriculum is crucial, with advanced and expert levels reserved for enthusiasts through elective courses and higher education.

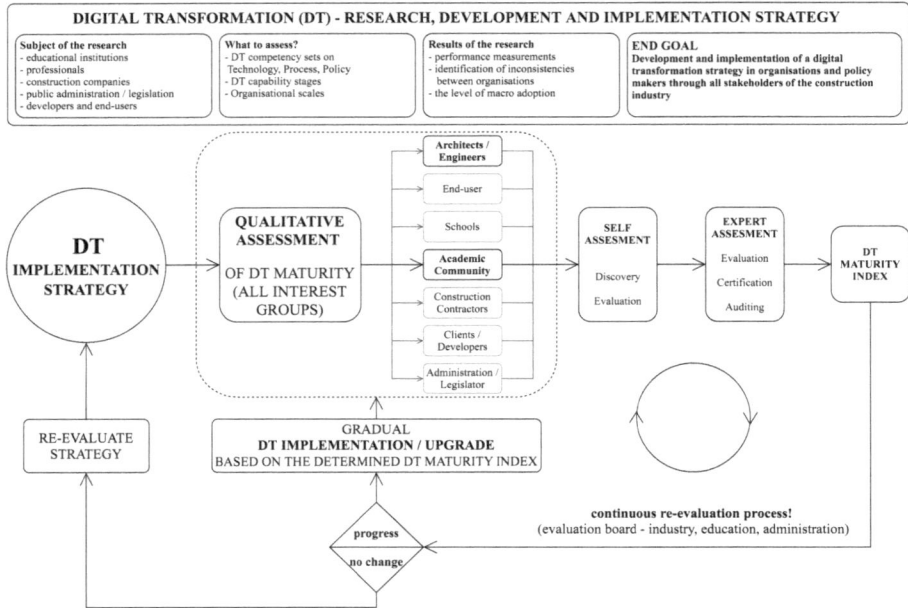

Fig. 3. Digital Transformation – Research, Development and Implementation Strategy, 2023, (by Damir Mance, previously unpublished). The diagram is based on BIM – Research, Development and Implementation strategy [11] and Building Information Modelling maturity matrix [13] and adapted for digital transformation of construction industry in general.

With the digital standards in the architectural sector being relatively low and lack of time frequently mentioned as a significant barrier to adopting new tools, identifying and addressing areas for improvement is essential. Key areas to focus on include:

Active Engagement with Technological Advancements: Tailoring the adoption of new technologies to match the current maturity level of users is crucial [13], which is especially important in integrating new digital technologies such as BIM or AI into the design studio at architectural schools [11]. That ensures a gradual and sustainable transition, enabling users to implement new tools into their workflows effectively (Fig. 3).

Use Case Analysis: It is vital to evaluate the suitability of each tool for specific business objectives, assessing how well a technology addresses the needs of a particular use case. This process not only aids in selecting the right tools but also provides guides on how they contribute to achieving the set objectives (illustrated in Fig. 4).

Enhanced Communication Strategies: Developing platforms dedicated to e-learning and the experiences exchange can foster a collaborative learning environment, ensuring that knowledge is shared efficiently across the board.

Integration of Advanced Project Management Techniques: Planning for the integration of digital technologies in both educational and professional settings requires advanced project management strategies. Emphasis should be placed on effective time management to ensure a smooth transition.

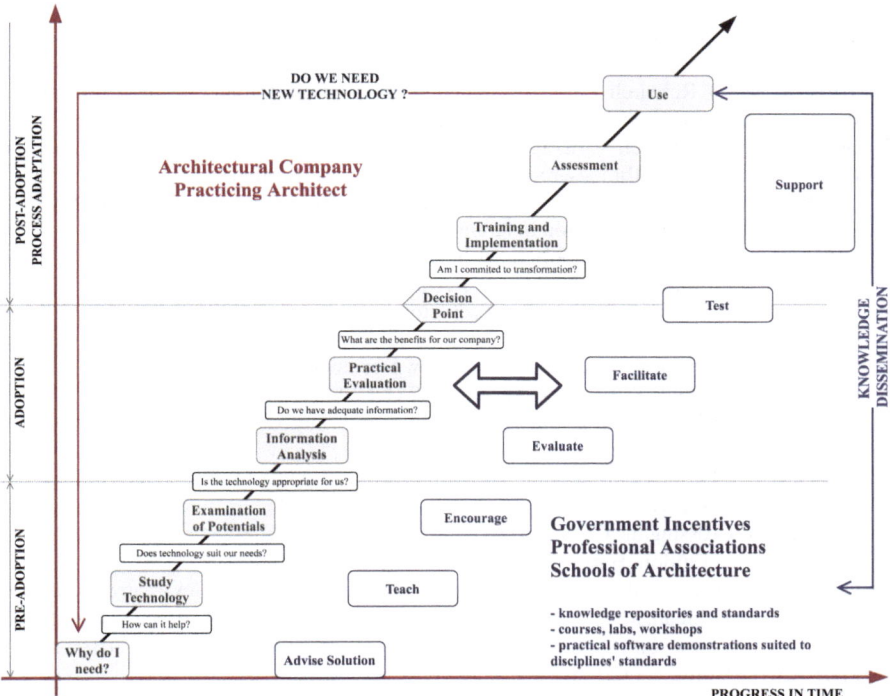

Fig. 4. Adoption of Digital Technologies decision-making process diagram, 2023, (by Damir Mance, previously unpublished). The diagram is based on Construction Technology Adoption Framework (CTAP) [14] and adapted for the architecture sector. The CTAP is a framework that delineates the phases of the process that customer organizations use when deciding to adopt a new digital technology as well as the parallel vendor activities [15]. In this adaptation, instead of vendors, government incentives, along with professional associations and schools of architecture, are positioned as facilitators in the transformation.

Extracurricular and Summer Learning Opportunities: Introducing students to new tools and digital processes early in their undergraduate studies is essential. By doing so, we can prevent studio sessions from becoming mere software training classes, ensuring a holistic educational experience in compliance with the Bologna process.

Leveraging Government Incentives: Proactively working towards digital technology adoption is critical, and government incentives can play a crucial role in facilitating this transformation. Encouraging policies and funding opportunities can provide the necessary support for educational institutions and small architectural practices.

In future research, I intend to refine the survey to engage more respondents and critically examine digital transformation, including AI. While the limited sample size of the survey necessitates caution in interpreting the results, the implications are significant if non-participation reflects a disinterest in digital change.

References

1. Mirza & Nacey Research Ltd (2023) Digitalisation within the architectural profession in Europe: Extract from the ACE Sector Studies 2018–2022. Brussels
2. Beetz, J., Brandenburger, Y., Krapp, S., et al.: BIM für Architekten Digitale Planung in der Hochschulausbildung. Bundesarchitektenkammer - BAK, Berlin (2022)
3. Compiled by Octavian Economics & Frank Hughes Architects PDrAR and SEMG& CE, The Economic Benefits of Regulation in Architectural Services. Brussels (2022)
4. Baker, K., Chu, J., Riskus, J.: American Institute of Architects, The business of architecture : AIA firm survey. American Institute of Architects (2006)
5. Mirza & Nacey Research Ltd., ACE Observatory/Observatoire du CAE. Brussels (2023)
6. Giattino, C., Mathieu, E., Roser, M.: The exponential growth in AI computation (2022). https://www.visualcapitalist.com/wp-content/uploads/2023/09/01.-CP_AI-Computation-History_Full-Sized.html. Accessed 27 Oct 2023
7. Mikulić, S.: AEC AI Hub. https://stjepanmikulic.notion.site/AEC-AI-Hub-b6e6eebe88094e0e9b4995da38e96768. Accessed 29 Oct 2023
8. NXT DEV 2023, Panel discussion: The future of AI in AEC. In: NXT DEV 2023 (2023). https://nxtdev.build/view-on-demand/. Accessed 27 Oct 2023
9. Moore, G.A: Crossing the Chasm: Marketing and Selling Disruptive Products to Mainstream Customers, 3rd Edition. Harper Collins e-books (2014)
10. Darmawan, A.H., Azizah, S.: Resistance to change: causes and strategies as an organizational challenge. In: Othman, M.Y., In'am, A., Shafrin Ahmad, N., et al. (eds.) Proceedings of the 5th ASEAN Conference on Psychology, Counselling, and Humanities (ACPCH 2019). Atlantis Press SARL, Dordrecht (2020)
11. Mance, D., Bačić, D.: Integrated building design approach in architectural students' curricula. In: Pračić, F., Barišić Marenić, Z., Bobovec, B., Arbutina, D. (eds.) Challenges of Recovery and Resilience: ArhiBau.hr 2022 scientific conference proceedings. Zagreb Society of Architects, Zagreb, pp. 78–89 (2023)
12. VUCA – Volatility, Uncertainty, Complexity and Ambiguity. In: PMI Disciplined Agile Online (2022). https://www.pmi.org/disciplined-agile/vuca-volatility-uncertainty-complexity-and-ambiguity. Accessed 28 Oct 2023
13. Succar, B.: Building information modelling maturity matrix. In: Underwood, J., Isikdag, U. (eds.) Handbook of research on Building Information Modelling and construction informatics: concepts and technologies, pp. 65–103 (2010)
14. Sepasgozar, S.M.E., Davis, S.: Digital construction technology and job-site equipment demonstration: modelling relationship strategies for technology adoption. Buildings **9** (2019). https://doi.org/10.3390/BUILDINGS9070158
15. Sepasgozar, S.M.E., Davis, S.: Construction technology adoption cube: an investigation on process, factors, barriers, drivers and decision makers using NVivo and AHP analysis. Buildings **8** (2018). https://doi.org/10.3390/buildings8060074

Aligning the Pedagogy of Postgraduate Professional Practice Courses to Develop the Meta-competencies Required of Architects Today

Claire Mullally[✉] and Catherine Brown-Molloy

South East Technological University, SETU Waterford, Cork Road, Waterford City X91 K0EK, Ireland
claire.mullally@setu.ie

Abstract. The architectural profession can be seen as a developing and multidisciplinary career, which has evolved from the Vitruvian sole master to the need for specialisation and collaboration in multidisciplinary teams. As a regulated profession, postgraduate professional practice courses in Ireland and the United Kingdom (UK) play a critical role in the registration process for Architects. However, there is little research published on the pedagogy of these courses nor their suitability to the evolving demands on the multifaceted role required of Architects today. As a result, there is a need to reassess professional practice courses considering the competencies required in Architect's diverse ways of practice. Over the past two decades, the number of Irish institutions offering Bachelor and Master of Architecture courses has increased significantly. Therefore, the demand for developing a new postgraduate course to meet the increased number of graduates is acknowledged. This presents an opportunity to address the research gap identified and to explore innovative approaches to curriculum design, delivery and assessment that can enhance learning while adapting to the changing societal, environmental, technological and professional challenges of architectural practice. In a desktop study, courses in Ireland and the UK were systematically examined to reveal fundamental similarities, with some significant variances. The study highlights the importance of reflective practice and multidisciplinary learning in preparation for the global challenges of the built environment. The complex nature of the architecture profession requires a diverse range of skill sets, knowledge and competencies as well as meta-competencies.

Keywords: Professional Practice · Postgraduate · Architecture · Education · Meta-competencies · Multidisciplinary

1   Introduction

Successful completion of postgraduate professional practice courses is a component of the common pathway for the registration of the title of Architect in Ireland and the United Kingdom (UK). The popularity of the architecture profession is increasing with

M. Barosio et al. (Eds.): EAAE AC 2023, SSDI 47, pp. 362–374, 2025.
https://doi.org/10.1007/978-3-031-71959-2_39

the number of Architects in Europe rising to 100,000, and Ireland's number of registered Architects rising by 25% in the last ten years to 3300 [1]. Despite this, "professional practice" is an under-researched area, within the context of architectural education [2]. In the past, Master Builders such as Michelangelo and Vitruvius had extensive knowledge of all aspects of the design and construction of buildings and worked independently to a high level of expertise and mastery. With significant societal, scientific and technological developments, the building process gradually became more complex and resulted in the requirements of specialised fields and expertise. As Rifaat [3] points out, it became unattainable for Architects to "effectively emulate the performance of the old masters." Today there is a requirement for Architects to oversee the process and to act as coordinators [4] and mediators [5] between different professionals. A considerable amount of literature has argued that the ability to collaborate across and communicate with professionals and workers from different disciplines is indispensable for the role of an Architect today [2, 5, 6]. The 21st-century Architect is understood to lead less and facilitate more and they are vitally important as part of a collaborative team [7].

2 Challenges of Professional Practice

In the pursuit of effective pedagogical approaches for architectural education, Borucka and Macikowski [4] assert that it is imperative to, first, understand the contemporary role that the Architect holds today. Similarly, Legény et al. [8] remind us of the evolving role of the Architect and argues that educational institutions need to react to the constant change in the architectural profession. Architects have to respond to an increasing level of complexity in the design and construction process. There is consensus in the literature that Architects have to navigate uncertain and changing situations that lead to diverse and evolving demands [9]. In particular, there is greater pressure on the role due to societal challenges, such as the recent pandemic and future inevitable forces that are unknown [7]. There are other challenges in the form of volatile economic forces [10], increased regulatory requirements [11], technological advances, including the evolution of digital tools, artificial intelligence and immersive technologies [12, 13]. However, one of the most pressing issues is the requirement to respond to climate change [14], and meet the requirements for climate neutrality by 2050 [15]. In addition, there is more emphasis placed on the social responsibility to design inclusive, sustainable spaces that are responsive to the needs of the community [16]. The perceived definition of 'ethics' within the architectural profession has expanded to create the requirement to design and build for the benefit of the public good and not just the good of the client [6]. Furthermore, it is acknowledged that the vast number of different specialists and stakeholders involved in the design and construction process today adds to the perplexity of the process. Bourka and Macikowski [4] point out, that the complexity is also concerned with the communications among different participants and the diverse fields of knowledge required as a result. Yet it is noted that unwillingness towards engagement with other construction professionals is a weakness in the profession [17].

These challenges will require changes in the working environment and in the professional landscape [7]; moreover, interdisciplinary cooperation and teamwork are proposed as key to dealing with these problems [18, 19]. Architects are required to possess proficiency in implementing theoretical knowledge to complex situations and a holistic knowledge of their own profession. However, MacLaren & Thompson [6] point out that they also need to possess an understanding of their own role in relation to the broader team, be able to communicate with these stakeholders and have an implicit understanding of the other roles within the team. While the importance of 'collaboration' within the construction industry has been extensively highlighted, criticism of the "singular discipline mentality" that exists within the industry has been recognised [20]. It has been argued that the need for effective collaboration in all areas of work will assume greater significance and be essential in order to meet the requirement to achieve Net Zero Carbon by 2050 [21]. It is argued that these challenges require more flexible and adaptable ways of practice [9, 18] and a positive attitude towards engagement for successful sustainable interdisciplinary collaboration [22]. Furthermore, Samuel [10] proposes reforms in architectural education to enable more socially aware professionals. Similarly, Scott [24] highlights the need for emerging ways of practice to be more effectively addressed in education and argues that there are shifts required in learning paradigms to tackle the complex global challenges [25]. Analogously, the Education Policies and Standards produced by the Royal Institute of Architects of Ireland (RIAI) [23] propose that to address the ever-expanding scope and complexity of the Architect's role, more effective collaboration within the profession and with allied professions is required. Consequently, the priority placed on collaboration by the RIAI is further evident in the theme and title of the RIAI Annual Conference 2023 entitled 'Collaboration'.

3   Education of the Profession

Traditional teaching approaches in architectural and engineering courses are primarily founded on passive learning and as such have received a significant amount of criticism. Stump et al. [26] argue that these traditional methods are unsuccessful in preparing students for professional practices. Oliveira et al. [27] suggest that the siloed discipline-based structure of architecture and engineering courses is the reason for failure to meet the current or future needs of the industry [27]. Furthermore, it is argued that the tendency within architectural education to give priority to traditional design knowledge rather than collaborative and critical learning is unfavourable and leads to disconnected experiences [27]. Notwithstanding this criticism, there are some notable examples of interdisciplinary learning present in the history of architectural education, specifically the London County Council (LCC) School of Building in Brixton (1904-1970s) and the Bauhaus Art School Dessau (1925–1932). In the LCC School of Building in Brixton, the emphasis was placed on instructional setting and Building Trades and Architects were taught under the same roof [28] (Fig. 1a and b).

Fig. 1. a and b. School of Building: Carpentry and School of Building: Stone Masonry Class by unknown, photograph © London Metropolitan Archives (City of London).

While at the Bauhaus, Walter Gropius' aimed to create a comprehensive artwork that would eventually bring all the arts together. The Bauhaus curriculum included elementary form and basic studies of materials taught over the first three years (See Fig. 2 and 3). Only after completing the fundamental courses were the best students allowed to progress to the core architecture course.

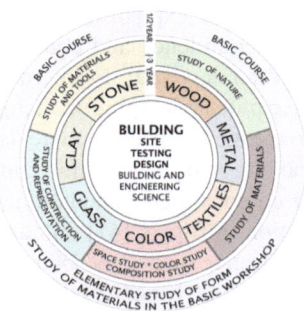

Fig. 2. Gropius, Walter, (1883–1969). (1925–1926, Image: 05/15/2003). Bauhaus, exterior, workshop wing, view from south. [Art schools]. Retrieved from https://library.artstor.org

Fig. 3. Diagram of the Bauhaus Curriculum, CC BY-SA3.0, https://creati vecommons.org/licenses/by-sa/3.0, via Wikimedia Commons

Although it is acknowledged that courses such as the Bauhaus and LCC School of Building had a historical influence on architectural education, there were few other examples of multidisciplinary or interdisciplinary learning in architectural education prior to 2000 [29]. The education sector has been much slower than industry to focus on integrating multidisciplinary collaboration and teamwork. There is a relatively small body of literature concerned with collaboration between design and construction disciplines within an educational context [30]. The majority of these studies have been qualitative and have focused on interdisciplinary design collaboration and teamwork using digital modalities [31]. A frequently observed theme in recent studies published on interdisciplinary learning within architecture courses is the presence of an overarching aim of sustainable building practices [30, 32, 33]. However, it should be noted that the preponderance of recent studies pertains to experiences and projects that are relatively short (commonly only lasting between four days and five weeks) [34].

There is consensus that the vast majority of architecture students' time is spent working alone with limited opportunities to collaborate with students from other disciplines. As a consequence, it is argued that architecture students lack the ability to effectively communicate and collaborate [5]. In this respect, the Architects' Council of Europe advise that teamwork and collective intelligence should be prioritised over the rigid categorisation of disciplines in a new systemic condition of teaching [15]. In addition, they advise developing relevant skills to support research in practice [35].

4 Pedagogy

The principal approach of architectural teaching focuses on the Design Studio. This learning paradigm supports integrative learning to lead to creative exploration and critical discourse. The Design Studio was proposed by Schön [36] as an exemplar for other professional courses. Paradoxically, other architecture modules (including professional practice) are generally taught in a more didactic, disintegrative way [37].

The scholarship of teaching and learning provides empirical support for the efficacy of peer learning in architectural education. The conversation theory framework offers a way to explain how interactions lead to knowing [38]. As well as bolstering students' confidence and competence, peer learning is also credited with alleviating students' anxiety [39]. Furthermore, in drawing from fields such as situated learning, interdisciplinary learning is recognised as emphasising the collaborative construction of knowledge in specific contexts and its benefits are not contested. Moreover, social anthropology aligns with this concept of learning as a social process in which knowledge is co-constructed and is situated in a specific social and physical environment [40]. In addition to Bigg's [41] outcome-based theory of constructive alignment, pedagogical approaches such as problem-based and participatory learning are more effective in bridging the gap between education and practice than the universal approach of imparting facts [9, 42, 43], challenge-based (team) learning [44] and the reflective process of assessing potential solutions [36] have been espoused as important for architecture students.

By establishing pedagogical approaches that promote and utilise the combined skills of Architects and Engineers, mutual respect and understanding of their own and the other disciplines will be possible [7]. Architects often do not apprehend and defend their unique skill sets, which are predominantly idea or solution-orientated. To enable their viable skill sets to adapt and respond to challenges and have new ideas for the future realised and acknowledged, reflective practice is important [25]. Reflective practice enables the ability to recognise one's limitations and ensure subsequent life-long learning. Facilitating this, for instance, by referencing frameworks such as the conscious competence model as a way of studying experiential learning, is beneficial for professional practice [45]. In addition, an interdisciplinary learning approach aligns with the evolving landscape of architectural ways of practice. This recognises the need for meta-competencies that facilitate the development of collaboration and communication in a multi-disciplinary environment. By aligning curriculum with real-world ways of practice, students are enabled to develop the necessary meta-competencies that underscore competency development, to operate effectively in complex and uncertain environments.

5 Professional Practice Courses

Demonstration of professional knowledge, skills and competencies are required for eligibility for Architectural Registration. These requirements are based on the elements listed under Article 46 of the EU Qualifications Directive (2005/36/EC). In the UK, registration for Architects is through the Architects Registration Board (ARB) whereas in Ireland, the RIAI [46] are the registration body and competent authority. Their Standard of Knowledge, Skill and Competence for Practice as an Architect, at a professional level, is currently under review and the revision awaits publication. The current eligibility requirements include knowledge of regulation, context, professionalism, management, technology and procurement and ability in design and communication. In Ireland and the UK, a candidate for registration usually completes the professional practice lecture course and examinations prior to registration. In Ireland the courses are set against the EU Directive 2005 36 - Article 46 [47] At present, the RIAI route 'C' sets the criteria for courses until the publication of a new competency framework.

The UK Courses are delivered and assessed in accordance with requirements of the Royal Institute of British Architects (RIBA)/Architects Registration Board (ARB) shared Professional Criteria for Part 3 [48]. These criteria include Professionalism (PC1), client users and delivery of services (PC2), legal framework and processes (PC3), Practice and management (PC4), and building procurement (PC5). In the UK the review of Architects Competencies Report [54] stated findings relating to the increasing importance of the Architects' roles regarding climate change/sustainability and management of health and safety. This aligns with the RIBAs' three specific mandatory competencies for attaining and maintaining chartered status; health and life safety, climate literacy, and ethical practice, with a possibility of research literacy being added [49]. This refinement of competencies is clear and direct and demonstrates a move away from a focus on cognitive and design skills to practical and interpersonal competencies required for professional practice.

6   Comparison of Irish and UK Professional Practice Courses

Ireland and UK postgraduate professional practice courses prepare architectural graduates for registered status as an Architect, knowledge and ability against a set of key criteria. In order to sit the examinations, a minimum of 24 months of relevant practical experience is required. As a result, candidates generally undertake their professional practice course part-time while working full-time in practice. Consequently, it can be an onerous undertaking for the candidate and the practice in which they are working.

The comparison of 20 Irish and UK Professional Practice courses revealed differences in European Credit Transfer and Accumulation System (ECTs) awarded. Ireland's courses award 30 ECTs and the UK courses most commonly award 60 ECTs for very similar curricula and assessments. The typical course duration is predominantly one year (part-time) however pathways do range from six months (London Metropolitan University) to 5 years flexible study option (University of Greenwich; Technological University Dublin). The majority of courses are delivered in person with exceptions including predominantly online, (University College London) and fully online courses (University of Bath). While in Ireland, University College Dublin (UCD) offers a choice of online and in-person modes. Although a small number of institutions simultaneously offer their professional practice modules to professionals in the built environment who may register for individual modules as part of their continuous professional development (CPD), there is a lack of institutions specifying or highlighting that their course is multidisciplinary. Despite the call for interdisciplinary learning within the architectural industry and in education, this does not seem to be provided for within the current course offerings.

In both Ireland and the UK, assessment components generally include documentary submissions relating to practical experience, as well as procedural, legal, professional and managerial themes. The practical experience submissions generally take the form of a Career Appraisal Report that focuses on professional practice as well as a case study with an average requirement of 8000 words. The word count of the case study ranges from 5000 (Architectural Association School) to 10,000–12000 (UCD). The other topics are generally assessed via essays and written examinations as well as a final interview. However, personal reflection on practice and learning is promoted as coursework by a few institutions including London Metropolitan University, London South Bank University, Newcastle University, and the University of Greenwich.

In the UK, The Architects Registration Board, report the need for modernisation to the structure of parts 1, 2 and 3 education of Architects [62]. A rethinking of course content and structure to include formative reflective practice and meta-competency development could be more conducive to lifelong learning and the development of reflective and critical practitioners.

7   Demand for Postgraduate Professional Practice Courses

There are currently five qualifications in architecture that are legally recognised for access to the Register of Architects in the Republic of Ireland. Table 1 lists the Higher Education institutes (HEI) delivering those courses along with the number of graduates on each, for four consecutive academic years. In addition to these courses, Atlantic Technological University Sligo has also been awarded provisional approval as a prescribed course. With 55 new entrants in the 2022–2023 academic year, it is expected that this will add significantly to the total number of graduates in the country. In addition to the pressure that this will put on the demand for postgraduate professional practice examination courses, there is further pressure from graduates coming from abroad and requiring registration in Ireland. As a result, the RIAI has stated that the provision of 140–150 places on professional practice examination courses may not meet the demand [50].

Table 2 lists the only two professional practice examination courses that are accredited by the RIAI and specified under the Building Control Act 2007 for access to the register of Architects in the Republic of Ireland. Although the RIAI have noted that each can enrol 70–75 students, the number of graduates shown in Table 2 reveals a significant disparity between enrolment and graduate numbers on both courses. Owing to attempts to help students complete the course and delays during the COVID-19 pandemic, UCD did not accept any new entrants in the 2023–2024 academic year. Despite being listed as part-time one-year courses, on both prospectuses, it is thought that the disparity between enrollment and graduate figures is because it is not uncommon for students to take two, three or more years to successfully complete the course. This limits the ability of these courses to meet the need and infers a demand for a new course.

Table 1. Graduates by HEI, course name, academic year (Higher Education Authority Statistics Unit. (2024). Personal communication)

Higher Education Institution	Course Name	2019–2020	2020–2021	2021–2022	2022–2023
Munster Technological University	Master of Architecture	10	10	15	20
South East Technological University	Bachelor of Architecture	15	15	20	15
Technological University Dublin	Master of Architecture	45	30	50	35
University College Dublin	Master of Architecture	45	30	50	35
University of Limerick	Bachelor of Architecture	20	20	20	35
Total		**135**	**105**	**155**	**140**

Table 2. Professional Practice Graduates by HEI, course name, academic year (Higher Education Authority Statistics Unit. (2024). Personal communication)

Higher Education Institution	Course Name	2019–2020	2020–2021	2021–2022	2022–2023
Technological University Dublin	Professional Diploma in Architectural Practice	15	15	20	15
University College Dublin	Professional Diploma in Architecture	40	35	30	30
University College Dublin	Professional Diploma in Architecture (online)	25	20	10	15
Total		**80**	**70**	**60**	**60**

8 Meta-competencies

Professional Practice is "the embodiment, indeed the expression, of the practitioner's everyday knowledge" [51]. The recruitment process in architectural practices has shifted focus from cognitive to soft/professional skills to include teamwork, leadership, negotiation and critical thinking [52]. In addition, interpersonal, communication, responsibility, and a positive attitude are also becoming integral to meeting the demands of the profession [53]. Meta-competencies and soft/professional skills are inextricably linked and interdependent with a range of skills that are often included with the term 21st-century skills [52]. Meta-competencies have appeared in professional competency models that incorporate reflection [36, 55, 56], yet are not included in most educational and professional frameworks. Meta-competencies enable people to become flexible and should be fostered in higher education to allow future adaptation in the workforce [12]. In addition, they can increase students' entrepreneurial mindset and readiness for innovation by enabling complex thinking and reflection [26]. The literature is varied as to its categorisations and definitions of the terms; competencies, competences and meta-competencies [25]. The term competence is understood as a combination of an Architect's knowledge, skills and experience [5]. Meta-competencies are higher-order abilities, which facilitate skillful, meaningful learning, thinking and adapting in diverse contexts, required for the activation of all other skills and competencies that help prepare people for future change [18, 25, 55, 57, 58].

Cheetham and Chivers [55] proposed a holistic model of professional competency by combining Schons' reflective practitioner [36] approach with meta-competencies, explained as the ability to cope with uncertainty, as well as with learning and reflection. Le Deist and Winterton explain cognitive competence (knowledge), functional competence (skills) and social competence (attributes) are required in order to be effective at work. They prioritise meta-competencies relating to learning to learn and align them with individual effectiveness, and that of social competence [25]. Meta-competencies such as volition, self-regulation and action competence can control the development of professional skills [59]. In addition, Bates et al. [60] prioritised meta-competencies of inter-relation, intrapersonal (self-management and self-reflection), domain-specific and normative (moral and ethical judgement) competence. Inter-relation skills enable the crossing of disciplinary and cultural boundaries, which harnesses empathy, communication and collaboration, to connect with other people. This was revealed as unique to address complex real-world problems aligned with the United Nations (UN) Sustainable Development Goals (SDGs) they focused upon. Furthermore, system and temporal thinking, interpersonal and ethical literacy, as well as creativity are five meta-competencies proposed for addressing the UN SDGs. This infers educational approaches to include case studies, guided inquiry with peer-to-peer learning, reflection essays along with self-assessment exercises and instruments [61].

Meta-competencies such as an open and creative mind, leadership, ability to prioritise, self-awareness, self-directed growth and self-reflection are also highlighted as important in the architecture profession [18]. While it has long since been acknowledged that these skills are required by Architects, the prevailing view in the literature is that meta-competencies are needed by graduates for future adaptation, employability, and success in the workplace [18, 60].

9 Conclusion

The role of Architects within the construction industry is demanding and evolving due to increased challenges and complexities. The recruitment process of Architects has shifted towards emphasising soft and professional skills rather than cognitive abilities, underlining the significance of teamwork, leadership, negotiation, and critical thinking. Skills such as interpersonal communication, accountability, and a positive attitude are also considered crucial for professional effectiveness. Despite this, professional criteria for Architects' registration typically relate to professional, procedural, legal, and managerial competencies. A comparison of Irish and UK postgraduate professional practice courses for Architects revealed similarities in duration, curriculum and assessment, yet significant variances exist concerning awarded credits in the qualifications. In considering how professional practice courses respond or adapt to complex global challenges and opportunities present today and in the future, interdisciplinary collaboration and meta-competency development are regarded as fundamental. In addition, fostering reflective practice as a learning outcome-based approach is considered beneficial for Architects as they journey through lifelong learning and navigate the diverse ways of practice.

References

1. ACE 2022 Sector Study: ACE. (n.d.). Accessed 16 October 2023. https://www.ace-cae.eu/activities/publications/ace-2022-sector-study/
2. Karakaya, A.F., Şenyapılı, B.: Rehearsal of professional practice: impacts of web-based collaborative learning on the future encounter of different disciplines. Int. J. Technol. Des. Educ. **18**(1), 101–117 (2008)
3. Rifaat, S.I.: The multidisciplinary approach to architectural education: bridging the gap between academic education and the complexities of professional practice. IOP Conf. Ser. Mater. Sci. Eng. **471**(8), 082067 (2019)
4. Borucka, J., Macikowski, B.: Teaching architecture – contemporary challenges and threats in the complexity of built environment. IOP Conf. Ser. Mater. Sci. Eng. **245**, 082058 (2017)
5. Jutraz, A., Zupancic, T.: The role of architect in interdisciplinary collaborative design studios. Igra Ustvarjalnosti – Creat. Game **2014**, 034–042 (2014). https://doi.org/10.15292/IU-CG.2014.02.034-042

6. MacLaren, A., Thompson, N.: 'Portfolio professionals' in the digitised built environment= education+ skills+ commercial environment+ communications network. In: Professional Practices in the Built Environment Conference, pp. 80–92 (2017)
7. Sanderson, L., Stone, S.: Emerging Practices in Architectural Pedagogy. Accommodating an Uncertain Future (1st ed.). Routledge, London (2021)
8. Legény, J., Špaček, R., Morgenstein, P.: Binding architectural practice with education. Glob. J. Eng. Educ. **20**, 6–14 (2018)
9. Luck, R.: Participatory design in architectural practice: changing practices in future making in uncertain times. Des. Stud. **59**, 139–157 (2018)
10. Samuel, F.: Why Architects Matter: Evidencing and Communicating the Value of Architects (1st ed.). Routledge, New York (2018)
11. Imrie, R.: The Interrelationships between building regulations and architects' practices. Environ. Plann. B. Plann. Des. **34**(5), 925–943 (2007). https://doi.org/10.1068/b33024
12. Rangaswamy, U.S.: Industry 4.0 disruption: Assessing the need for adaptive capability. Dev. Learn. Organ. Int. J. **35**(6), 7–10 (2021). https://doi.org/10.1108/DLO-01-2021-0003
13. Stals, A., Caldas, L.: State of XR research in architecture with focus on professional practice – a systematic literature review. Archit. Sci. Rev. **65**(2), 138–146 (2022). https://doi.org/10.1080/00038628.2020.1838258
14. Andrić, I., Le Corre, O., Lacarrière, B., Ferrão, P., Al-Ghamdi, S.G.: Initial approximation of the implications for architecture due to climate change. Adv. Build. Energy Res. **15**(3), 337–367 (2021)
15. Architects Council of Europe ACE (2023) Upskilling to deliver high-quality architecture for a beautiful, sustainable and inclusive environment. https://www.ace-cae.eu/services/news
16. Proenca, et al.: The role of universities on forming social inclusive and sustainable environments (2022). https://re.public.polimi.it/handle/11311/1232688
17. Joliffe, E., Crosby, P.: Architect.The Evolving Story of a Profession (1st ed.). RIBA Publishing, London (2023)
18. Senova, M.: Meta-skills are the key to human potential. J. Behav. Econ. Soc. Syst. 133–137 (2021). https://doi.org/10.5278/OJS.BESS.V2I1.6463
19. Ely, A.: Professional Behaviours: Being a Professional, being Professional. In: Defining Contemporary Professionalism. RIBA Publishing, London (2019)
20. Becerik-Gerber, B., Gerber, D.J., Ku, K.: The pace of technological innovation in architecture, engineering, and construction education: integrating recent trends into the curricula (2011)
21. Morrell, P.: Collaboration for change: The edge commission report on the future of professionalism. The Edge Commission, Edge Debate, London (2015)
22. Kasali, A., Nersessian, N.J.: Architects in interdisciplinary contexts: representational practices in healthcare design. Des. Stud. **41**, 205–223 (2015)
23. RIAI, Education Policies and Standards (2016)
24. Scott, C.L.: the futures of learning 3: what kind of pedagogies for the 21st century (2015). https://unesdoc.unesco.org/ark:/48223/pf0000243126
25. Le Deist, F.D., Winterton, J.: What Is competence? Hum. Resour. Dev. Int. **8**(1), 27–46 (2005)
26. Stump, G.S., Hilpert, J.C., Husman, J., Chung, W., Kim, W.: Collaborative learning in engineering students: gender and achievement. J. Eng. Educ. **100**(3), 475–497 (2011). https://doi.org/10.1002/j.2168-9830.2011.tb00023.x
27. Oliveira, S., Olsen, L., Malki-Epshtein, L., Mumovic, D., D'Ayala, D.: Transcending disciplines in architecture, structural and building services engineering: a new multidisciplinary educational approach. Int. J. Technol. Des. Educ. **32**(2), 1247–1265 (2022)
28. Crinson, M., Lubbock, J.: Architecture—Art Or Profession?: Three Hundred Years of Architectural Education in Britain. Manchester University Press, Manchester (1994)
29. Elzarka, H.: Teaching value engineering effectively: an interdisciplinary approach (1998)

30. Holley, P.W., Dagg, C.: Development of expanded multidisciplinary collaborative experiences across construction and design curricula. Int. J. Constr. Educ. Res. **2**(2), 91–111 (2006)
31. Idi, D.B., Khaidzir, K.A.M.: Critical perspective of design collaboration: a review. Front. Archit. Res. **7**(4), 544–560 (2018). https://doi.org/10.1016/j.foar.2018.10.002
32. Gil-Mastalerczyk, J.: Architectural education in the formation of the built environment with sustainable features (2020)
33. O'Dwyer, S., Gwilliam, J.: Ways of choosing: the role of school design culture in promoting particular design paradigms in irish architectural education. In: EAAE Annual Conference Proceedings, pp. 44–65 (2020)
34. Septelka, D.: The design-build charrette–an educational model for teaching multidiscipline team collaboration. In: ASC Proceedings of the 38th Annual Conference Virginia Polytechnic Institute and State University-Blacksburg, Virginia, pp. 85–96 (2002)
35. Hay, R., Samuel, F., Farrelly, L.: Demonstrating the value of design through research in architecture practice. University of Reading, Architects' Council of Europe (2020)
36. Schön, D.A.: The architectural studio as an exemplar of education for reflection-in-action. J. Archit. Educ. **38**(1), 2–9 (1984)
37. De Graaf, E., Cowdroy, R.: Theory and practice of educational innovation through the introduction of problem-based learning in architecture. Int. J. Eng. Educ. **13**, 166–174 (1997)
38. Weimer, M.: Learner-Centered Teaching: Five Key Changes to Practice. Wiley, Hoboken (2013)
39. Rodarte-Luna, B., Sherry, A.: Sex differences in the relation between statistics anxiety and cognitive/learning strategies. Contemp. Educ. Psychol. **33**(2), 327–344 (2008)
40. Lave, J., Wenger, E.: Situated Learning: Legitimate Peripheral Participation. Cambridge University Press, Cambridge (1991)
41. Biggs, J.: Constructive Alignment. Constructive Alignment (1996). https://www.johnbiggs.com.au/academic/constructive-alignment/
42. Hatleskog, E.K.: Reflection, participation and production of ideas through architectural design practice. Reflect. Pract. **15**(2), 144–159 (2014)
43. Weisz, C.: Resilient design: 'systems thinking' as a response to climate change. Archit. Des. **88**(1), 24–31 (2018). https://doi.org/10.1002/ad.2255
44. Nichols, M., Cator, K., Torres, M.: Challenge based learning guide (2016). https://www.researchgate.net/publication/337029776_Challenge_Based_Learning_Guide/citation/download
45. Cannon, H.M, Feinstein, A.H., Friesen, D.P.: Managing complexity: applying the conscious-competence model to experiential learning. developments in business simulation and experiential learning. In: Proceedings of the Annual ABSEL Conference (2010)
46. RIAI, Standard of Knowledge, Skill and Competence for Practice as an Architect (2009)
47. Directorate-General for Internal Market, Industry, Entrepreneurship and SMEs (European Commission). User guide, Directive 2005/36/EC: All you need to know about recognition of professional qualifications. Publications Office of the European Union (2020). https://data.europa.eu/doi/10.2873/49563
48. Architects Registration Board, Prescription of Qualifications: ARB Criteria at Part 3. Architects Registration Board (2012)
49. RIBA, Mandatory Competences (2021). https://www.architecture.com/knowledge-and-resources/resources-landing-page/mandatory-competences
50. RIAI, Professional Practice Examinations Update, 10 August 2023
51. Cuff, D.: Architecture: The Story of Practice. MIT Press, Cambridge (1992)
52. Khodeir, L.M., Nessim, A.A.: Changing skills for architecture students employability: analysis of job market versus architecture education in Egypt. Ain Shams Eng. J. **11**(3), 811–821 (2020)
53. Salleh, S.M., Zahari, M., Said, N., Ali, S.: The influence of work motivation on organizational commitment in the workplace. J. Appl. Environ. Biolog. Sci. **6**, 139–143 (2016)

54. SQW, Review of Architects Competences Report for the Architects Registration Board (2021)
55. Cheetham, G., Chivers, G.: The reflective (and competent) practitioner: a model of professional competence which seeks to harmonise the reflective practitioner and competence-based approaches. J. Eur. Ind. Train. **22**(7), 267–276 (1998)
56. Uzunkaya, A., Paker Kahvecioğlu, N.: Deciphering architectural knowledge as reflective practice: revealing map. Reflective Pract. **21**(4), 499–519 (2020). https://doi.org/10.1080/146 23943.2020.1779685
57. Ustav, S., Venesaar, U.: Bridging metacompetencies and entrepreneurship education. Educ. + Training **60**(7/8), 674–695 (2018). https://doi.org/10.1108/ET-08-2017-0117
58. Brown, R.B., McCartney, S.: Competence is not enough: meta-competence and accounting education. Acc. Educ. **4**(1), 43–53 (1995)
59. Bach, C., Suliková, R.: Competence development in theory and practice: competence, meta-competence, Transfer competence and competence development in their systematic context. Management (18544223) **14**(4), 289–304 (2019). https://doi.org/10.26493/1854-4231. 14.289-304
60. Bates, R., Brenner, B., Schmid, E., Steiner, G., Vogel, S.: Towards meta–competences in higher education for tackling complex real–world problems – a cross disciplinary review. Int. J. Sustain. High. Educ. **23**(8), 290–308 (2022). https://doi.org/10.1108/IJSHE-06-2021-0243
61. Bartlett, P.W., Popov, M., Ruppert, J.: Integrating core sustainability meta-competencies and SDGS across the silos in curriculum and professional development. In: Nhamo, G., Mjimba, V. (eds.) Sustainable Development Goals and Institutions of Higher Education. Sustainable Development Goals Series, pp. 71–85. Springer, Cham (2020). https://doi.org/10.1007/978-3-030-26157-3_6
62. Architects Registration Board, ARB Annual Report and Financial Statement (2022). https:// arb.org.uk/wp-content/uploads/ARB-Annual-Report-and-Financial-Statement-2022-publis hed.pdf

Design Institutes and Design Studios

Cases of Permeability Between Teaching and Practice (Including Research)

Marco Trisciuoglio[1,3](✉) and Bao Li[2,3]

[1] Politecnico di Torino, Torino, Italy
marco.trisciuoglio@polito.it
[2] Southeast University, Nanjing, People's Republic of China
[3] "Transitional Morphologies" Joint Research Unit SEU/PoliTo, Nanjing, China

Abstract. The contribution describes an urban design teaching activity developed in the last eight years (2016–2023) at the top School of Architecture in China. Several Design Studios "Urban morphology, architectural typology, contemporary settlement patterns" has been held at the School of Architecture, in Southeast University Nanjing, China.

One of the main features of the teaching activity has been the strong connection with the practice activities developed within the Design Institute of the same University, such as the Urban Architectural Lab, founded in 2006 as part of the historic Architects & Engineers Co. Ltd. of the same university.

The role of the Design Institutes is specific of the Chinese context, where those public structures are the legacy of the process of collectivization of the professions promoted during the Fifties. The strategic role of the Design Institutes located in the universities allows nowadays not only high quality in design productions, but also the opportunity for students to face real topics of great complexity and to improve their competencies: in design as well as in socio-economic management.

The aim of the contribution is also quoting some urban regeneration projects developed in Nanjing historical urban tissues where the connection between Design Institute and Design Studios was fruitful and strong, from the choice of the topic and the surveys to the exams involving stakeholders within the final jury.

Keywords: Design Studios · Design Institutes · China · Urban regeneration · Urban morphology

1 The Role Played by Design Studios in Chinese Urban and Architectural Design Education

Architectural education in China dates back to a specific date, 1927. In that year, in the Capital city of Nanjing, the first university course aimed at teaching the practice of architectural design was activated.

The institution that takes charge of this initiative was the National Central University (later Nanjing Institute of Technology), divided from 1952 in two different main

M. Barosio et al. (Eds.): EAAE AC 2023, SSDI 47, pp. 375–385, 2025.
https://doi.org/10.1007/978-3-031-71959-2_40

high education institutes: the "generalist" Nanjing University (NJU) and the "technical" Southeast University (SEU). This last one, with its School of Architecture SEU-Arch, is considered as the heir of that tradition of the first School of Architecture based in 1927 within the National Central University.

Four are considered the founding fathers of modern Chinese architecture: Liang Sicheng, Liu Dunzhen, Yang Tingbao and Tong Jun. If Liang Sicheng (1901–1972) was the "inventor" (overall operative in Beijing) of historical Chinese architecture as field of knowledge, Liu Dunzhen (1897–1968) founded, as Japanese trained scholar founding the earliest architectural departments in China in 1920s at Suzhou Technical School, Tong Jun (1900–1982) became the leading expert on Chinese garden art, Yang Tingbao (1901–1982) was probably the greatest architect of all of them, crossing the twentieth century with a precise attention to the relationship between tradition and innovation. The last three are directly involved in the establishment of the Nanjing school, while the first one, Liang Sicheng, had a decisive role in the foundation of the Beijing School (at the Tsinghua University), as well as representing an intellectual figure who more than others had the merit, starting from the 1920s, of revealing to the world the existence of a historical Chinese architectural culture of a level and importance equal to the Western classical one.

The Chinese pedagogical model in architecture is imported, in a sort of global circle of references [1]. Many of the Chinese architects who trained in the first half of the 1920s did their studies either in Japan or in the United States. There, in particular, in Philadelphia, at Pennsylvania University, a French architect and professor, Paul Philippe Cret (1876–1945), trained at the Ecole des Beaux Arts in Lyon, was active. Most of the younger professor of late 1920s studied at "Penn", where that French Master had brought the Fine Arts way to teach design. Thus, the Western method of teaching architectural design in Design Studios arrived in China from the US, at the foundation of the Chinese Schools of Architecture in late Twenties on the basis of a European eclectic model [2].

For all these reasons, the pedagogical system on which the Chinese architecture school is built is precisely the "beaux arts" one: an atelier in which a few students (a dozen at most) refer by imitation to a Master who guides them in dealing with precise and given design themes, strong in conspicuous collections of repertoires and catalogs from which to draw ("copying", so exercising the main action of nineteenth-century art, or in the most extraordinary cases reinterpreting, through minimal scraps of minimal emancipation gain from a given model).

Nowadays, Design Studios still exist. In the work of students, the imitation of the Master's design work in Fine Arts ateliers has been replaced by forms of discussion with teachers upon design works and the attempt of investigating more and more design questions (for example the urban contexts) by collecting data is one of the main phases of the work. The design process is defined as an incremental step-based process, which involves different stages: diagnosis of the context, envisioning exercise and development of a number of design options, selection of a design alternative and implementation.

2 The Role Played by Design Institutes in Chinese Professional Practice Ecosystem in Urban and Architectural Design

The Design Institutes are Chinese bodies governed by public law, responsible for the design of works, neighborhoods and urban settlements. After 1949, with the establishment of the People's Republic of China, collectivism radically transformed the work system, no longer oriented towards the market and capitalist profit, but towards effectiveness and efficiency, with respect to the functions to be performed. The old liberal professions are obviously overwhelmed by this revolution. The two main professions (those to which the European Union still today recognizes a special status today due to their necessity for people's lives), those linked to the practice of medicine and architecture, are interpreted in the pivotal role of public utility and therefore it is decided that they be carried out within specific structures. In this sense it can be said that, at least initially, the design institutes (first located within local, municipal or provincial government structures, then also within public universities) are for architects, engineers and planners, at least in their conception and in their functioning, what the hospitals or clinics are for doctors and surgeons [3].

Following the first Five Years Plan (towards a collectivistic transformation of China), in 1952 the East China Industry Bureau Architectural Design Company, as the first Shanghai's state-owned Design Institute, was established. From that moment, the Design Institutes will become the main actors of the architectural and urban transformation until nowadays in China Mainland.

Nowadays, even after the Chinese economic and commercial reforms of the 1990s, design institutes continue to occupy a predominant role in the panorama of Chinese professional practice. There are, as in the whole world, large and powerful private design companies, some of which are multinational in nature, and there are also small studios that offer a sort of brand of their products (with a large circle of real architects/artists with personalities relevant, often at the level of the great international star-architects), but the power of the public design institutes remains unshakeable: they are reliable, have important tools, human resources and skills, a great ability to deal with the public sector of which they are part, often (from within the universities where they are located) have the opportunity to experiment [4].

In China, Design Institutes (within universities or within municipalities) are the dominant subject in the professional environment, where they are the key between local government and developers (as, for example, in the key area of Yuzui CBD of Hexi New City with the interplay among urban infrastructures, ecological resources and high-rises in vertical dimension) [5].

The general framework described here should also be considered as a possible operational horizon to which a Chinese architecture student today aspires: the average student expects to work in a large company (design and development companies carry out frequent enrollment sessions within schools, directed at final year bachelor students), while he/she dreams of doing an apprenticeship that will one day allow him/her to open a business (his own or in a small group) as an independent designer. However, also considering the great difficulties of the Chinese national exam for the qualification to practice the profession which very few are destined to pass and which confers an almost purely notarial seal in the project validation/approval process, the best, most prepared and most disciplined students are immediately involved in the design institutes and saw operational careers of some interest opening up.

3   "Urban Morphology, Architectural Typology, Contemporary Settlement Patterns" (SEU Nanjing, from 2015)

The Design Studios "Urban morphology, architectural typology, contemporary settlement patterns" held at the School of Architecture in Southeast University (Nanjing, China) are working, since almost a decade (2015–2024), in strong connection with the Design Institute of the same University «Urban Architectural Lab» founded in 2006 as a part of the historic Architects & Engineers Co. Ltd. of the same university.

Design studios have not changed much in the century that has now passed since the founding of the architecture school in China, at least in a high-ranking public university like Southeast University and many others. Above all, the pedagogical structure has remained unchanged. There is always a very small number of students (from 6 to 12) who present themselves as a team. The teacher prefers to exercise his authority with a guiding role (through his projects and/or through its methodological approach), rather than becoming a trainer of students in a series/sequence of practices that can be used in a professional context. The type of project training is still firmly anchored to the evaluation of the formal, constructive and functional outcome of the project action, rather than to the enhancement of the student's educational path read in the form of a design process (Fig. 1).

However, the themes are very current and similar to a lot of architecture schools around the world: the use of innovative technologies and construction systems, the dialogue with traditional forms and techniques also in terms of sustainability, the use of performance satisfaction metrics, the comparison with an elderly and weak society, the search for design solutions capable of developing the opportunities offered by digital devices, the question of heritage as element able to switch on fruitful connections between tourism and marketing.

The themes of urban regeneration have had a certain importance, especially in the last ten years. The innovative design studio named "Urban morphology, architectural typology, contemporary settlement patterns" was experimented in Nanjing from 2015. There, a favorable connection between the Chinese Southeast University and the Italian Politecnico di Torino created the conditions for joint teaching actions, based on a simple mission: using the standard morphological-based method that was characteristic of the Italian school of urban analysis and urban design to teach students how to read the settlement forms, the spaces and the urban objects of the Chinese city (including urban fabrics) and also to design accordingly [6].

This has allowed the conceptual tool represented by the Italian typo-morphological tradition to update itself by dealing with a new theme such as that of the Asian city, and the question of the urban regeneration of the Chinese city to find new possible approaches, not necessarily based on new urbanism practices. or of pushed gentrification, but ultimately oriented (as we will see) on protocols of innovative participatory forms directly played on aspects of urban form.

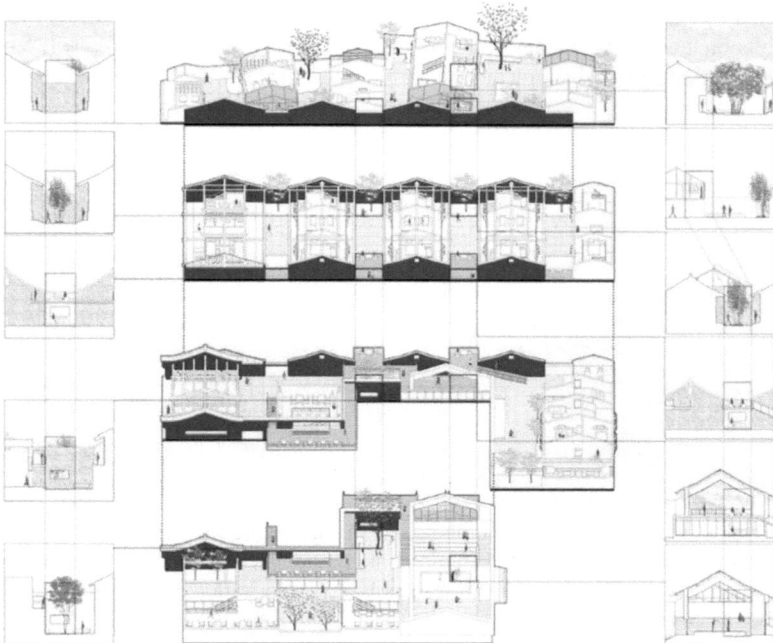

Fig. 1. Studies on the transitional morphologies of the XiaoXiHu block in Nanjing (Qinhuai District), from the Design Studio "Urban morphology, architectural typology, contemporary settlement patterns" at SEUArch 2018 (Professors Bao Li and Marco Trisciuoglio), traditional typologies and their renovation (from Archive "Transitional Morphologies" Joint Research Unit)

Thus, the first aim of the Design Studios is practicing fundaments of urban morphology and buildings typology in order to read the urban spaces and artefacts. The second aim of the Design Studios is using that reading activity in order to look for innovative design solutions for the contemporary city. The teaching activity is based on the reading of the Chinese city of nowadays and the work in the Design Studios is organized by weekly collective discussions about design development.

A series of more or less extensive areas (in any case at the scale of the urban project, between 1:200 and 1:1000, with in-depth analyzes at 1:100 scale), located in Nanjing or in other Chinese areas, have been the subject of the attention of teachers and students.

Without necessarily distorting the traditional pedagogical structure of the Design Studio, this experiment makes use of at least two important innovations. The first innovation consists in addressing real and not hypothetical issues, having them suggested directly by the Design Institute of the same university (therefore with the involvement of stakeholders both at the level of developers and at the level of politicians and managers directly operational on urban regeneration practices). The second innovation consists in hiring, alongside the usual dozen undergraduate students, about half of master students with the organization of three-person working groups, made up of two undergraduates and one master students, where mentoring by the older students becomes fundamental.

Based on the concepts of TECTONICS, TYPOLOGY and TOPOGRAPHY, the Design Studio lets bachelor and master students together investigate on the interplay between tradition and modernity (through design activities as surveys, sketches, models, diagram) [7] (Fig. 2).

Fig. 2. Studies on the transitional morphologies of the XiaoXiHu block in Nanjing (Qinhuai District), from the Design Studio "Urban morphology, architectural typology, contemporary settlement patterns" at SEUArch 2018 (Professors Bao Li and Marco Trisciuoglio), detail of the entrance (from Archive "Transitional Morphologies" Joint Research Unit)

The master students are not only more adults. It in a pedagogical system like the Chinese one (which is American-style 1+4+2, i.e. one preparatory year, four undergraduate years and two optional master's years), they are often the ones already involved in work as an internship within the school and departments, able to develop, with the relevant professors, projects in preliminary stages intended for the design institutes, when not directly projects already being developed within the design institutes.

4   The «Architectural Lab» Within the Context of the Historical Architects & Engineers Co. Ltd (SEU Nanjing, from 2006)

It should not be thought at all that design institutes are anonymous professional bodies, capable of providing a low-quality service that is in no way comparable to that provided by large design companies or celebrated star architects. In the Chinese system, the reciprocal roles of architecture schools, renowned designer architects, the world of communication that revolves around design, public developers and local governments are very different from those in Europe (and, above all, the university has a pre-eminence of position which is still relevant today).

In this context, it is not uncommon for important practitioners, who are also professors, to set up their own professional studio within the same design institute. One of the most interesting realities on the Chinese professional scene in recent years has been the Urban Architectural Lab of the Southeast University of Nanjing.

The Urban Architectural Lab (UAL), founded in 2006, is based at and part of the Architects & Engineers Co. Ltd., the Design Institute of Southeast University (around 580 employees versus the 800 at the Design Institute at Tsinghua University).

The core members of UAL began the team's professional activities in 2000 and after around 20 years of development, the team has now more than 50 members: 5 architects and faculty, 15 full-time architects, 30 among PhD students and master students.

It is a separate structure, more streamlined than the great Design Institute, and directed today by Han Dongqing, former Dean of the School of Architecture, well known Master of urban and architectural design [8].

The Design Institutes signed at 2020 around 70 projects (mostly in Nanjing), some of them are very important works, published on international journals as demonstrative. One interesting case for the SEU Design Institute UAL is the reconstruction in Nanjing, with a great symbolic intention, of the Jinling Da Bao En Temple (financed in 2015 by the investor Wang Jianlin of Dalia Wanda Group), the former Porcelain Tower, described in 17th century as one of the Seven Wonders in the World [9].

Being part of an academic environment (or the bridge between the academic world and the practice world), the activity of the Design Institute is often not only focused on design, but also on methodological investigations. In the last three years, for example, UAL made great efforts in linking ownership's data with typo-morphological map in order to improve innovative participatory models, for the for the implementation of large and innovative urban regeneration projects.

5   The Southern Part of the Walled Center of Nanjing Between Real Estate and Urban Regeneration

Nanjing today still retains much of its Ming-era city wall (when it was Capital City of Chinese Empire). The southern part of the city in particular, called Qinhuai District and crossed from west to east by a navigable canal, still shows large parts of the ancient city, with its urban fabric made up of courtyard houses. Of course, much of the extension of this sector (which still constitutes a fifth of the surface area of historic Nanjing enclosed by the walls) is also irremediably compromised by the presence of functionalist building types built between the 1960s and the 1980s, by the presence of a series of road infrastructures that do not respect historical routes, by the looming heavy gentrification, especially of a commercial nature, which has significantly altered the urban spaces of the traditional city [10, 11].

For a first-level Chinese city, capital of the province of Jiangsu, the most technologically advanced of all the provinces of the People's Republic, therefore with prospects of a lively city of the future, the theme of urban regeneration is urgent. Furthermore, the general conditions of the historical spaces and buildings aren't at a level so sufficient to let inhabitants live in a safe and healthy way. However, the sense of community is very strong: every house, every small courtyard, every person tells stories about the places and the vivid daily life into them [12].

Over the last twenty years the local government has sequentially experimented with three different regeneration methods. A first phase, with an exquisitely commercial imprint, has reconstructed the urban fabric around the Confucius Temple (Fuzimiao) through an operation dictated by a pop culture of tourism and tradition. A second phase, suggested by unbridled real estate practices, razed part of the historical fabric to the ground (in Laomendong), moving the resident population to much more functional suburbs and creating very expensive and refined urban villas in the choice of materials, all immediately sold but all left irremediably uninhabited. A third phase, based on the direct involvement of citizens, has in fact invented almost from scratch a participation system which has borne excellent and clearly extraordinary results (for example in the case of the Xiaoxihu block) (Fig. 3).

Fig. 3. View of the regeneration project at Xiaoxihu, Nanjing 2023 (photo by Author).

The stubbornly sought connection between the activities of the Design Institute and the activities of the Design Studio contributed significantly to the development of this third phase, after the local government asked SEU a help to find different design processes after the experiences of Fuzimiao and Laomendong.

In fact, for the first time, starting around six/seven years ago, the involvement of stakeholders, developers and local decision makers has intensified in the discussion of the outcomes (even partial) of the training activity conducted by the Design Studios. At the same time, the Design Institute entrusted entire sections of blocks to be redeveloped to some of the teachers directly operating in the same project areas [13].

In this way, on the one hand the approach of the Design Studio took strictly into consideration professional opportunities gradually proposed to the Design Institute, on the other hand the work carried out with the students immediately found a testing ground in entire passages of the real city. In short, an incredible virtuous circularity has been created between operators and students, under the guidance of designers/professors capable of keeping the world of study and that of the profession closely together.

6   The Regeneration of the Block XiaoXiHu as a Living Lab for Students' Design Investigation. Coincidences

Within the context of a series of Design Studios, students analysed the pilot block of Xiaoxihu (very close either to Laomendong or to Fuzimiao) in order to demonstrate a more careful approach to what still exists (and what was existed) of the old town.

The design activity within the Design Studios (attended by mixed groups of Bachelor and Master Students) gave some first important guidelines and suggestions for the future uses of buildings and spaces and for the image of Xiaoxihu.

The final results of Design Studios were an important pre-figuration of the possible processes to reactivate ways of living in the block, with its internal spaces and its paths, so as to recreate the typical porosity of traditional fabric and daily life.

The system of images produced within the Design Studios and the results of the physical and social surveys became shared element of critical discussion in the context of the Design Institute, in a harmonious relationship of reciprocity where the two institutions, the didactic one and the professional one, have worked "shoulder to shoulder" (Fig. 4).

Fig. 4. View of the regeneration project at Xiaoxihu, Nanjing 2023 (photo by Author).

A first result was a real urban regeneration project nowadays almost completed, which won the 2022 UNESCO Asia Pacific Award for Cultural Heritage Conservation.

A second result was the improvement of the competences of students, thanks to the opportunity to work on real project in the connection Design Studio + Design Institute.

The more important result was the regeneration and the improvement of the urban social daily life of the Xiaoxihu block, through a mix of technical skills and investigation for innovation (Design Institute) and braveness, imagination and attention in design (Design Studios).

Xiaoxihu is today an important demonstration project in China.

The shared activities between Design Studios and the Design Institute helped a lot the scientific research: papers, books, seminars, international conferences, PhD dissertations were and are promoted, also deeply supporting the activities of the "Transitional Morphologies" Joint Research Unit (established in 2018 between Southeast University and Politecnico di Torino) [14].

Generally speaking, in China, the permeability between Design Studios and Design Institutes was until now not only advantageous from the point of view of both project training and the choice of specific design solutions to the detriment of others. Actually, it has allowed us to identify and develop important lines of research. One of these concerned, for example in Nanjing, the possibility of creating "augmented" urban typological maps with property and land value data deduced from the intersection of old land registers with a current survey of building structures and also of housing conditions. Another line of research prefigured, tested and then verified an innovative participation system based in Nanjing on the so called "diagram of the five actors" (local government, developers, designers, insiders and outsiders) and on the possibility of alienating part of the families' assets in favor of activities of microeconomics to be achieved in the most complete respect of building types and settlement morphologies. The most recent line of research concerns, always in Nanjing, the monetary valorization of urban spaces and objects involved in participatory negotiation, imagining connections between urban morphology and urban economics.

All three of these shortly described three lines of research used, as a case study, the Xiaoxihu block, located in the Qinhuai District, characterized by the presence of some structures from the historical era and also by a very high number of modern compromises. For some years now, the block has been the subject of redevelopment actions which aim not only at the mere protection of the buildings placed under protection, but also and above all at identifying new, "transitional" roads to prefigure the urban settlement of the future.

As mentioned above, Xiaoxihu's project won the 2022 UNESCO Asia Pacific Award for Cultural Heritage Conservation. At the same time, the same block became the subject of some Design Studios also held in Italy, at the Politecnico di Torino. One of the purposes of the connections established between the Politecnico di Torino and Southeast University is in fact to mutually exchange experiences and solutions, again in pedagogy, research, urban design practice.

References

1. Li, X., Chong, K.H.: Implications of Chinese architectural education in contemporary Chinese architecture. J. Archit. **8**(3), 303 (2003)
2. Cody, J.W., Steinhardt, N.S., Atkin, T.: Chinese Architecture and the Beaux-Arts. University of Hawaii Press, Honolulu (2011)

3. Trisciuoglio, M., Lei, J.: L'ombra della pagoda. Note sul progetto di architettura tra mestiere e scuola nella Cina contemporanea. In: Giovanni Rocco Cellini. La domanda di architettura. Le risposte del progetto. Atti del VI Forum della Società scientifica nazionale del progetto. Docenti ICAR 14-15-16 ProArch, pp. 130–133. Roma, (2018)

4. Wu, F., Xu, J., Gar-On Yeh, A.: Urban development in post reform China. State, market, and space. Routledge, New York, Oxon (2007)

5. Santi, E.: Il 'dispositivo' dell'architettura sperimentale cinese. Identità e soft power nell'era del sogno cinese. Territorio 76 (2016)

6. Bao, L., et al.: Typological Permanencies and Urban Permutations. Design Studio of Regeneration in Hehua Tang Area. SEU Press, Nanjing (2017)

7. Trisciuoglio, M., Bao, L.: Capire le Città Cinesi. Ri-Disegnare gli Strumenti Italiani/Understanding Chinese Cities. Redesigning Italian Methods. Agathón. In: International Journal of Architecture, Art and Design 3 (Didattica e Progetti nelle Scuole di Architettura. Teaching and Projects of Architecture Schools), pp. 123–132 (2018)

8. Han, D., et al.: Multiple dimensions of urban design development from a practice perspective: a case study of an institute in Nanjing. Front. Archit. Res. **10**, 79–91 (2021)

9. Anonymous: The Porcelain Tower of Nanjing: History and Legacy of One of the China's Most Famous Buildings. Charles River Editors, Amazon Kindle (2017)

10. Chen, F., Thwaites, K.: Chinese Urban Design. The Typomorphological Approach. Ashgate, Farnham Burlington (2013)

11. Jiang, L.: Morphological research of the historical urban boundary. In: The Inner Fringe Area of Nanjing, Nanjing/Torino. Transitional Morphologies" Joint Research Unit/Southeast University – Politecnico di Torino, Nanjing/Torino (2019)

12. Tang, L., Li, Q., Ding, W.: The role of urban design in urban regeneration process – an urban design research in Nanjing, China. In: 9th IFOU Conference - International Forum on Urbanism. The 9th International Conference of the International Forum on Urbanism (IFoU) 2016 Buenos Aires/UBA, FADU (2016). https://www.researchgate.net/publication/316787 487_THE_ROLE_OF_URBAN_DESIGN_IN_URBAN_REGENERATION_PROCESS_-AN_URBAN_DESIGN_RESEARCH_IN_NANJING_CHINA. Accessed 18 May 2024

13. Dong, Y., Han, D., Trisciuoglio, M.: A graphical method of presenting property rights, building types, and residential behaviors: a case study of Xiaoxihu historic area, Nanjing. Front. Archit. Res. **11**, 1077–1091 (2022)

14. Trisciuoglio, M., et al.: Transitional morphologies and urban forms: generation and regeneration processes - an agenda. Sustainability **13**, 6233 (2021)

Ways of Architecture(s)

A New Form of Practice: *La Rivoluzione delle Seppie*

Rita Elvira Adamo(✉)

La Rivoluzione delle Seppie, Amantea, Italy
ritaelvira.adamo@gmail.com

Abstract. The paper describes the work of *La Rivoluzione delle Seppie* in Belmonte Calabro, where they have regenerated places and promoted new forms of community and social cohesion. The tested and implemented strategies aim to make these so-called marginal areas competitive and attractive within local and territorial systems and globally. The experimental nature of this practice is characterized by a methodology for designing and constructing placed-based interventions tailored to the territories and carried out through the involvement and protagonism of academics, students, institutions and local communities. A methodology that enhances the local skills implemented by knowledge of global stakeholders and it uses critical issues as opportunities to generate elements of innovation necessary to qualify the relaunch and the development of the area concerned. The action of *La Rivoluzione delle Seppie* started in Belmonte Calabro. The group acted as an agent of change and, above all, as a facilitator between the old and the new inhabitants. Not only Belmonte, but also *BelMondo*!

Keywords: marginal area · conviviality · learning by doing · agent of change · selfconstruction · commons · The *BelMondo* case study in Belmonte Calabro

1 The *BelMondo* Case Study in Belmonte Calabro

(See Fig. 1).

The development that characterized industrial growth and intense urban aggregation processes in Italy during the last century generated contradictions and imbalances at a territorial and social level. In the last decades, faced with large public debt, the margins of policies aimed at great investments or welfare-type interventions have been increasingly shrinking. However, the criticality marked by the abandonment and depopulation today for these territories can be seized as an opportunity. It is not a question of soliciting new oppositions concerning those that have marked the historical events of the last century, nor of cultivating anti-urban imagery or fueling settlement dispersion, but, on the contrary, of shifting attention to the territorial organization as a whole. There is a need to imagine a territory that, thanks to the rich articulation of its settlement models, continues to hold together weak and robust areas with a dense network of increasingly intangible flows; a territory that in its entirety knows specific processes of ecological transition and redevelopment of fixed capital and the relationships of proximity. In this

© The Author(s) 2025
M. Barosio et al. (Eds.): EAAE AC 2023, SSDI 47, pp. 389–393, 2025.
https://doi.org/10.1007/978-3-031-71959-2_41

Fig. 1. *Questa non è Campagna*, workshop in collaboration with Cheap, Crossings 2022, Belmonte, Photo by Nicola Barbuto, July 2022.

framework, the question arises of a new social pact that can guarantee new levels of integration and social inclusion. For this to happen, a more innovative function of public infrastructure is needed, but also a greater protagonism of the agents that on the territory can be facilitators of the processes of satisfaction of new needs and social interests (Fig. 2).

Fig. 2. Happy Lab, Ceramic Workshop with refugee children, *Casa di BelMondo*, Belmonte Calabro, Photo by Giulia Rosco, October 2023.

The La *Rivoluzione delle Seppie*'s practice conceived and tested intentions to go in this direction, fuelled by a strong focus on planning and implementing the imagined design. The intention is not to photograph local systems as they are now but to stimulate and prefigure a decisive transformation based on global and collective influences.

It may be helpful to point out three traits that characterize its work and to help grasp its meaning and specificity. The first is related to combining a specific radical vision and an effort of concreteness in action. This radical vision seeks to make room for itself with proposals that outline measured, tactically and concretely actionable efforts. A second important aspect is that the actions carried out decline a form of multifunctional planning, which aims with the same investment to obtain multiple benefits.

The objective is not only related to the traditional renovation of a building or a space but also to develop civic research for each step, making proposals in a collective form, combining visions of the future and concreteness of action on different levels. A third aspect is that this set of actions, developed by many hands, not only by professionals but also by scholars, students, locals and migrants from different locations, proposes to focus on a precise way of understanding a diverse approach to academic research and architectural practice (Fig. 3).

Fig. 3. Collaboration Rooms, Design by Orizzontale, Crossings 2019, Belmonte Calabro, Photo by Antonio D'Agostino, July 2019.

The main question on which La Rivoluzione delle Seppie's work is: "How can we build our communities around places?".

By strengthening the connection between people and places, Le Seppie's work is based on a collaborative process through which the public sphere can be shaped to maximize shared value. With community-based participation, both local and temporary, an effective place-making process that capitalizes on the resources, inspiration, and potential of the above-mentioned heterogeneous group results in the development of the *BelMondo* process.

This process has generated a series of actions since 2016 that contribute to both cultural welfare development and the people's well-being. The action started in Belmonte Calabro, South of Italy, and has progressively taken shape through the annual editions of *Crossings*, a series of residencies, symposia and workshops: an inter-weaving of activities that have generated a network of collaborators and constant moments of exchange between different actors for a common goal: the construction of *BelMondo*.

The physical place, the *Casa di BelMondo*, is the space where manual and artistic activities take shape, the non-physical place, *BelMondo Altlas*, is the virtual space where collective sessions at a distance take shape, where debates, seminars, and confrontations alternate with radio projects, experiential atlases, and communication projects. In this context, La Rivoluzione delle Seppie aims to redefine the village as a living architecture, a set of places and non-places that can accommodate new living forms and offer collectively usable hybrid spaces. This allows communities to appropriate them according to collective and social needs based on the current cultural and resource-sharing approach. A solution is not proposed but a dynamic process.

Fig. 4. Baywatch movable structure part of the Market Temporary Square in Belmonte Marina, Design by Orizzontale, Crossing 2022, Belmonte Calabro, Photo by Giulia Rosco, July 2022.

The objectives are twofold: the first is the elaboration of a diverse model of living and working collectively, as opposed to competitive living and hyper-specialized work culture; the second is the exchange of knowledge to inhabit a place temporarily but constantly so that experimental, conceptual but not ideological forms can be manifested, with a different conception of a participatory approach to public living. To achieve these objectives, *Glocal Tools* have been developed and deciphered: eight tools that characterize the operational approach of La Rivoluzione delle Seppie, derived precisely from the know-how accumulated so far, which defines shared values that can be adapted according to the opportunities and skills in a given socio-cultural and territorial context (Fig. 4).

Everyone Belongs to Everyone Else

Giacomo Ardesio[✉], Alessandro Bonizzoni, Nicola Campri, Veronica Caprino, and Claudia Mainardi

Fosbury Architecture, Milan, Italy
info@fosburyarchitecture.com

Abstract. The scarcity of resources in a context of permanent crisis is not only an opportunity but also the only direction in which it makes sense to practice. In our view, there is a generation that has already accepted the challenge and is trying every day to develop antibodies to disillusionment. The Italian Pavilion at the 18th International Architecture Exhibition – La Biennale di Venezia invited these practitioners to recognize themselves as a "movement." Titled *Spaziale*, it referred to an expanded notion of the discipline where the built artifact was not seen as an ultimate goal but as one of the possible tools for intervening in the fabric of relationships between people and places. Nine practices were invited to collaborate with nine advisors – from other fields of creativity – in nine Italian territories representative of conditions of fragility or transformation, with the support of as many local interlocutors. Each intervention represented a chapter in an incomplete agenda of urgent research themes for the national context and for architecture: open questions, traceable to the scenario of transition – not only ecological – that we are facing in these years.

Keywords: La Biennale di Venezia · Italian Pavilion · permacrisis · critical spatal practices

We represent a generation that grew up in a state of permacrisis. The neologism, elected Word of the Year in 2022 by Collins Dictionary, expresses a recurring dimension of catastrophic occurrences that perfectly describes the unfolding of events of the last twenty years. After the near miss of the Y2K bug, the century began with the 9/11 attack on the Twin Towers: the crisis of the West being screened on live television. We enrolled at university during the 2007–2008 financial crisis and, once we graduated, we started to look for work in the smoking ruins of the market. Today an energy and geopolitical crisis, tomorrow the environmental crisis; and this is only a partial representation of reality.

The global consequences have produced obvious repercussions on our profession and, while the age of architectural exuberance came to an end with the 2008 crisis, the pandemic has broadened that widespread awareness of the total depletion of resources [1]. As Rory Hyde suggests: "All crises have spatial consequences that architects are well prepared to deal with, yet instead of diving into them, we seem to be experiencing our own crisis: one of relevance" [2]. The clear risk is that the umpteenth internal discourse within the discipline makes us lose sight of how architecture, rather than providing solutions,

© The Author(s) 2025
M. Barosio et al. (Eds.): EAAE AC 2023, SSDI 47, pp. 394–399, 2025.
https://doi.org/10.1007/978-3-031-71959-2_42

is often part of the problem: on the one hand, the construction sector is one of the main causes of the environmental crisis; on the other hand (especially in Italy), the increase in land consumption does not correspond to the involvement of architects in transformation, neither their involvement in such process. An inversely proportional relationship between growth and development that lays the foundations for an unprecedented alliance between

Fig. 1. Italian Pavilion, La Biennale di Venezia, 2023. Photo © Delfino Sisto Legnani.

Fig. 2. Italian Pavilion, La Biennale di Venezia, 2023. Photo © Delfino Sisto Legnani.

the environment and professionals in the sector. An opportunity for architecture to evolve, albeit only opportunistically, in order to survive (Figs. 1, 2, 3 and 4).

Fig. 3. *Siren Land*. Authors: BB (Fabrizio Ballabio + Alessandro Bava) + Terraforma. Location: Baia di Ieranto (Napoli), Campania. Photo © Piercarlo Quecchia/DSL Studio.

There is a generation of sustainable natives who, in our view, have already accepted the challenge and seek to develop antibodies to disillusionment in their daily practice. In our opinion these critical spatial practices, term coined by Jane Randell in 2003 to indicate those practices "working across public and private, art and architecture" [3], are those who use the codified tools of design to question the social conditions of the places where they intervene. Accustomed by training to operating within a regime of scarcity, these practices foster transdisciplinarity as a means of pushing back the limits of architecture to hitherto little-explored fields.

We have intended our appointment as curators of the Italian Pavilion at the La Biennale 2023, titled 'Spaziale. Everyone Belongs to Everyone Else', as an unprecedented occasion to present to a broader public a series of Italian critical spatial practices and an invite to probe their diverse attitude confronting them with real design occasions, using curatorial practice as a powerful tool to shift from archiving to action. Necessarily recognising an ethical dimension to this role, we decided to use the Italian Pavilion as

Fig. 4. *Uccellaccio.* Authors: HPO + Claudia Durastanti. Location: Ripa Teatina, Abruzzo. Photo © HPO.

a pretext to activate pioneering projects, concrete actions that go beyond the six-month duration of the Biennale. Moreover, ethics also lies at the heart of the discourse when questioning the meaning and impact of temporary events of this scale: be they exhibitions, fashion shows, concerts, sporting events or fairs, they are all extractive processes that dissipate a great deal of energy and resources. In order to continue to celebrate moments of confrontation and contamination in a sustainable manner, it is now urgent to drastically rethink formats, flows and temporalities. To convert consumption into investment and the end into a beginning, Spaziale employed a substantial part of the public funds earmarked by the Italian Ministry of Culture to initiate and realize those pioneering projects in the months prior to the inauguration of the Venice Biennale 2023.

Each project tackled an agenda of urgent research topics for the national context and for the discipline as a whole: open questions that may be traced back to the transition scenario – and not only the ecological one – that we have been dealing with over recent years. An incomplete list of 'impossible' challenges that have been up for debate for decades, yet which – on the scale of the micro-histories of local contexts – are capable of producing tangible results. The definition of the themes guided the selection of designers under forty, who in their daily practice develop independent research in line with our curatorial proposals. Identified on the basis of the approach with which they operate, the territories in which they intervene, the means they use, the questions they raise and the answers they put forward, they were called upon to develop site-specific actions for the Italian Pavilion. Each practice was associated with an Advisor: nine professionals supporting the designers, from various fields across the creative industry, capable of informing and enhancing the ongoing projects, making them an undoubtedly transdisciplinary product. The installations were implemented in sites representative of conditions of fragility or transformation of our country: nine Stations narrating an unprecedented Italian landscape, a series of symbolic destinations on a renewed Italian Journey. Each collaboration was supported by one or more local interlocutors: public or private institutions which, as Incubators, helped to root the projects in the selected Stations.

What was shown in the Italian Pavilion has not to be intended as an exhibition but rather as a visual and formal synthesis of the multiple design processes and approaches embodied by the participants. The ultimate objective of the exhibition being the manifestation of a new disciplinary attitude that sees architectural interventions not as a goal, but rather as an one of the possible instruments to act in the space, or a network of relationships between communities and places; the basis of any architectural project.

Fig. 5. *Concrete Jungle.* Authors: Parasite 2.0 + Elia Fornari (Brain Dead). Location: Marghera, Veneto. Photo © Melania Dalle Grave/DSL Studio

Fig. 6. *La Casa Tappeto*. Authors: Studio Ossidiana + Adelita Husni Bey. Location: Librino (Catania), Sicily. Photo © Piercarlo Quecchia/DSL Studio.

References

1. Durastanti, C.: Tutto esaurito. Vulcano **4**, 34–39 (2022)
2. Hyde, R.: Future Practice: Conversations from the Edge of Architecture, p. 17. Routledge, New York (2013)
3. Rendell, J.: A place between art, architecture and critical theory. In: Place and Location, Proceedings of the Estonian Academy of Arts, pp. 221–233. Tallinn (2003)

Assemblage and Rituals

Giovanni Glorialanza[✉]

False Mirror Office, Genova, Italy
falsemirroroffice@gmail.com

Abstract. False Mirror Office analyses its own architectural practice, exposing the connections between some of its recent projects, research activities and publications. The following text is based on the presentation led by Giovanni Glorialanza and Filippo Fanciotti (False Mirror Office).

Keywords: Collage · Assemblage · Rituals · Domestic Environment · UFO

False Mirror Office's (FMO) work, being as multidisciplinary as somewhat devious, involves refining techniques for combining heterogeneous elements into assemblages in architecture, using pop-art's method of transfiguration of form and meaning. FMO's members conduct a rigorous initial phase of individual research before each design opportunity, uncovering any underlying themes that such project could imply in order to further explore them way after the assignment is completed. The exploration spectrum of those researches often extends beyond the disciplinary field of architecture, landing into cross-cutting issues; this is why in its research journals the complex representations of James R. Thompson Center's, designed by the architect Helmut Jahn, Chicago 1980, easily appear next to Frances Glessner Lee's dollhouses or the voyeuristic scenes in Hitchcock's Rear Window (1954). By juxtaposing and combining diverse elements, the natural mutability of the ever-changing referent in Architecture is further amplified: whereas the Doric capital caricatured at the Allen Memorial Art Museum, by Venturi and Scott-Brown (Ohio, 1977), still refer to elements disciplinarily belonging to architecture, it was precisely the encroachment of everyday-objects that demonstrated that 'Alles Ist Architektur' [1]. This oscillation between highbrow and inevitably lowbrow references makes the outcome of these assemblages distinctly autobiographical.

A first example of the application of these techniques can be observed in the project for a villa in the Roman countryside: a purely speculative exercise consisted of a three-part development. The first act consisted in grafting the floor plans of Roman villas with those typical of modern construction in the Roman countryside of today (Fig. 1), followed by a collage of places and characters associated with everyday-life in the Roman countryside; the third document presents a model of the villa, revealing the original forms of both architectures and some prominent representatives of the referential park from the collage, presented in a literal primordial broth, displayed in an elegant tureen.

© The Author(s) 2025
M. Barosio et al. (Eds.): EAAE AC 2023, SSDI 47, pp. 400–404, 2025.
https://doi.org/10.1007/978-3-031-71959-2_43

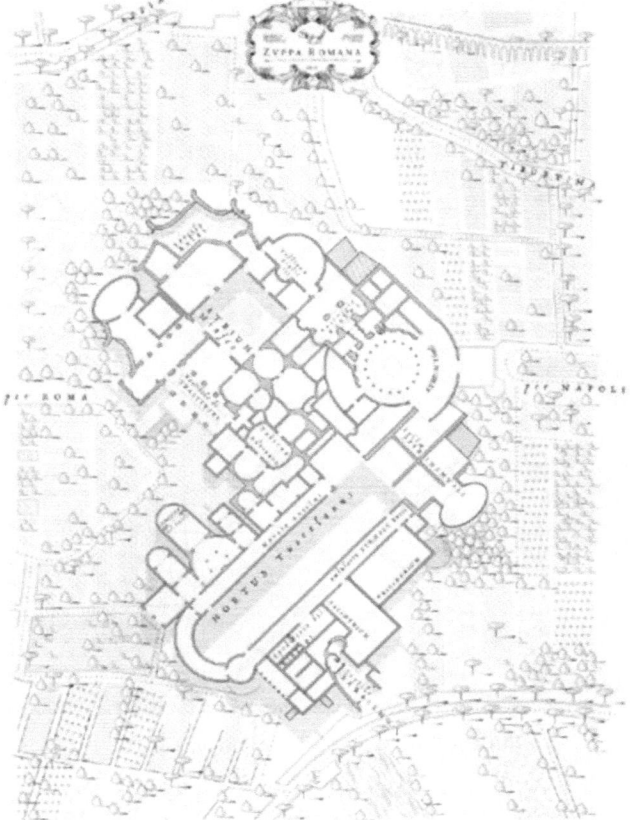

Fig. 1. *Zuppa Romana* (grafting), False Mirror Office, [2017].

A second attempt in architectural assemblages happened throughout the project to transform a neglected warehouse in the port of Trondheim into a public food-hall; the new exterior facade of the building was purely designed as a scenic backdrop, featuring the tail of a whale, a small boat, a mysterious chimney, and the profile of a riverfront *bryggen*. The proposal for a decommissioned building in Cuneo included irreverently positioned elements such as the golden palms from the *Austrian Travel Agency*, designed by Hans Hollein in Vienna (1978), and colossal versions of *LEGO pine trees*. Additionally, a grand-scale theft from the *National Collegiate Football Hall of Fame*, by Venturi and Scott Brown, in New Brunswick (1967) was incorporated, reduced to a mere form, ready for new functions (Fig. 2).

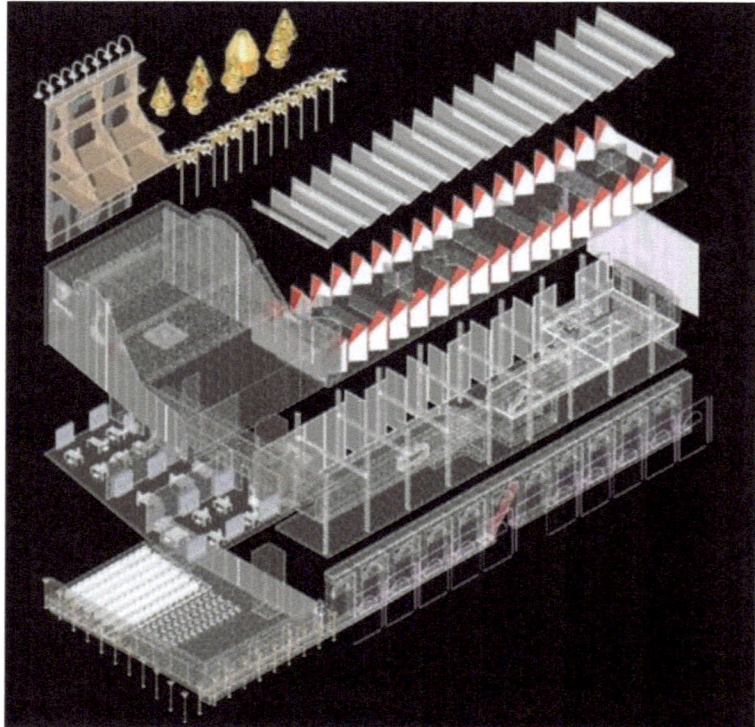

Fig. 2. *Frigo*, False Mirror Office, [2020].

Studying those cases in which the technique of *assemblage* has been applied in architecture, it is impossible do overlook the handful of projects realized by the radical group *UFO* (Lapo Binazzi, Riccardo Foresi, Titti Maschietto, Carlo Bachi, Patrizia Cammeo), active in Florence between '68 and '78. FMO's interest in the UFO's work, born out of genuine curiosity, first took shape in the publication *L'assemblaggio come testo figurativo per l'architettura* [2] and was then further explored in the monograph *UNIDENTIFIED FLYING OBJECT for contemporary architecture*, published by ACTAR and financed by the Italian Art Council of 2021.

While UFO performances aimed to subvert interactions between people and public space, at the domestic scale such rituals only intensify, further defining rigid rules in the division of spaces associated with living; the increasingly common practice of smart working ultimately introduces rituals related to *negotium* into the domestic space. These assumption inspired the project for a *New post-pandemic habitat* [3]; designed to accommodate multiple inhabitants, the habitat features semi-public devices provided for shared activities between small groups of inhabitants, while individual living units are designed to keep fundamental activities separate into *poche* obtained digging the perimetrical wall (Fig. 3).

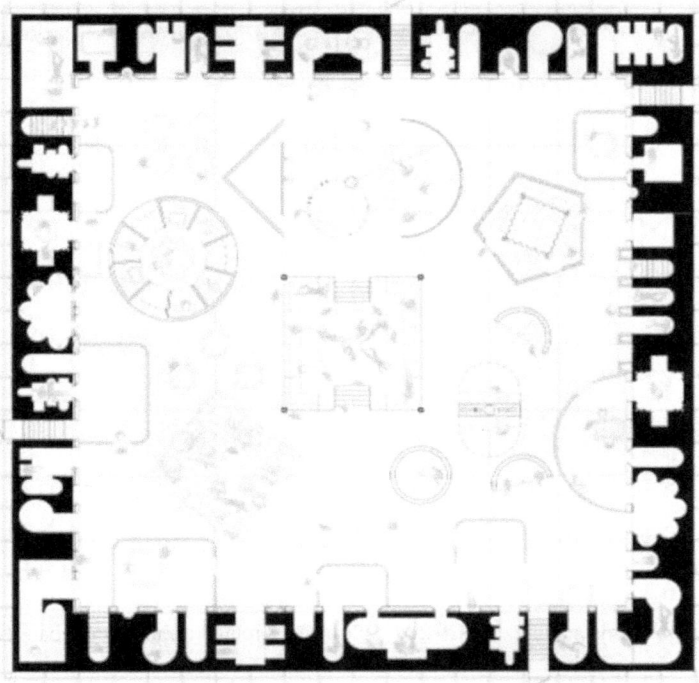

Fig. 3. *Nuovo habitat post-pandemico*, False Mirror Office, [2020].

Starting from these assumptions takes form also a device for the cult of *otium* within the domestic space, exposed first for the exhibition *Italy: The New Domestic Landscape. New York 1972 / Venice 2020*, held in Venice (2020) and then in Milan (*Otiarum*, Super-attico, 2022) together with five additional votive temples, each dedicated to a distinct aspect of idleness in the domestic space.

The same shrine-like form finally became the object of production of a workshop called *Station to Station* (Monesiglio, 2021); in that occasion participants selected stories from the small village in Alta Langa, artfully misrepresented them, and elevated them to the status of myth by representing them within a diorama. At the climax of the workshop, the participants staged a procession through the streets of Monesiglio; by stopping at various significant spots in the village to display their dioramas, they intentionally enacted a profoundly lay version of the stations of the Cross (Fig. 4).

Fig. 4. *Station to station*, Monesiglio, False Mirror Office, [2021].

References

1. Hollein, H.: Alles ist architektur. In: Bau: Schrift für Architektur und Städtebau, 1:20, pp. 1–2 (1968)
2. Office, F.M.: L'assemblaggio come testo figurativo per l'architettura. Un dialogo tra UFO e False Mirror Office. Piano B. Arti E Cult. Visive **4**(2), 88–118 (2019)
3. Anselmo, A., Hamzeian, B., Office, F.M.: Abitare oltre la pandemia: verso un nuovo habitat domestico. GUD **2**, 28–35 (2020)

New Territorial Narratives

Viviana Rubbo[✉] and Alessandro Guida

Paesaggi Sensibili, Turin, Italy
info@paesaggisensibili.org

Abstract. Spatial and territorial complexity require new and multiple forms of representation. Photography becomes the privileged tool to reflect on contemporary shifting landscapes adopting a critical look at our society. *Paesaggisensibili* is an independent observatory on contemporary landscape, a space for research and experimentation for the elaboration of territorial narratives based on interdisciplinary and multi-actor engagement.

Keywords: landscape photography · cultural landscape · narrative · community engagement · transdisciplinary approach

1   Introduction

Alessandro Guida and Viviana Rubbo graduated at the Polytechnic of Turin, Italy, and, from there, their paths have been very different for quite a long time. Alessandro worked as an architect, 3D visual designer and photographer. Viviana worked for several years in social planning and community-based development projects until when she found herself dealing with urban and territorial dynamics in Europe and around the world. Those two paths began to intertwine along the years because of a common interest in the observation of the territorial shifting and its spatial and societal implications. In 2016 they embarked on a new common adventure with the creation, together with other photographers, of the collective *Urban Reports* which has led –few years later– to the duo format of *Paesaggisensibili*. This practice was born to develop new forms of territorial narrative through the use of different languages (mainly photography, video and texts) as a method of engagement of a large spectrum of disciplines and knowledge sectors aimed at the involvement of a wider public in the discussion around landscape's transformations.

2   Photography as a Cultural Tool to Delve into the Landscapes (the Act of Looking as an Instrument for a Critical Observation of Our Society)

In a moment when we are confronted with an unprecedented complexity of territorial systems, an increasing acceleration of the processes of transformation and global phenomena (climate crisis, massive migration flows and pandemic events to name a few),

© The Author(s) 2025
M. Barosio et al. (Eds.): EAAE AC 2023, SSDI 47, pp. 405–409, 2025.
https://doi.org/10.1007/978-3-031-71959-2_44

the landscapes, and our perception of them, are subject to change, questioning the systems of values that, as individuals and as a society, we assign to them. These changes are happening faster than before and society doesn't have the time to assimilate them.

Already thirty years ago geographer Eugenio Turri proposed the metaphor of 'the landscape as a theatre' to express the urge to reflect on the way we transform our spaces of life; "man's relationship with the territory does not only concern his accomplishment as a doer, who operates and transforms nature, but also, if not above all, his being an observer [...] who understands the measure of his footprint: that is, the reflection on himself, the awareness of his own action" [1]. In this sense, landscapes are not just a mere spatial concept, but a cultural construction, made up of images that we inherit from the past, and which is enriched, day by day, by observing them.

Landscapes are like a text made of stories, memories and signs and, as such, they must be deciphered, read and interpreted.

Photography is a medium that can well respond to this challenge. In fact, it expands the possibilities of critical analysis of the space because it requires the immersion in the territory, the physical and mental experience of the places, introducing the subjective dimension in the interpretation of the landscape. At the same time it is also a powerful means of communication, because of its empathy and subjectivity, capable of dealing with the existing imagination, updating or replacing it, and thus giving life to territorial narratives capable of reaching out a large variety of actors, such as technicians, specialists and policy-makers, but also the general public with the aim to encourage more inclusive decision-making processes. This is why the privileged target of *Pesaggisensibili* are landscapes in transition, places in abeyance or going through a process of change. The goal is to draw attention to hurdles and issues, as well as hidden potentials and values to be rediscovered, and meanings to be reassigned. The sense of a place shouldn't be lost as Juan Noguè, the Catalan geographer recalls, "when the landscape loses its imagination and one is unable to replace it, in that very moment, the landscape dies" [2].

For this paper, three projects were selected to display a number of possibilities offered by this tool.

In the first one, the perceptive dimension of the space was introduced to nourish a research-led design process. The occasion was the project called *Arcipelago*, proposed by Mario Cucinella, the curator of the Italian Pavilion at the XVI International Architecture Biennale in Venice in 2018. The goal was to experiment new paradigms based on interdisciplinary and multi-actor working methods stimulating the role of architecture as change-maker in the reactivation of the inland areas of the country.

In this context photography:

1. Proved to be a concrete knowledge tool supporting the design process, providing a new layer of understanding of the areas of interest (the photographic research was not limited to the single intervention site but had explored a wider region);
2. Offered insights and unexpected connections between the sites when seen all together (territorial analysis and observation). Seeking to provide a larger vision of the sites, each photographer has adopted his/her own point of view, capturing the diversity and the richness of each area;

3. Gave an organic vision, a new representation of these areas (from recomposing the photographers' individual investigation into a collective narrative), showing the photographers' view on the reality explored.

The photographic campaign saw the photographers travel along the Apennine ridge and develop a vision of the places that was intended to encourage the dialogue with the architects, being able to broaden their cognitive horizons, "an interpretative space that (has been) a resource and inspiration, to grasp the invisible reality, the silent words of the landscape" [3]. The areas of investigation chosen by the curator were: the Casentinesi Forests in the Emilia-Romagna region, the town of Camerino in the Marche region, the Basento Valley in Basilicata, Gibellina in Sicily and Ottana in the heart of Barbagia, in Sardinia. This work has produced a new level of interpretation for each area and therefore, as a whole, a visual synthesis of the identity of these territories.

The second project shows the use of the photographic narrative as a research instrument for territorial analysis and a method for communities engagement to raise awareness around the topic of the Ecosystem Services (ESs).

L.U.I.G.I stands for *Linking Urban and Inner Alpine Green Infrastructure*s and was the title of a European project funded by the Alpine Space Program aimed at recognize, analyze, map and enhance ESs. The Habitats Directive (Council Directive 92/43/EEC) was adopted in 1992. It requires all EU Member States to establish a strict protection regime for species endangered with the aim of halting and reversing the loss of biodiversity and ecosystem services.

The study case chosen by the Turin Metropolitan City was the five lakes area of Ivrea, characterized by the presence of a majestic geological structure of glacial origins: the Morainic Amphitheater of Ivrea (AMI). The area is today a site of community importance and a *Special Conservation Zone* under the European Union Habitat Directive embracing six municipalities. Through a series of individual interviews, small group meetings (with the representatives of the productive and economic sectors, and with the administrators) and public gatherings, the process of listening has made possible to renew, and in some cases, to establish new connections between the territory and the representatives of the Metropolitan City and, at the inter-municipal level, between the public administrations, the associative world and the productive sectors which operate locally. The outcomes included a visual representation of this particular *milieu* which was brought at the citizens' attention during a series of exhibitions in the public realm. In addition, an open-call was organized inviting the population to bring their own point of view with respect to the elements of the landscape that today require more care and attention.

Photography was therefore used, on the one hand, as a 'cultural tool' for the technical and sector-specific discussion between the Metropolitan City, the local administrators and the experts responsible for the identification and mapping of the ESs; on the other hand, as a 'narrative language' to address the theme of the ESs in dialogue with the territory with the aim to start a process of recognition of the naturalistic and environmental values of the site.

Finally, a project where photography has become the means for a group of inhabitants to explore and rediscover their own territory, assigning new, unexpected and, perhaps, forgotten meanings.

Découvrir pour promouvoir Saint-Vincent was a pilot project developed by the *Associazione Poetica del Territorio* together with *Paesaggisensibili* in collaboration with the Municipality of Saint-Vincent in the Italian Alps. The programme was funded by the European Union Interreg Alcotra Program. The Alpine town has been for more than two-thousand years at a crossroad of international influences, playing a strategic role in the socio-economic, cultural and urban development of the region.

Today the town is represented by a binary narrative: its casino and the thermal baths. The goal of the two year plan of activities was to steer a process of engagement of the local community in the recognition and re-appropriation of the cultural heritage and values of the municipal territory. A first step included the identification of a wide spectrum of disciplines to develop the contents necessary to deepen the knowledge of the site from a geological, archeological, artistic, historical and anthropological perspective. The outcomes were presented to the community during two public events. A second step centered on a workshop activity where participants were asked to elaborate their own vision of the local material and immaterial values supported by site visits and based on the awareness given by the direct experience of the place. The individual works were discussed collectively leading to a new multi-voice representation which seemed much more realistic and close to the complexity of the present reality. The results became a public show where an actress in duet with a musician interpreted and gave voice and atmosphere to the visual narrative with the aim to share the emerged vision with the larger community.

We are facing times which require to address an extraordinary number of territorial challenges starting from retrieving our ability to observe and interpret the space that surrounds us, what French geographer Jean-Marc Besse defines 'la nécessité du paysage' [4], possibly allowing different forms of representations to express such dense territorial conditions.

References

1. Turri, E.: Il Paesaggio Come Teatro: Dal Territorio Vissuto al Territorio Rappresentato, p. 16. Venezia, Marsilio (1998)
2. Zagari, F.: Questo è Paesaggio: 48 Definizioni, p. 221. Mancosu (2006)
3. Cucinella, M.: Progetti per il futuro dei territori interni del paese: arcipelago Italia. In: Urban Reports, L'altra Italia, Racconto Per Immagini Delle Aree Interne del Paese, pp. 8–9. Johan&Levi, Milano (2018)
4. Besse, J.M.: La Necessite du Paysage. Editions Parenthèses, Marseille (2018)

Author Index

© The Editor(s) (if applicable) and The Author(s) 2025
M. Barosio et al. (Eds.): EAAE AC 2023, SSDI 47, pp. 411–412, 2025.
https://doi.org/10.1007/978-3-031-71959-2